Harry Rosenbusch

Microscopical Physiography of the Rock-Making Minerals

Harry Rosenbusch

Microscopical Physiography of the Rock-Making Minerals

ISBN/EAN: 9783744686259

Printed in Europe, USA, Canada, Australia, Japan

Cover: Foto ©berggeist007 / pixelio.de

More available books at **www.hansebooks.com**

MICROSCOPICAL

PHYSIOGRAPHY

OF THE

ROCK-MAKING MINERALS:

AN AID TO THE

MICROSCOPICAL STUDY OF ROCKS.

BY

H. ROSENBUSCH.

TRANSLATED AND ABRIDGED FOR USE IN SCHOOLS AND
COLLEGES

BY

JOSEPH P. IDDINGS.

Illustrated by 121 Wood-cuts and 26 Plates of Photomicrographs.

NEW YORK:

JOHN WILEY & SONS,

15 ASTOR PLACE.

1888.

TRANSLATOR'S PREFACE.

In preparing an English translation and abridgment of Professor Rosenbusch's "Mikroskopische Physiographie der petrographisch wichtigen Mineralien," with his permission, it has been my desire to present to English-speaking students the essential features of this valuable work, which contains all that is necessary for an accurate and complete determination of the rock-making minerals; hoping that in so doing I may not only meet the wants of those who take up unaided the study of rocks, but may assist those who are teaching this important branch of geology by providing them with a reference-book containing the diagnostic characters of these minerals. It is also hoped that it may lead to a more general interest in and a more accurate knowledge of microscopical petrography in this country, and may increase the number of those who, by exact and patient study, shall add to the store of established facts, and thus advance the science of lithology.

In abridging the book, I have endeavored to retain all that appeared to be essential to a fair, general comprehension of the subject, omitting what seemed to be refinements beyond the need of the average student, and for which the advanced student is referred to the original work. Thus most of the historical portions have been omitted, as well as the elaborate treatment of the optical anomalies of certain minerals, and many notes on European localities; while a number of notes on American occurrences have been inserted.

In two instances I have taken the liberty of departing from the original in the use of names, the grounds for which Professor Rosenbusch will undoubtedly appreciate. The term *Sphärokrystal* has been rendered *spherulite*, as the latter has become well established in English petrographical literature, and is not open to the objection which might be raised against the new term. *Liparite* has been rendered *rhyolite*, also for the reason that it is in such general use in this country and in England that to supplant it would lead to great confusion. With these exceptions I trust Professor Rosenbusch may find

Bedeutung gerade der optischen Eigenschaften für die Erkennung
der Mineralien unter dem Mikroskope und die oft gemachte Er-
fahrung, wie sehr dieselben von den Studirenden zu ihrem eigenen
grössten Schaden vernachlässigt werden. Allenthalben, wo es nöthig
schien, habe ich die besprochenen Verhältnisse durch schematische
Zeichnungen zu erläutern gesucht; dieselben sollen nur die An-
schauung erleichtern und machen auf strenge Winkelgenauigkeit
keinen Anspruch. — Die ganze Anlage und der Zweck des Buches,
welches ja kein Lehrbuch der Mineral-Optik sein soll, dürften es
wohl hinreichend erklären, dass nicht eine strengere Form für die
Besprechung dieser Verhältnisse gewählt wurde. Aehnliche päda-
gogische Erwägungen leiteten mich auch, wenn ich z. B. die Er-
läuterung der optischen Erscheinungen in dünnen, doppeltbrechenden
Minerallamellen im polarisirten Lichte der Erklärung der gleichen
Phänomene in dickeren Krystallplatten vorausgehen liess, obwohl
ja die im ersten Fall auftretenden Farben nur der centrale Theil
des Bildes sind, welches wir im zweiten Fall erhalten.

Aus der reichlichen Benutzung fremder Arbeiten wird mir um
so weniger ein Vorwurf erwachsen können, als ich die Resultate
derselben, wo es mir nur irgend möglich war, stets an eigenhändig
gefertigten Präparaten geprüft habe. Die wenigen Fälle, wo das
nicht geschehen konnte, wird man beim Lesen des Textes leicht
herausfinden.

Ich war lange schwankend, ob nicht auch ein Abschnitt über
die Technik des Mikroskopes hätte aufgenommen werden sollen;
schliesslich war der Umstand entscheidend, dass ein solcher, wenn
er nutzbringend sein sollte, das Buch unverhältnissmässig vergrössert
und also vertheuert haben würde. Hierfür muss demnach auf die
einschlägigen Werke, besonders das von Harting, verwiesen werden.

Nach meinen Erfahrungen wird der Werth mikroskopischer Beschreibungen wesentlich durch bildliche Darstellung erhöht; es war
daher mein Bestreben, diese in möglichster Reichhaltigkeit, theils
als Holzschnitte im Text, theils in den Farbentafeln zu geben. Dass
bei den ersteren manche fremde Zeichnung mit Angabe der Quelle
benutzt wurde, bedarf wohl keiner Entschuldigung. Bei Anfertigung
der Tafeln, auf denen sich nur eigene Zeichnungen finden, habe ich
mit Fernhaltung Alles dessen, was man Schematisirung derselben
nennen könnte, stets eine absolut objective Wiedergabe des mikroskopischen Bildes angestrebt. — Dass auf den Farbentafeln die
Mineralien der späteren Systeme gegenüber den amorphen und re-
gulären etwas stiefmütterlich behandelt worden sind, hat seinen

Grund darin, dass ursprünglich mehr Tafeln in Aussicht genommen waren. Doch beliefen sich die Kosten für Anfertigung derselben so hoch, dass ihre Zahl auf 10 beschränkt werden musste, wenn nicht allzu weit über das billige Maass hinausgehende Anforderungen an die dankbar anzuerkennende Opferwilligkeit des Herrn Verlegers gestellt werden sollten. — Von den Zeichnungen solcher mikroskopischer Verhältnisse, die schon in leicht zugänglichen Specialarbeiten eine graphische Darstellung gefunden hatten, konnte Abstand genommen werden. Ferner wurde darauf gesehen, solches Material als Object zu den Zeichnungen zu wählen, welches unschwer für Jeden zu beschaffen ist, damit der Lernende an selbstangefertigten Präparaten nach Anleitung des Buches seine Beobachtungen und Studien machen könne. Denn das muss man nicht vergessen: mit dem blossen Lesen und Studiren ist es nicht gethan; — wer mikroskopische Mineralogie lernen will, muss an den Schleiftisch und an das Mikroskop.

Die genaue und gewissenhafte Angabe der Literatur bei jedem Gegenstande, sowie die Zusammenstellung derselben am Schlusse des Buches, dürfte auch dem Fachmann nicht ganz unwillkommen sein und ist besonders darauf berechnet, dem Anfänger Gelegenheit zu geben, sich in die historische Entwicklung der Wissenschaft einzuleben. Eine eingehende Kenntniss der Geschichte der Wissenschaft scheint mir durchaus nothwendig, um den organischen Zusammenhang des Individuums mit der Gesammtheit herzustellen, durch welchen allein die fördernde Einheit und das klare Bewusstsein der anzustrebenden Ziele in die wissenschaftliche Entwicklung kommt. Ferner aber kann nur durch die historische Kenntniss seiner Wissenschaft jedem Studirenden das Seiende als ein Gewordenes erscheinen und ihn erkennen lassen, wie

> Alles sich zum Ganzen webt.
> Eins in dem Andern wirkt und lebt.

Sollte hie und da eine nennenswerthe Arbeit unerwähnt geblieben sein, so bitte ich das im Hinblick darauf zu entschuldigen, dass ja dem Einzelnen nicht alle Bücher und alle Zeitschriften zugänglich sind. Für Belehrung und Unterstützung in dieser Richtung würde ich in ganz besonderem Grade dankbar sein.

Im Ganzen und Grossen glaube ich, gestützt auf die Erfahrungen, die ich im akademischen Vortrage des Gegenstandes dieses Buches zu sammeln Gelegenheit hatte, einen nicht durchaus falschen Weg eingeschlagen zu haben, bescheide mich aber gern gegenüber dem Urtheil erfahrenerer Forscher, deren sachliche Kritik mir in hohem Grade willkommen sein wird.

Trotz aller angewandten Sorgfalt sind im Text einige Druck-fehler stehen geblieben, deren Verzeichniss angeheftet ist und die ich zu corrigiren bitte.

Schliesslich fühle ich mich gedrungen, meinem Freunde, Herrn Professor H. FISCHER in Freiburg, den aufrichtigsten Dank für die unermüdliche Bereitwilligkeit auszusprechen, womit er durch die Erlaubniss zur Benutzung des akademischen Cabinets, seiner Privat-bibliothek und seiner reichen Sammlung mikroskopischer Präparate meine Arbeit freundlichst gefördert hat.

Freiburg i. B. im Mai 1873.

H. ROSENBUSCH.

VORWORT ZUR ZWEITEN AUFLAGE.

Die gewaltigen Fortschritte, welche die mikroskopische Mineral-
diagnose seit dem Jahre 1873 gemacht hat, bedingten eine voll-
kommene Umarbeitung der ersten Auflage dieses Buches. Ander-
weitige Beschäftigungen und eine angestrengte Lehrthätigkeit ver-
zögerten die Erfüllung dieser Aufgabe fast über Gebühr.

Ob es mir gelungen ist, eine dem heutigen Standpunkt unserer
Wissenschaft entsprechende Darstellung des Gegenstandes zu geben,
muss ich der Beurtheilung berufener Fachgenossen überlassen. Die
hohe Vervollkommnung der Methoden und Instrumente, sowie eine
gewisse Verschiebung der Ziele und Gesichtspunkte bei mikroskopischen Mineraluntersuchungen, welche in dem letzten Decennium
sich vollzogen hat, bedingten manche durchgreifende Aenderung in
dem Plane, nach welchem die erste Auflage dieses Buchs be-
arbeitet war. Das rein Descriptive in derselben musste auf das
unumgänglich nothwendige Maass beschränkt, das Hauptgewicht auf
die Anleitung zu einer möglichst exacten mikroskopischen Bestimmung
der Mineralien gelegt werden. Immerhin durfte und sollte dieses
Buch kein Lehrbuch der Krystalloptik werden, es sollte ein Hülfs-
buch bei petrographischen Untersuchungen bleiben. Das be-
dingte in manchen Punkten eine Abweichung von den strengen
Methoden der Optik, über deren Berechtigung man verschiedener
Ansicht sein kann und wird. Ich habe mich bei der Behandlung
des Gegenstandes durch die Erfahrungen leiten lassen, welche ich
in dem Zusammenarbeiten mit lieben Schülern seit nunmehr 16
Jahren habe sammeln können.

Die auf mikroskopische Mineralogie bezügliche Literatur ist
eine so zahlreiche geworden und eine ihrem inneren Werthe nach
so verschiedene, dass eine volle Berücksichtigung derselben, wie
im Jahre 1873, nicht möglich war. Ich habe mich jedoch bestrebt,
dieselbe in solcher Vollständigkeit bei jedem Gegenstande zu geben,
dass der Leser ohne Schwierigkeit die historische Entwicklung der
Erkenntniss desselben verfolgen kann. Vor jedem Capitel steht das
allgemein Wichtige aus derselben, in Fussnoten das nur oder mehr

einzelfällig Bedeutsame. Nicht aufgeführt wurden die Lehrbücher
von F. Fouqué und A. Michel-Lévy, von Ern. Mallard, G. Tschermak und J. Verdet, die ich ausgiebig benutzt habe. Ihr hoher
Werth bedarf nicht meiner besonderen Anerkennung. — Einen nach
Möglichkeit vollständigen Literaturnachweis habe ich dem Buche
auf vielseitig mir ausgesprochenen Wunsch angehängt; ich folge
darin dem Rathe von lieben Fachgenossen, auf deren Urtheil ich
sehr grosses Gewicht lege, fast gegen meinen Willen.

An die Stelle der 10 chromolithographischen Tafeln der ersten
Auflage sind 26 Tafeln in Lichtdruck getreten; dieselben machten
viele lange Beschreibungen entbehrlich. Herr Professor Cohen in
Greifswald gestattete freundlich die weitestgehende Benutzung seiner schönen Mikrophotographieen. Auch zur Vervollständigung des
Literaturnachweises half er mir in bereitwilligster Weise. Ebenso
erfreute ich mich der liebenswürdigen Unterstützung und des be-
währten Rathes des Herrn Professor Klein in Göttingen Beiden
lieben Freunden danke ich auch an dieser Stelle nochmals herzlich.

Die Herstellung der Newton'schen Farbenskala war nicht ohne
Schwierigkeit; es erwies sich unmöglich, dieselbe mit andern, als
mit Anilinfarben auszuführen. Man wolle sie also thunlichst vor
Sonnenlicht schützen. Dieselbe wird hoffentlich die Benutzung der
Interferenz-Erscheinungen zur Mineralbestimmung dem Anfänger
wesentlich erleichtern.

Die Beschreibung des neuen Fuess'schen Mikroskops, welches
mir erst vor wenigen Tagen bekannt wurde, glaubte ich in einem
Nachtrage geben zu sollen.

Der Herr Verleger ist mit solcher freundlichen Bereitwilligkeit auf jeden meiner Vorschläge zur Ausrüstung dieses Buches,
ohne Rücksicht auf die dadurch erwachsenden Kosten und Mühen
eingegangen, dass ich mich ihm in hohem Grade verpflichtet fühle.

Allen meinen lieben Schülern und Freunden endlich, nah und
fern, die hier die Früchte ihres Fleisses verwerthet finden, möge
dieses Buch eine Erinnerung sein an die frohen Tage gemeinschaft-
licher Arbeit und Anregung!

Einige sinnstörende Druckfehler wolle man vor der Benutzung
des Buches corrigiren.

Heidelberg, September 1885.

H. Rosenbusch.

TABLE OF CONTENTS.

SPECIAL PART.

ABBREVIATIONS.

N. J. B. = Jahrbuch, or Neues Jahrbuch für Mineralogie, Geologie und Paläontologie. Stuttgart.

Z. D. G. G. = Zeitschrift der deutschen geologischen Gesellschaft. Berlin.

P. A. = POGGENDORF's Annalen für Physik und Chemie. Leipzig.

A. M. = Annales des mines. Paris.

S. W. A. = Sitzungsberichte der K. K. Akademie der Wissenschaften zu Wien.

B. M. = Monatsbericht der K. Akademie der Wissenschaften zu Berlin.

S. B. A = Sitzungsberichte der K. Akademie der Wissenschaften zu Berlin.

S. M. A. = Sitzungsberichte der K. bayrischen Akademie der Wissenschaften zu München.

T. M. M. Mineralogische Mittheilungen ges. von G. TSCHERMAK. Wien.

T. M. P. M. = Mineralogische und petrographische Mittheilungen ges. von G. TSCHERMAK. Wien.

A. Ch. Ph. = Annales de Chimie et de Physique. Paris.

A. Ch. Pharm. = Annalen der Chemie und Pharmacie. Leipzig.

Z. X. = Zeitschrift für Krystallographie und Mineralogie, her. von P. GROTH. Leipzig.

C. R. = Comptes rendus hebdomadaires de l'Académie française. Paris.

F. K. = Földtani Közlöny. Budapest.

Min. Mag. = Mineralogical Magazine. London.

Geol. Mag. = Geological Magazine, etc. London.

Q. J. G. S. = Quarterly Journal of the Geological Society. London.

B. S. M. and Bull. Soc. Min. Fr. = Bulletin de la Société minéralogique de France. Paris.

Bull. Soc. géol. Fr = Bulletin de la Société géologique de France. Paris.

G. F. i Stockh. Förhdl. = Geologiska Föreningens i Stockholm Förhandlingar. Stockholm.

MICROSCOPICAL PHYSIOGRAPHY

OF THE

ROCK-MAKING MINERALS.

The Microscopical Physiography of rock-making minerals describes the characteristics by which these minerals may be determined in thin section or in grains by transmitted light under the microscope.

It may be divided into two parts: a *general part*, in which the three great classes of mineral characteristics, the morphological, physical, and chemical, are applied to microscopical diagnosis; and a *special part*, which contains the particular description of each mineral species as it appears under the microscope.

PREPARATION OF MATERIAL.

Literature.

J. G. und L. G. BORNEMANN, Ueber eine Schleifmaschine zur Herstellung mikroskopischer Gesteinsdünnschliffe. Z. D. G. G. 1873. XXV. 367–374.

H. CLIFTON SORBY, On the microscopical character of sands and clays. Monthly Microscop. Journ. 1877. February 7.

G. STEINMANN, Eine verbesserte Steinschneidemaschine. N. J. B. 1882. II. 46–54.

J. THOULET, Note sur un nouveau procédé d'étude au microscope des minéraux en grains très-fins. Bull. Soc. Minér. Fr. 1879. II. 188.

H. VOGELSANG, Philosophie der Geologie und mikroskopische Gesteinsstudien. Bonn 1867, 225–228.

F. ZIRKEL, Mikroskopische Gesteinsstudien. S. W. A. 1863. XLVII. 227–229

The microscopical investigation of minerals or mineral aggregates is carried on by observing them by transmitted light, either in thin plates with parallel faces, called thin sections, or in the form of powders. In general, for optical diagnosis thin sections are the most convenient; while for microchemical determination a thin section is to be preferred in some cases, mineral powder in others.

The manner of preparing a thin section may be modified in many ways. In cases where it is desired to prepare thin sections in a particular direction through a mineral or rock, thin plates are cut by means of a stone-cutting machine, using a metal disk set with diamond-dust or emery.   If direction is of no consequence, it is better to chip with a hammer thin splinters or flakes from the material to be studied. These chips should not be less than half an inch in diameter, and without cracks or flaws.   One side of the chip or of the thin plate is ground plane and smooth.   The grindstone may be a fixed cast-iron plate, emery-stone, sandstone, or whetstone, with or without emery; but it is more convenient to use a small grinding-machine having a vertical axis, on which may be screwed horizontal grindstones of different coarseness.   Having prepared a plane surface which should extend across the whole chip, it is then polished, carefully washed with a stiff brush, and dried.

The chip is then fastened or cemented to a thick object-glass by means of Canada balsam.   The object-glass should not be too thin, as it will bend under the pressure of the fingers, and the edges of the rock section will grind away faster than the middle, producing a lenticular-shaped section.   The Canada balsam may be used in a viscous form, being handled with a glass rod, or after it has been hardened by evaporation.   An excellent cement, which is to be preferred to Canada balsam, is obtained by slowly melting together a mixture of 16 parts by weight of viscous Canada balsam and 50 parts of shellac, and keeping them heated for some time.   The mass, before it completely cools, may be drawn out in strings and rolled between the hands into convenient rods about 1 cm. thick and 20–30 cm. long.

The chip is cemented in the following manner: Spread over the thoroughly cleaned and gently heated object-glass a continuous coat of cement,—which should not be too thin,—at the same time heating the chip with the polished side up to drive off any moisture, and then lay it with the polished side on the balsam.   The object-glass is then gradually heated till the balsam loses most of its turpentine, care being taken that no bubbles form.   On cooling, press the chip in place.   When the whole is thoroughly cold, the exposed surface is ground down as before.   The grindstone is used as long as possible with safety to the section, which in most cases becomes nearly or quite transparent.   It is then ground on the whetstone until it is completely transparent.   In place of the whetstone a smooth glass plate may be used with water, together with the finest possible emery-dust floated off from previously used emery.

The thinness required in a particular case depends on the object of the study, and may be determined by examining the section from time to time with the microscope during the final process of grinding, the section being first moistened with water. When the grinding is finished the superfluous balsam around the rock section is removed with a heated knife-blade, and the section is thoroughly washed with a stiff brush and alcohol, rinsed quickly in water, dried with a linen cloth, and then brushed.

A new object-glass having been thoroughly cleaned, a drop of Canada balsam is put upon it and gently warmed so that it spreads slightly. The old object-glass is then gradually heated till the balsam is melted, when the rock section is slid on to the new glass and the balsam heated so that the section adheres to it. Another drop of balsam is put upon the section, and the thinnest possible glass cover placed over it. The whole is gradually heated and the glass cover pressed closely down. The superfluous balsam is removed by a warm knife-blade and alcohol, rinsing thoroughly with water.

In many cases, where the material to be studied is in the condition of small particles (sand, volcanic ashes, etc.), or where its composition does not permit of the preparation of a thin section (clay, mud, etc.), or, finally, where the mineral elements of a rock have been separated for individual study, the choice of methods for the preparation of material for observation depends on whether the outward form or the internal structure is the special object of investigation. In either case it is advisable to use powder of very nearly equal grain. This is easily obtained by freeing the entire powder of dust by repeated washing in water and decantation, and then passing it through a graduated series of sieves.

If the outward form of the powder is to be studied, it is best to place it in a fluid whose index of refraction is considerably lower than that of the solid body. Water is therefore used, care being taken not to suspend too much powder in the water. The whole is covered with a glass to give it a plane surface.

On the other hand, if the internal structure is to be studied, one must avoid the total reflection from the surface of the solid which occurs with water, and employ a medium whose index of refraction is as near that of the solid body as possible. The substances most frequently used are glycerine ($n = 1.46$), almond oil ($n = 1.47$), cassia oil ($n = 1.606$), or a concentrated solution of iodide of potassium and mercury ($n = 1.733$). Canada balsam ($n = 1.54$) is not so satisfactory, for the powder is apt to crowd together when the balsam is melted. This may be avoided by spreading grains of powder over a thin film of cold balsam, heating slightly so the grains will adhere, and covering all with balsam dissolved in ether or chloroform, and with a glass cover.

GENERAL PART.

MORPHOLOGICAL CHARACTERS.

I. Crystals and Crystal Sections.

Literature.

G. Wertheim, Ueber eine am zusammengesetzten Mikroskop angebrachte Vorrichtung zum Zweck der Messung in der Tiefenrichtung und eine hierauf gegründete neue Methode der Krystallbestimmung S. W. A. Math.-naturw. Classe. 2. Abth. Bd. XLV. 157–170. 1862.

J. Thoulet, Procédé pour mesurer les angles solides des cristaux microscopiques. Bull. Soc. min. Fr. 1878. I. 68.

E. Bertrand, De la mesure des angles dièdres des cristaux microscopiques. C. R. LXXXV. 1175. 1877.

— De l'application du microscope à l'étude de la minéralogie. Bull. Soc. min. Fr. 1878. I. 22.

The extraordinary importance of the morphological characters of minerals for their macroscopical determination is greatly reduced for their determination microscopically.

Complete crystal bodies are only seen under the microscope in particular cases, as in isolated material with very small crystals and as so-called individualized interpositions; in other cases only the cross-sections of crystals have to be considered, and as the position of these sections bears no regular relation to the crystal axes, it is readily seen that, with the endless number of possibilities for the cutting plane, neither the outline of the cross-section nor the relation of its angles are of any absolute value in determining the crystal.

If the direction in which the section cuts a crystal were known, it would not be difficult to calculate the form of its outline and its angles; but in most cases there is no basis for such a calculation. Where the optical phenomena establish the position of the optical constants in a crystal section, a more or less definite conclusion can be drawn as to the position of the section, by a proper combination of these with the outlines of the cross-sections and other characters, such as the course of the cleavage; and this conclusion may be so far substantiated by angle measurements on the cross-sections, that it may be considered as practically established.

Moreover, a statistical method of procedure frequently furnishes valuable criteria; for although there may be an endless number of section planes, yet the probability of the occurrence of all of these is not the same for each, but is essentially dependent on the relative dimensions of the crystals and on the arrangement of them in the rock. So if those cross-sections which occur most frequently are properly combined, the form of the crystal may be determined without difficulty.

If, for example, quadratic and six-sided, colorless sections of a mineral should be found to preponderate over all others, the interpretation would have an exceedingly wide scope. But if, on investigation in polarized light, it should be found that both kinds of sections remained dark in every position between crossed nicols, one would rightly conclude that they must belong to an isotropic or isometric mineral which crystallized in the form of rhombic dodecahedrons—most likely a mineral of the garnet or the haüyne groups.

If, however, the tetragonal sections (parallel OP) remained dark between crossed nicols, while the hexagonal ones (parallel to the principal axis through a combination ∞P . P) were generally light, they would belong to a mineral crystallizing in the quadratic system—possibly to the scapolite group.

On the other hand, they would belong to the hexagonal crystal system, if the hexagonal sections (parallel OP) remained dark between crossed nicols, while the quadratic ones (parallel to the principal axis through a combination ∞P . OP) were generally light, and the mineral would probably be nepheline or apatite.

If, finally, both quadratic and hexagonal sections were for the most part light between crossed nicols, they might probably belong to an orthoclastic feldspar, which had been so cut that the section in one case passed about parallel to the orthopinacoid, and in the other was so inclined to the base as to pass through the anterior and posterior prism faces.

It need scarcely be mentioned that the correctness of these conclusions in many cases may be placed beyond doubt by further optical determinations, by measurements of angles, by comparison of the microstructure of the questionable cross-sections with those of known material, and finally by microchemical tests.

The measurements which may be made with the microscope relate either to linear extension, to plane angles, or to solid angles. Linear extension is measured by means of an ocular micrometer. This consists of a glass plate on which a sufficiently fine scale has been engraved with a diamond. Generally a millimetre is divided into 10 parts, whole millimetres being separated by long marks, half ones by medium lines, and tenths by short ones. With an ocular micrometer one does not measures the object itself, but its image. In order to determine the actual value of a division of the ocular micrometer for a particular system of objectives, a glass plate with fine divisions (object micrometer) is placed under the objective, and the relation between the two scales is established. If the ocular micrometer is divided into tenths of a millimetre, and the object micrometer into hundredths of a millimetre, and three divisions of the former cover one division of the latter, then with this system of lenses one division of the ocular micrometer will correspond to an actual extent in the object of 0.0033 mm.

The measurement of a plane angle is made by placing the apex of the angle to be measured on the centre of the cross-wires in the ocular; and since the stage of a petrographical microscope is made to rotate accurately about the optical axis of the microscope, the sides of the plane angle are covered in turn by the cross-wires, and the amount of rotation is read off on the graduated circle of the stage.

The measurement of solid angles, which is fully treated in the German edition and by the authors already cited, is here omitted.

II. NORMAL AND ABNORMAL CRYSTALLIZATION.

a. The External Form.

Literature.

H. BEHRENS, Die Krystalliten. Mikroskopische Studien über verzögerte Krystallbildung. Kiel. 1874.

M. L. FRANKENHEIM, Ueber das Entstehen und das Wachsthum der Krystalle nach mikroskopischen Beobachtungen. Pogg. Ann. CXI. 1860.

O. LEHMANN, Ueber physikalische Isomerie. Z. X. 1877. I. 97–131.

— Ueber das Wachsthum der Krystalle. Z. X. 1877. I. 453–496 (auch als Beilage zum Programm des Gymnasium zu Freiburg i. B. 1877).

F. LEYDOLT, Ueber die Krystallbildung im gewöhnlichen Glase und in den verschiedenen Glasflüssen. S. W. A. 1852. Math.-naturw. Classe VIII. 261–277.

LINK, Ueber die Bildung der festen Körper. Berlin. 1841.

H. VOGELSANG, Ueber die mikroskopische Structur der Schlacken und über die Beziehungen der Mikrostructur zur Genesis der krystallinischen Gesteine. Pogg. Ann. 1864. CXXI. 101–125.

— Philosophie der Geologie. Bonn. 1867.

— Sur les crystallites. Arch. Néerland. V. 1870; VI. 1871; VII. 1872.

— Die Krystalliten. Nach dem Tode des Verfassers herausgegeben von F. ZIRKEL. Bonn. 1875.

IF an aqueous, molten, or gaseous solution contains crystallizable compounds under conditions (saturation) which make their separation or secretion possible, the development of crystals will begin when there is sufficient mobility of the molecules, and will continue as long as the conditions are favorable for their separation and regular grouping. Their number and size will depend on the number of centres of crystallization and the quantity of the material contributing to the growing crystal.

Every growing crystal will exert an attracting and directing influence upon those molecules in the solution which are capable of entering into the composition of the crystal, and are within the sphere of its molecular forces; and it will grow through their accession and arrangement. In this way there arises about every growing crystal a mantle of solution which is poorer in matter pertaining to the crystal, and to which crystallizable molecules are constantly being supplied through diffusion out of the saturated mother-liquor, and from which they are being withdrawn by their incessant addition to the crystal.

So long as this process continues normally, the growing crystals will be bounded in every stage of their growth by continuous plane faces. If we transfer this process to a molten solution, and imagine that the condition of saturation ceases with respect to a substance separating out, and that at about the same time the movement of the molecules is gradually hindered by the increasing viscosity of the solution, then at some particular moment the diffusion of the crystallizable compound from the mother-liquor into the space of crystallization (Krystallisationshof) (Pl. I. Fig. 1) would cease; yet in the immediate vicinity of the crystal, that is, within this space, so much heat is liberated by the passage of the accessory molecules into a solid state, that crystallizable molecules within this space can attach themselves to the crystal from the space of crystallization.

Consequently, after the complete cessation of crystallization the space of crystallization will be noticeably poorer in the crystallizable compound than the mother-liquor. If there was a tendency in this compound to color the mother-liquor, then after the solidification of the whole the crystal will be surrounded by a space which is lighter colored than the mother-liquor. This phenomenon is often observed in porphyritic rocks, and Pl. II. Fig. 6 exhibits the same around augite in the obsidian of Hammarsfjord. If the centres of crystallization are sufficiently far apart, and the process of crystallization ends while there is mother-liquor still present, then the boundaries of the crystals formed will be essentially determined by their proper laws of formation (morphology); if one or both of the above conditions are not fulfilled, then the perfecting of each single individual will be hindered and disturbed by those lying next to it, and there will result a more or less irregular crystalline aggregate.

If the accession of matter through the diffusion currents into the space of crystallization is very abundant and accelerated, then certain parts of the crystal, particularly those to which a greater portion of this space is tributary, will grow faster than other parts. O. Lehmann has called attention to the fact that in the growth of a crystal the

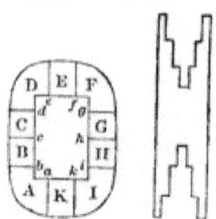

edges and corners would have an advantage over the same-sized part of the faces, as is illustrated in Fig. 1, in which *ab*, *bc*, *cd*, *de*, *ef*, etc., represent equal-sized portions of the faces of a growing crystal, and *A*, *B*, *C*, etc., the part of the space of crystallization tributary to each portion of the faces.

Fig. 1   Fig. 1a   If the growth of the crystal ceases during the exuberant growth of the points and edges, then its outline will be in the form of a ruin or of steps, or will be indented, as Fig. 1a shows; but each boundary element will be parallel to every other equivalent boundary element. Such forms are quite frequent phenomena among the feldspars, augites, hornblendes, olivines, etc., of porphyritic eruptive rocks.

The growth of a crystal takes place in essentially the same manner, as soon as any of the above-mentioned normal conditions of growth are in any way disturbed. If the necessary mobility of the crystallizable molecules in the solution is wanting because of the too rapid evaporation of the solvent, or because of its too great viscosity, its too strong adhesion to the containing walls (object-glass and cover-glasses, for instance), or from any other circumstances; or if the mobility ceases too soon, or if the necessary saturation with the crystallizable compound is lacking in any or in all parts of the space of crystallization,—then there must occur disturbed crystallizations, and forms arise which are designated in general as *forms of growth*.

In porphyritic and glassy eruptive rocks, forms of growth which have been produced by too great viscosity of the magma are of frequent occurrence. The explanation and most of the nomenclature of these extremely variable structures are derived from the studies of H. Vogelsang, and have been elaborated by the work of O. Lehmann.

If a solution of sulphur in carbon bisulphide, which has been thickened with Canada balsam, is spread on an object-glass, in a short time larger and smaller spherules separate out, which are strongly refracting and are saturated drops of the sulphur solution. By evaporation these lose their solvent, and finally become solid. Vogelsang saw in these amorphous, round, drop-like forms the elementary bodies of crystals, and called them *Globulites* (Pl. I. **Figs.** 1 to 4). If the solution dries up about the time the globulites are formed, they suffer no further change. But if the solution preserves sufficient mobility for some time, currents set in, by which the globulites change their place, and are sometimes aggregated in quite irregular heaps, *Cumulites* (Pl. I. Fig. 5), sometimes in more or less regular structures. Frequently they arrange themselves in rows like strings of pearls (Pl. I. Fig. 3), which **Vogelsang** called *Margarites.*

Moreover, globulites increase in volume by coalescing with one another (Pl. I. Fig. 2). So long as the resistance of the solvent is not too great the enlarged globulite retains the spherical form; otherwise there arise cylindrical, disk-like, or sharply conical and crooked forms, which are classed together as *Longulites* (Pl. I. Fig. 3). Globulites and longulites, as well as their manifold aggregations with one another, do not possess the characteristics of crystals. **Vogelsang** named them collectively *Crystallites*, and found them singly refracting so long as their elements preserved the globulitic form, or their more complex forms did not exceed the stage of globulitic aggregation.

With sufficient mobility of the solution the **supersaturated** drop or globulite does not solidify as such, but takes the form of the orthorhombic sulphur pyramid at the **instant** of solidification. This is especially noticeable when the globulites have been driven by the currents on to normally developed sulphur crystals or forms of growth, with which they have grown into skeleton crystals with several axes. Upon the loss of the globulitic form and the accession of a crystallographic boundary double refraction regularly appears.

The researches of H. **Vogelsang** and his successors explain those appearances in rocks which are so closely related to the artificial productions. These products of incomplete crystallization naturally occur in the more or less basic porphyritic eruptive rocks which have not reached a holocrystalline development. Indeed in many of these rocks the residuum of crystallization, called the base, is entirely made up of these kind of crystal structures. The globulites (Pl. I. Fig. 4), which in rocks poor in silica are generally strongly colored or opaque, and in the siliceous rocks are usually clear and transparent, occur uniformly disseminated, or strung together into margarites (Pl. II. Fig. 1). In place of the round or disk-shaped globulites, or beside them, are longulites; and the closer study of many obsidians and vitrophyres reveals an endless variety of all imaginable intermediate forms between the loosely strung margarites and crystal needles. The crowding together of globulites, and irregularly massed or more or less regularly arranged groups, are confined principally to the most acid siliceous rocks (quartz-porphyries and rhyolites). Cumulites in which there is a radial arrangement of the single globulites are called *Globospherites* (Pl. I. Fig. 6). Since in such globospherites there is a constant diminution in density from the centre toward the periphery, interference phenomena appear in certain instances in polarized light which are analogous to those in spherulites (sphärokrystalle).

Those crystalline structures, or forms of growth, which have developed beyond the stage of globulitic aggregates but have not attained complete perfection of form, are exceedingly manifold. With all their variation in appearance, they agree in that they are not composed of elementary bodies, and in that they often possess the physical characters of crystals in a recognizable manner, or permit them to be conjectured from their form. Their most important forms may be characterized as *microlitic* structures.

Trichites (θρίξ = hair), according to Zirkel, are hair-like crystals whose length greatly exceeds their breadth; they are often more or less twisted, and even bent in loops. They have a great tendency to arrange themselves in many-armed groups about a central grain (Pl. II. Fig. 2). They usually appear opaque, because of their small diameter and the consequent total reflection of transmitted light.

Spherulites (*sphärokrystalle*) form another group of incipient forms of growth closely related to trichites. They include a great part of the spherulites which occur in different porphyritic rocks. They are homogeneous spherical crystal structures, which are radially fibrous, in some cases with a rough surface, in others with a more or less smooth one (Pl. II. Fig. 4). The needles composing a spherulite are not always simple crystal needles, but are sometimes many-branched forms arising from the repeated splitting of a simple needle into two or more slightly diverging arms, which are in turn split up. Pl. III. Fig. 4 shows a variety of spherulites of feldspar in a trachytic rock from the Caucasus.

While the trichitic structure appears in general to be confined to the more basic rock constituents rich in iron, and therefore to relatively older periods in the development of a rock, the spherulitic structure belongs to the more acid, feldspathic, or feldspar-like constituents, poor in iron, and to comparatively late periods of the rock formation.

Skeleton crystals, strictly speaking, are those crystallizations which have not produced entire and complete individuals, but have led to crystallographically parallel or symmetrical aggregates of small individuals; the latter may be arranged throughout their whole extent as a single individual or as a twinned one. Pl. III. Fig. 2 shows skeleton crystals of magnetite, and Pl. III. Fig. 3 those of augite. Pl. III. Fig. 1 shows an intermediate form between a crystal and a skeleton crystal of olivine.

The name *microlite* ($\mu\iota\kappa\rho\sigma\varsigma$ = small; $\lambda\iota\theta\sigma\varsigma$ = stone) may be applied to more or less completely defined crystals, without reference to their habit or to their optical behavior, and which are only recognizable microscopically, and cannot be specifically determined. If their nature is determinable, they are called by the specific name, together with an expression indicating their habit; as, for example, lath-shaped feldspar, augite prisms, mica plates, perofskite octahedrons, etc. Microlites are true crystals, as is proven by their form.

Besides the deformation of crystals which has been produced by conditions attending their growth, the microscope reveals a number of others which have affected completed individuals and are due to mechanical and chemical processes.

To the *mechanical deformation* of crystals belong the cracking and breaking apart of the older secretions, porphyritic crystals, which occurs so frequently in porphyritic rocks. It is recognized by the irregular broken outline of the mineral section across the surface of fracture. Elastic minerals like mica exhibit bending instead of breaking. Pl. III. Fig. 5 shows a broken feldspar crystal; Pl. III. Fig. 6 a bent mica plate.

Another group of deformations of rock constituents due to mechanical processes is met with, especially in greatly faulted and uplifted mountain masses, and is evidently occasioned by the dynamical processes of mountain-building. From the fact that these deformations take place in solid rock under pressures exerted on all sides, they do not generally appear as great alterations of the outward form, but more as internal displacement of parts of a crystal with respect to one another. These deformations, therefore, are often first recognized in polarized light by the greater or less variation of the optical orientation of the parts of the crystal.

Examples of this kind of deformation are found in the bending of the twin lamellæ of triclinic feldspar (Pl. IV. Fig. 6); in the varying position of the axes of elasticity in particular parts of feldspar and quartz crystals, which shows itself by the shadowy and rapidly shifting extinction over the section during its rotation between crossed nicols. Through greater pressure the crystal may be more or less broken up

(Pl. IV. Figs. 1, 2), or the crystal outline disappear altogether (Pl. IV. Figs. 3, 4).

It has been very frequently observed that a rock constituent which in its normal condition is optically uniaxial, becomes biaxial after undergoing great pressure. The natural occurrence of pressure figures in mica plates may also be referred to this cause (Pl. IV. Fig. 5).

Chemical deformations appear in many ways among the older secretions of porphyritic eruptive rocks. It lies in the conception of a crystal and in the conditions under which porphyritic secretions are formed, that they should possess regular crystallographic boundaries. When therefore the normal outward boundary is wanting, it must have been lost through secondary action. Since the production of porphyritic secretions belongs to an early stage in the history of a rock, and follows the laws which obtain for crystallization from a mixed solution, it is possible to imagine that through changes in the chemical composition or physical condition of the mother-liquor (rock magma) the older secretions are no longer able to exist, but must dissolve again to make room for other crystallizations. The older secretions are therefore melted again, and if the process of resorption is interrupted by further changes or by the solidification of the magma before their complete fusion, there result rounded grains in place of the sharp-edged crystals. Often this corrosion in its earliest stages appears to have been one-sided, as is shown for quartz (Pl. V. Fig. 1), and for nosean (Pl. V. Fig. 2).

If the crystal substance which is corroded and dissolved by the magma is converted immediately into new crystalline forms, as is so often the case with the micas and hornblendes of eruptive rocks, there arise no properly corroded crystals, but pseudomorphs after the dissolving crystals which grow from the border inward, and which will be described in another place.

b. The Internal Structure, or Homogeneity.

Literature.

DAVID BREWSTER, On the existence of two new fluids, etc. Transactions of the Royal Society of Edinb. T. X. 1, Auszug daraus Edinb. Phil. Journ. vol. IX. 94 u. 268. — Sehr vollständiger Auszug, z. Th. Uebersetzung in Pogg. Ann. VII. 1826. 469.

HUMPHRY DAVY, On the state of water and aëriform matter in cavities, found in certain crystals. Philos. Transactions 1822, in französischer Uebersetzung in Annales de chimie et de phys. T. XXI. 1822. 132.

TH. ERHARD und ALFR. STELZNER, Ein Beitrag zur Kenntniss der Flüssigkeitseinschlüsse im Topas. T. M. P. M. I. 1878. 450–458.

C. W. Gümpel, Ueber die mit einer Flüssigkeit erfüllten Chalcedonmandeln (Enhydros) von Uruguay. S. M. A. 1880. II. Math.-phys. Classe. 241–254.
— Nachträge zu den Mittheilungen über die Wassersteine (Enhydros) von Uruguay und über einige süd- und mittelamerikanische sog. Andesite. ibidem 1881. I. 321–268.
W. N. Hartley, On the presence of liquid carbon dioxide in mineral cavities. Journal of the Chemical Society. London. 1876. I. 137–143.
— On variations in the critical point of carbon dioxide in minerals and deductions from these and other parts. ibidem 1876. II. 237–250.
— Observations on fluid cavities. ibidem 1877. I. 241–249.
— On attraction and repulsion of bubbles by heat. Proceed. Roy. Soc. XXVI. 137. 1878.
— On the constant vibration of minute bubbles. ibidem XXVI. 150. 1878.
G. W. Hawes, On liquid carbon dioxide in smoky quartz. Amer. Journ. 1881. XXI. 203–209.
Al. A. Julien, On the examination of carbon dioxide in the fluid cavities of Topaz. Journ. of the Amer. Chem. Soc. III.
W. Prinz, Les enclaves du saphir, du rubis et du spinelle. Ann. de la Soc. belg. de microscopie. 1882.
H. Cl. Sorby, On the microscopical structure of crystals etc. Quart. Journ. of the Geol. Soc London 1858. Nov. vol. XIV. 453–500 und andere Arbeiten desselben Verfassers, cf. Literaturnachweis.
H. Vogelsang und Geissler, Ueber die Natur der Flüssigkeitseinschlüsse in gewissen Mineralien. Pogg. Ann. vol. CXXXVII. 1869. 56 und Nachtrag zu dieser Abhandlung von Vogelsang. ibid. 257.
Arth. W. Wright, On the gaseous substances contained in the smoky quartz of Branchville, Conn. Amer. Journ. 1881. XXI. 209–216.

Theoretically, the substance of a crystal should be of unbroken continuity and perfectly homogeneous; but these qualities seldom exist together in nature.

Zonal Structure.—From the fact that the growth of a crystal from a solution is not always a single continuous act, but is at times interrupted by longer or shorter intervals of inaction, there arises a shelly structure, which in cross-section produces the appearance called zonal structure.

This is particularly well shown in many zircons, frequently in the feldspars of trachytes and andesites, and in the nepheline and leucite of the more basic lavas (Pl. V. Fig. 3).

Where the shelly individuals are minerals which may be regarded as isomorphous mixtures of several molecular groups (garnet, tourmaline, pyroxene, amphibole, mica, etc.), then the successive shells often differ in chemical composition. If the isomorphous, laminated compounds are colored, the variation in composition is frequently recognized by the different colors of the separate zones (Pl. V. Figs. 4 and 5). The form of the different shells naturally depends on the manner of growth in the crystal.

The lines of zonal structure are usually parallel to the outlines of the crystal, or to certain of its outlines. Consequently, if the outer form of a crystal has been destroyed by resolution, its character may be reasonably inferred from the zonal structure. Yet cases occur in which this parallelism is not present, and the zonal structure indicates another crystal form from that shown by the outline of the individual.

Since the optical behavior of a substance, independent of pressure and temperature, is a function of its molecular composition, it follows that the value and position of the axes of elasticity and of the optical axes, as well as the pleochroic relations, may differ in the various shells of a crystal with isomorphous lamination.

Inclusions.—The term *discontinuity of crystal substance* may be applied to a group of phenomena which arise from the fact that the space occupied by a crystal is not entirely filled by the crystal substance, but in part by bodies foreign to it. All these so-called foreign bodies are classed as inclusions or interpositions, and may be divided into *unindividualized* inclusions and *individualized* inclusions, according as the inclusion consists of amorphous substances in any state of aggregation whatever, or of crystallized bodies. The first are produced by the growing crystal taking up particles of the magma or gases and fluids contained in it; the second arise through the inclusion by the growing crystal of pre-existing crystallizations, or of those which were being secreted at the same time from the magma. Experience with the artificial production of crystals has shown that interpositions are taken up more abundantly by growing crystals as their growth is more rapid.

According to the character of unindividualized inclusions they are divided into *gas inclusions, fluid inclusions,* and *glass inclusions.*

Gas Inclusions.—Gas inclusions are also called gas and vapor cavities: they are recognized chiefly by their outward appearance. Because of the great differences in the indices of refraction of solid and gaseous bodies, these inclusions must appear glistening by incident light, but in transmitted light like small spots with large dark borders (Pl. VI. Fig. 1). This phenomenon, which results from the total reflection of the rays of incident light, may be observed on any gas bubbles, as those which rise in soda-water or champagne, or occur too frequently in the Canada balsam of thin sections.

The form of gas inclusions varies greatly, but round and elliptical shapes predominate, along with which occur irregularly jagged, bay-shaped, branching, and other forms. Less frequently they appear as negative crystal cavities, that is, with a polygonal boundary corresponding to the crystal form of their host (*Wirth*). Gas cavities seldom occur isolated, but are generally grouped in rows and planes through the substance of a mineral, and with low magnifying power appear as a local clouding of the mineral.

The secretion of crystals can take place from aqueous solutions, and from those which are fluid when melted, or by sublimation, and in all three ways they may acquire gas cavities. Their formation in crystals resulting from sublimation needs no special explanation. It is well known that water at different temperatures absorbs different amounts of various gases; from this the enclosure of primary gas inclusions follows naturally.

Secondary gas cavities can occur in certain hydrogenous crystals, if original fluid inclusions evaporate, as not infrequently happens in the case of minerals with very perfect cleavage.

The great capacity of melted fluids to dissolve gases is an established fact. As soon, however, as cooling sets in with the consequent solidification the absorbed gas must be liberated; hence the "sprouting" of silver, the porous structure of lavas, etc. This explains the presence of gas cavities in glassy, solidified fluids, like obsidian, and in pyrogenous minerals, like nepheline and others. There is little definite knowledge concerning the chemical nature of the gases filling such cavities; or whether they are always filled with gas, and are not sometimes in the case of glassy bodies simply contraction phenomena. Whether the cavities are filled with gas, and with what kind, depends also upon the permeability or impermeability of the walls of the cavities. The amethyst of Schemnitz has been found to be impermeable to gases, while calcite is always found to be permeable.

If the enclosed gases possess a certain tension, or possessed it at the time of their inclusion, while the enclosing body had a certain plasticity of substance, as with glasses and other amorphous bodies, then the pressure exerted by it would induce a molecular strain in the solid substance, which might lead to the phenomena of double refraction, which do not otherwise occur in amorphous bodies, but which may be produced in them artificially by the application of external pressure. A similar disturbance of the normal optical properties of a crystallized matrix also may result from the tension of enclosed gases.

Fluid Inclusions.—Fluid inclusions, like gas inclusions, occur mostly in groups, and like them are usually arranged in lines and along planes, which in some cases pass irregularly through the crystal, in others are arranged more or less closely in accord with the crystallographic constants.

The shape of fluid inclusions is extremely variable. Besides the round and elliptical forms, which are most frequent, are cylindrical, club-shaped, pear-shaped, quite irregular, and often branched forms. Not infrequently they have a plane polyhedral boundary, which is determined by the crystal form of their host. Thus the fluid inclusions in rock salt are cubical, in calcite often rhombohedral, in quartz dihexahedral, and so on (Pl. VI. Figs. 2 and 3). Their dimensions vary greatly, so that in the same crystal, besides fluid inclusions, which may be recognized as such with the naked eye, are those which when highly magnified appear only as a clouding of the mineral substance.

Since the index of refraction of fluids differs less from that of solid bodies than that of gases does, the dark border about fluid inclusions produced by the total reflection of the transmitted light will be generally narrower than that about gas inclusions. But sometimes the indices of refraction of the fluid and of the mineral containing it are very different; moreover, the breadth of the dark border depends not only on the relative indices of refraction of the two substances, but on the shape of the fluid inclusion, whose bounding plane, if greatly inclined to the line of vision, may produce a border as broad as that of a gas inclusion; therefore the distinction based on this character is not absolutely certain.

The fluid may either completely or partially fill the cavity in which it occurs, as shown in Figs. 2 and 3, Pl. VI. In the latter case the appearance differs with the ratio between the amount of fluid and the size of the cavity. If there is much fluid present, so that the cavity is nearly filled, then the remainder will be occupied by the vapor of the fluid, or by another gas in the form of a round bubble. The border of this bubble within the fluid is broad; that of the fluid within the solid, narrow. If, however, the volume of the fluid is quite small compared with that of the cavity containing it, and if the fluid does not wet the substance of the crystal (Pl. VI. Fig. 6), then the fluid forms a drop surrounded by an envelope of its vapor or of another gas. In this case the fluid apparently forms a bubble with dark border next to the vapor envelope, which in turn is bounded by a still broader margin. The presence of a bubble naturally prevents the confusion of fluid and gas inclusions.

The bubbles in fluid inclusions may arise from the contraction of the fluid after its enclosure in the crystal, or from the condensation of vapor after its inclusion, or be due to the fact that both vapor and fluid were imprisoned at the same time. In the first instance there should be a constant ratio between the volume of the bubble and the fluid in all the inclusions of one individual, which is seldom observed in nature.

The fluids included in crystals are almost always colorless; occasionally they have a yellowish color, and rarely an orange color. The bubbles occurring in fluid inclusions often show a spontaneous movement, in some cases swinging slowly back and forth, in others hurrying about in a lively dance. The mobility appears to be greater the smaller the bubble; large bubbles generally remaining stationary. A motion may be produced artificially in many cases by heating one end of the inclusion.

The mobility of the bubbles is a sufficient proof of the fluid condition of the inclusions in which they occur, and constitutes an important distinction between these and glass inclusions. It is not to be assumed that all fluid inclusions in minerals are primary, for it is easy to imagine that original gas inclusions, or secondary cavities produced by chemical action, may be filled with fluid through capillary crevices.

Chemical and physical investigation of the contents of fluid and gas inclusions has shown that they vary greatly, both as to the nature of the material and the tension under which it exists. The fluid is usually water, carrying more or less of other substances in solution; in some cases it is petroleum. The gas has sometimes the composition of ordinary air; is often carbon dioxide, nitrogen, or a mixture of gases. Instances are frequent, especially in the quartz of granites and crystalline schists, in rock crystals, topaz, beryl, etc., where the fluid inclusions contain double bubbles, one within the other. These have been shown to consist of liquid and gaseous carbon dioxide in water, the water wetting the walls of the cavity, and the liquid and gaseous carbon dioxide occupying the central part of the space; the liquid carbon dioxide envelops the gaseous when the amount of the former is relatively great, and both take the spheroidal form. On the other hand, the gaseous carbon dioxide separates the liquid carbon dioxide from the water when the relative proportions are reversed. The position of the broad and narrow borders produced by total reflection generally distinguishes these two cases from one another.

The presence of crystalline secretions of various kinds in the fluid inclusions of very different minerals has been confirmed by many observers. The strikingly widespread occurrence of cube-like crystals in the fluid inclusions of quartz of the greatest variety of crystalline rocks and in many other minerals is specially to be noted (Pl. VI. Fig. 5). These are probably sodium chloride in some cases, but they cannot always be referred to this mineral.

The conditions under which crystalline bodies separate out of fluid inclusions are quite analogous to those under which a glass inclusion is converted into a devitrified inclusion. In the case of fluid inclusions they may be briefly summarized as physical changes in the fluid which prevent its retaining the dissolved salts longer in solution. In exactly the same manner crystalline separation may take place out of gas cavities when the enclosed sublimation products cool.

Glass Inclusions.—Solidified portions of the once molten magma are often found enclosed in minerals which crystallized out of melted solutions. These are called glass inclusions (glass cavities: Sorby) when the solid is amorphous, and slag inclusions (stone cavities : Sorby) when it has a more or less crystalline development, whether this accompanied the consolidation of the inclusion or was subsequent to it. The shape of these glass and slag inclusions is just as irregular and manifold as that of gas and fluid inclusions, and they often possess the form of their host (Pl. VII. Figs. 1 and 2). The color of glass inclusions varies with their chemical composition, and especially with the iron percentage of the rock glass. They are usually colorless when they occur in the minerals of the acid eruptive rocks, but are very often colored yellow, red, or brown in those of basic rocks. Very frequently these glass inclusions contain one or more darkly margined bubbles, which are not moved by changes of temperature ; and often the glass particle is fairly riddled by a great number of bubbles. The immobility of the bubbles and the presence of several in one glass inclusion are the best distinctions between these and fluid inclusions (Pl. VII. Fig. 3). The occurrence of bubbles in glass inclusions arises from the presence of gases in the molten magma, which were enclosed along with the glass.

Individualized Inclusions.—The occurrence of individualized inclusions, inclusions of one mineral in another, was a well-known fact in the case of transparent minerals before the introduction of the microscope. Microscopical investigation has only demonstrated the very wide distribution of this kind of interpositions, and placed in a clear light their significance for the results of chemical analyses. Many optical phenomena also have been explained by their presence. as the schillerization of crystals, asterism, etc. Only those foreign crystals which are older than the enclosing mineral or are contemporaneous with its growth are called interpositions. Infiltrations in cracks and products of the decomposition and alteration of a mineral are not considered inclusions.

In many cases there is no particular relation between the arrangement of crystalline interpositions and their crystal host (Pl. VII. Fig. 4). Yet we know from the macroscopic parallel growth of many minerals (rutile with hematite, hematite with mica, etc.), as well as through the investigations of Frakenheim on crystallization, that a crystal can exert a directing influence on crystals of a different kind which grow upon it. There is also frequently recognized among microscopic crystalline interpositions a definite arrangement of these with respect to one another and their host (Pl. VII. Fig. 5).

Another kind of orderly arrangement of inclusions is determined, not by a crystallographically directing force, but by mechanical conditions, namely, the rate of growth. It is their accumulation in certain parts of the host while other parts of it are relatively or entirely free from them. This regularity applies to all varieties of inclusions. Three kinds of orderly arrangement are recognized: central (Pl. VII. Fig. 6), peripheral (Pl. VIII. Fig. 1), and zonal (Pl. VIII. Fig. 2). In the central arrangement the inner portion of the crystal is full of inclusions, the outer more or less free from them. In the peripheral the case is the reverse. In the zonal arrangement the inclusions lie on the surface of concentric shells of the crystal.

The amount of individualized inclusions in a crystal is often so great, that one may speak of a mutual penetration of two or more materially and morphologically different substances. Such a mutual penetration of quartz and acid feldspars is especially common: it presents a peculiar appearance, characteristic of certain members of the quartz-porphyry group, and is the so-called *micropegmatite* or *granophyre* structure (Pl. VIII. Fig. 3). The same intergrowth is frequently observed between different members of the feldspar group (microcline, albite, orthoclase) in the older massive rocks, where it is usually controlled by rigid mutual crystallographic relations. The basic massive rocks also exhibit similar phenomena, as, for example, when the larger porphyritic augites are so filled with apatite, magnetite, mica, nepheline, haüyne, etc., that the augite substance only forms a cement, as it were, for the different minerals (Pl. VIII. Fig. 6). This structure has been called *poicolitic*.

c. Twins.

The twinning of minerals is recognized microscopically either by the occurrence of reëntrant angles in the outline of the section or by optical phenomena. The occurrence of reëntrant angles only characterizes certain varieties of twinning, and then is only observed when the outlines of the crystals are regular. The optical phenomena in polarized light prove the presence of twinning in all cases, except in minerals of the isometric system and in certain twins with parallel axes. The discussion of the optical phenomena in twinned crystals will be found in a later part of the book.

d. Aggregates.

Under the term *aggregates* are here included only those mineral aggregations which are homogeneous, or which cannot be shown to be heterogeneous. They may consist of amorphous or of crystalline substances; but since their chief characteristic is their optical behavior, they cannot be properly described before the optical properties of minerals in thin section have been discussed. They will therefore be considered at the end of the chapter on that subject.

PHYSICAL PROPERTIES.

AMONG the physical properties of minerals their cohesion and behavior towards light are specially useful in microscopical studies.

I. PHENOMENA OF COHESION.

Cleavage.—Through the shattering consequent upon grinding, cracks and crevices are formed in many minerals, the sharpness and more or less continuous course of which depends on the greater or less perfection of the cleavage in the particular mineral, while their direction corresponds to the intersection of the cleavage planes with that of the section.

Cohen states that by heating thin sections to redness cleavage cracks sometimes arise, which did not make their appearance during the grinding. The more perfect the cleavage of a mineral is, the more closely crowded, uninterrupted, and sharp will be the cleavage cracks in its thin section. In less perfectly cleavable substances the cleavage cracks are less frequent, and it is highly characteristic of some minerals that the cracks often stop in the middle of a section and reappear in another parallel plane, while an irregular fissure connects the two cleavage cracks. The perfection of the cleavage cracks depends also on the angle at which the section cuts the plane of cleavage. They are sharpest when these directions are at right angles to one another. In an inclined position the cleavage cracks often appear broad and dark, because of the total reflection from the capillary layer of air between their walls; their margins are sometimes very finely indented.

Cleavage cracks, especially in colorless minerals, may often remain undetected by full illumination,—that is, in a strong light,—and first become evident with dull illumination, which is obtained by depressing the polarizer and its lens, or the condenser when using strongly convergent light.

The fracture of a mineral has no corresponding microscopical phenomenon: the irregular cracks and fissures in cleavable and uncleavable minerals either result from aggregation, following the outlines of the individuals composing the aggregate, or correspond to previously existing internal fractures, which in many cases have been developed by mechanical pressure while the minerals were part of a mountain mass; in other cases they have been produced by chemical processes, for example, the cracking of olivine through serpentinization, etc.

The course and the relative position of cleavage cracks depend on the direction in which the section cuts the mineral. A *pyramidal* cleavage always furnishes four systems of parallel cleavage cracks, which intersect at angles dependent on the position of the section. A good example is anatase. An exception to this rule would occur in the case of cleavage after a holohedral hexagonal pyramid, which is not met with among the petrographical minerals. *Prismatic* cleavage furnishes two (in the hexagonal system three) systems of parallel lines, which cross one another so long as the section lies at right angles or inclined to the axis of the prism, but which are all parallel when the section is parallel to this axis. *Pinacoidal* cleavage, the cleavage face being parallel to two axes, gives parallel lines in all sections, except in the case of the regular cube (the isometric system).

Cleavage parallel to several pinacoids would produce the same effect as prismatic or pyramidal cleavage, but would be distinguished from these by the unequal perfection of the cleavage cracks parallel to the different pinacoids. Pl. IX. Figs. 5 and 6, Pl. X. Figs. 1–6, and Pl. XI. Figs. 1–3 present different degrees of perfection and the mutual position of cleavage cracks.

It is evident that one can calculate the angle at which the cleavage planes must intersect when the position of the section and the normal cleavage angle are known. In the same manner, from the cleavage angle measured for a particular case the position of the section may be determined when the zone in which it lies is known, which is often recognized with approximate certainty by optical means.

Gliding planes (Gleitflächen) and *pressure planes* (Druckflächen) also give rise to cracks which from their appearance cannot be distinguished from cleavage cracks. Up to the present they have been recognized only in mica and cyanite (Pl. IV. Fig. 5); but it may be stated that they have a wider distribution, and that certain planes of parting (Absonderungsflächen) observed in the pyroxene, amphibole, and feldspar groups may be considered as pressure planes.

The investigation of chemical cohesion by means of etched figures in petrographical work is somewhat hindered by the dependence of the form of the etched figures on the position of the surface etched, and by the uncertain determination of the position of this plane in thin sections. Nevertheless it is serviceable in particular cases, which will be mentioned under the description of the chemical properties and in the second part of this book.

II. Optical Properties.

a. *Refraction and Index of Refraction in Isotropic Media.*

Literature.

BABINET, Ueber die optischen Kennzeichen der Mineralien. Comptes rendus 1837. I. 758 und Auszug in Pogg. Ann. XLI 115. 1837.

A. DES CLOIZEAUX, Mémoire sur l'emploi du microscope polarisant et sur l'étude des propriétés optiques biréfringentes propres à déterminer le système cristallin des cristaux naturels ou artificiels. Ann. des Mines VI. 1864 and Pogg. Ann. CXXVI. 1865.

— De l'emploi des propriétés optiques biréfringentes en minéralogie. Paris. 1857

— Nouvelles recherches sur les propriétés optiques des cristaux naturels ou artificiels et sur les variations que ces propriétés éprouvent sous l'influence de la chaleur. Mém. prés. à l'Institut impérial de France. T. XVIII. 1867.

Optics teaches us that light is transmitted in a straight line without change of direction in one and the same homogeneous medium as vibrations of particles of the luminiferous ether, which take place at right angles to the direction of transmission. There are media in which the velocity of transmission of the light is independent of the direction in which it is propagated : these are called *isotropic*. In other media the velocity of transmission changes with the direction : these are called *anisotropic*.

Gaseous, fluid, and amorphous (glassy) bodies, and those crystallizing in the regular system, are isotropic ; on the other hand, substances crystallizing in the quadratic, hexagonal, orthorhombic, monoclinic, and triclinic systems are anisotropic.

The independence of the rate of transmission of light of its direction in isotropic media leads to the conclusion that the distribution and elasticity of the luminiferous ether is the same in all directions throughout such media. The absolute magnitude of the elasticity of this ether in materially different media is different, and since the rate of transmission (velocity) of the light is proportional to the square root of the elasticity of the ether, light will be transmitted in different isotropic, homogeneous media at different rates (with different velocities). Thus there are optically denser and optically rarer media.

In consequence of these different optical densities, a ray of light is generally diverted from its former direction in passing out of one medium into another and always experience a change in its rate of transmission. This is termed the *refraction of light*.

The phenomena connected with the passage of a ray of light out of one homogeneous *isotropic medium* into another homogeneous *isotropic medium* of different optical density are the following (air being taken as the medium out of which the ray of light comes): Let ab be the bounding plane between the air and the second isotropic body (Fig. 2), fc the incident ray of light, and de the normal to ab at the point of incidence c. Then will the light reaching c from the direction fc in part pass into the second medium in a changed direction and with different velocity (the *refracted ray*), part of it will return in the first medium according to a definite law (the *reflected*

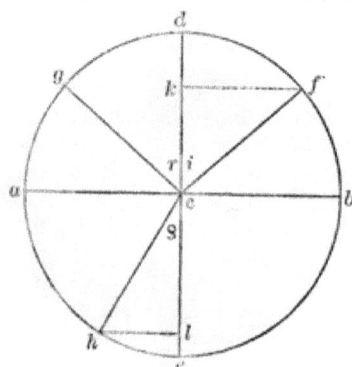

Fig. 2

ray), and part will be scattered irregularly in all directions or be diffused. Most of this diffusion of the light would not take place if ab were a mathematical plane, since it arises from an unevenness of the surface. If one calls the angle which the incident (cf), the reflected (cg), and the refracted ray (hc) make with the normal ecd, the angle of incidence $= i$, angle of reflection $= r$, and angle of refraction $= \rho$; and further, the planes through each of these rays and the normal, the plane of incidence, plane of reflection, and plane of refraction, respectively, then there exists between these quantities the following relations:

(1) The planes of incidence, reflection, and refraction fall together.

(2) The angle of incidence is equal to the angle of reflection; $i = r$.

(3) The angle of incidence and angle of refraction bear a constant relation to one another.

Describing about c a circle with cf as radius and letting fall from f and h (fk and hl) perpendicular to the normal dee, then $\sin i = kf$ and $\sin \rho = hl$. Then whatever be the direction of the incident ray, that of the corresponding refracted ray (the media remaining the same) is so conditioned that the quotient of the sine of the angle of refraction into the sine of the angle of incidence is a constant quantity (n or μ), which is called the *index of refraction* or coefficient of refraction. Thus the third relation may be precisely formulated:

$$\frac{\sin i}{\sin \rho} = n.$$

The index of refraction n is therefore a constant, which can be employed in the determination of a substance, just as the specific gravity or any other constant. By the term "index of refraction," as ordinarily used, is understood the index of refraction of an isotropic medium compared with air; and since the index of refraction of air compared with a vacuum varies with the thermometer and barometer, it is dependent on temperature and pressure; at 760 mm. pressure and 0° C. temperature, $n = 1.000294$.

The index of refraction of isotropic media compared with a vacuum is called their *absolute index of refraction*. In actual practice and under all the conditions of pressure and temperature found at the surface of the earth the index of refraction may be considered unchangeable. This index of refraction for almost all fluid and isotropic solid media lies between 1 and 2, and seldom exceeds the latter figure. For example, it is 1.336 for water, 1.498 for rock salt, 1.553 for glass, 1.435 for fluorite, 2.270 for diamond. In passing into an optically denser medium the incident ray is bent toward the perpendicular, when into an optically rarer medium it is bent from the perpendicular.

Finally, the amount of deflection upon the passage of a ray from air into another medium is dependent on the wave-length of the incident ray; it is consequently different for different-colored rays, and is inversely proportioned to the wave-length. Thus the index of refraction for blue rays is greater than for red, $n_v > n_\rho$.

This phenomenon is called the *dispersion of light*. Its amount is different for different media, and is measured by $\frac{n_\rho}{n_v}$. Moreover, the amount of difference in the dispersion for particular colored rays— yellow and green, for instance—holds no general relation to the total dispersion, but is different and characteristic for each part of the spectrum in each and every substance.

From the ratio $\frac{\sin i}{\sin \rho} = n$, when n and i are known, the direction of the refracted ray may be calculated. Among all possible values for i there are three of special importance, namely, $i = 0°$, $i = 90°$, and $\tan i = n$.

If $i = 0°$, the incident ray coincides with the perpendicular, and $\dfrac{0}{\sin \rho} = n$; that is, the angle of refraction $= 0°$, and the transmitted ray coincides with the perpendicular. Thus when the incident ray is perpendicular to the bounding plane there is no deflection of the transmitted ray, only a change in its rate of propagation.

If $i = 90°$ (grazing incidence), $\dfrac{1}{\sin \rho} = n$ or $\sin \rho = \dfrac{1}{n}$. This value of the angle of refraction is called the *limiting or critical angle;* for water this is 48° 35′, for flint glass 37° 36′, for diamond 23° 53′. From the general law that a motion follows the same way back as forth, a ray of light from a denser medium coming upon a rarer medium at the critical angle continues parallel to the bounding plane between both media, that is, at right angles to the normal.

If the light from a denser medium strikes a rarer one at a greater angle than the limiting angle it cannot pass into the latter, but will be reflected from the bounding plane. Since, in distinction to the previously mentioned reflection, no part of the light in this case enters the second medium, this latter reflection is called *total reflection*. This cannot take place on the passage of light from a rarer into a denser medium.

The above-mentioned relations explain certain phenomena which are very frequently observed in the microscopical investigation of minerals and rocks. If we imagine any particular substance enclosed in another of exactly the same color and index of refraction, then the boundaries of the enclosed substance against the surrounding one could not be observed at all. On the other hand, the enclosed substance would have the highest degree of transparency in all its parts. Therefore if it is desirable to see the outward form of a substance with the greatest possible sharpness it must be immersed in a medium with as different an index of refraction as possible (air or water). But if it is desired to observe particularly the internal characters of the substance, an envelope with as nearly the same index of refraction as possible should be chosen (oil and other strongly refracting fluids, or Canada balsam). If substances with various indices of refraction immersed in the same envelope of water, oil, or solid are studied simultaneously, the surface of one appears smooth and even, while that of another is rough and wrinkled. The latter are said to be shagreened. The surface of that substance will appear smooth whose index of refraction is smaller or equal to that of the envelope, for all of the rays coming out of it can pass through the surrounding substance. If, however, the enclosed body is more strongly refracting than its envelope, there will be many rays which will strike the rough surface, produced by incomplete polishing of the section, at angles greater than the limiting angle, and these will suffer total reflection, in consequence of which the surface of the substance is visible because of a diminution of the light. One and the same substance will therefore show a smooth surface in certain envelopes and a rough one in others, so when the index of refraction of the enclosing substance is known the refraction of the enclosed substance may be inferred.

Strongly refracting minerals appear more glaring or clearer in contrast to less refracting ones, because the amount of light striking any point of the former becomes concentrated into a smaller part of the

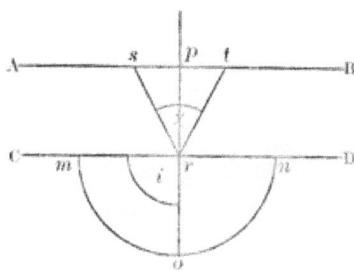

Fig. 3

surface. If there falls on the point *r* of the lamella *ABCD* (Fig. 3) a hemispherical bundle of rays *mon*, the same become within the lamella a cone of rays *srt*, the radius of whose base *pt* has the following relations:

$$\frac{\sin i}{\sin \frac{x}{2}} = n.$$

For $i = 90°$, $\sin \frac{x}{2} = \frac{1}{n}$. The circular base, therefore, is smaller in proportion as the index of refraction of the lamella is larger, and consequently the illumination becomes stronger, in fact, in proportion to the squares of the indices. On the other hand, the boundary of the more strongly refracting body against the less refracting must appear the darker in proportion as the difference between their indices of refraction is greater, because the critical angle becomes the smaller and the total reflection occurs so much the sooner.

For this reason gases enclosed in solid or liquid bodies have very broad total reflection borders, while those borders for fluid-inclusions, *cæteris paribus*, are smaller, and for inclusions of solids within solids still smaller. These relations are made use of in distinguishing gaseous, fluid, and solid inclusions in minerals from one another, and the breadth of the total reflection border of gas bubbles compared with the size of the clear centre of the same may be employed to determine the size of the index of refraction of the enclosing substance.

Among the various microscopical methods used for determining the index of refraction of isotropic media is the following: If one focuses the objective of a microscope exactly on any point, and then slips between this and the objective a refracting medium,—for example, a glass plate with parallel faces,—then the object which was distinctly seen before is no longer visible, or not distinctly so, and the objective of the microscope must be raised a certain amount in order to see the point as distinctly as before. The extent to which the point in question appears to be raised depends on the thickness of the inserted plate and its index of refraction. Let o (Fig. 4) be the point in air; if the lamella L be

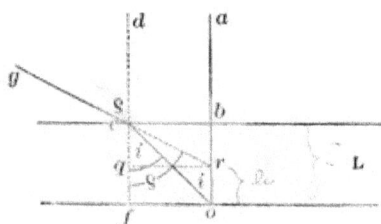

Fig. 4

placed over it, then a ray oba will reach the objective through the lamella with unaltered direction, but with altered velocity. A ray oc, on the other hand, will be deflected at c to cg, since it passes into air, and will appear to come from r. In r, where both rays intersect, the point o will apear to be; it is therefore raised a distance ro. Placing $ro = h$, $ob = D$, we have, if df is the perpendicular at c and rq is at right angles to df,

$$\tan i = \frac{cb}{D} \text{ and } \tan \rho = \frac{qr}{qc} = \frac{cb}{D-h};$$

therefore,

$$\frac{\tan i}{\tan \rho} = \frac{D-h}{D}.$$

And since for small angles, which are those met with in microscopical observation the tangents and sines may be interchanged, this equation becomes

$$\frac{\sin i}{\sin \rho} = n' = \frac{D-h}{D},$$

for the passage of light from the lamella into air. For the passage of light from air into the lamella, i and ρ are reversed, and we have

$$n = \frac{D}{D-h}$$

The process of measurement may be varied in a great variety of ways for particular cases. It must not be forgotten that the accuracy of the process increases with the closeness of the focusing; the sharpest possible test-objects should therefore be chosen, and the highest practicable magnifying power. The weak point of the method is the determination of the thickness of the lamella, because this is seldom the same for all parts of the lamella, and should be determined at the point where h is to be measured.

To determine the index of refraction in thin sections where the lamellæ lie between Canada balsam and glass, it is best to make use of the law readily derived from the foregoing, that the apparent thicknesses of two equally thick lamellæ are inversely proportional to their indices of refraction, $n_1 : n = D' : D_1'$, and $n_1 = n \cdot \dfrac{D'}{D_1'}$. There is placed on the same glass, alongside of the thin section of the substance to be investigated, a thin section of the same thickness of a substance whose coefficient of refraction (n) is known, or another known substance in the first thin section may be used, or finally the Canada balsam itself, if its coefficient be known. The apparent thickness D_1' of the substance in question is measured, and also the difference, d, between the focusing on the test-object seen through the lamella under investigation, and through the known lamella. And since this difference may be positive or negative according as the first lamella is more strongly or more weakly refracting than the known lamella, we have

$$n_1 = n \frac{D_1' \pm d}{D_1'},$$

in which n_1 alone is unknown.

In spite of the fact that the apparent thickness of a lamella is smaller the larger its index of refraction, strongly refracting substances in rock sections stand out in relief from the web of less refracting substances around them, and one can judge after a little practice of the relative indices of refraction of any two substances from their greater or less relief as compared with one another. This apparently contradictory phenomenon is the result of several circumstances. The more glaring illumination of the surface of strongly refracting lamellæ combined with the marginal total reflection causes their surface to appear nearer than that of less glaring lamellæ; moreover, the fact that their under surface appears more raised combined with the consciousness that both are of equal thickness increases the impression that the upper surface projects in relief.

Through the refraction of light upon its passage out of one isotropic medium into another, not only its direction and rate of transmission are changed, but another phenomenon shows itself to a greater or less degree, which is called the *polarization of light.* Since light is propagated as vibrations of ether particles at right angles to its direction, and since there are innumerable normals to a line in space, the particles of ether may vibrate in an endless number of planes during the propagation of ordinary light. On the other hand, polarized light is that which is propagated as vibrations of the ether in a single plane. The difference between these two kinds of light cannot be detected by the unaided eye. Polarized light may be recognized by its being in some cases completely absorbed by doubly-refracting, absorbing media, by its not being reflected from mirrors under certain conditions, nor resolved (analyzed) by doubly-refracting media under particular conditions.

The reflection and refraction in isotropic media are among the many processes through which ordinary light becomes polarized; that is, has all the oscillations of the vibrating luminiferous ether reduced to one azimuth.

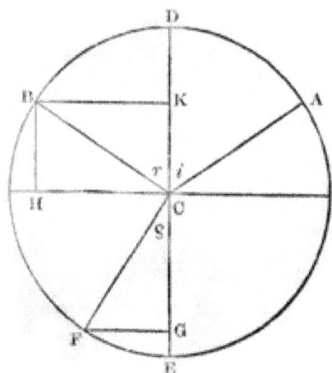

Fig. 5

A partial polarization of light takes place with every reflection and refraction. But when the reflected ray stands at right angles to the refracted one, both of these rays are polarized, the reflected one is completely polarized when the substance is transparent, and the planes of polarization are perpendicular to one another.

The angle of incidence for which the reflected and refracted rays are at right angles to one another is one peculiar to every substance, and naturally depends on its index of refraction.

Let AC (Fig. 5) be the incident ray and $\sphericalangle BCF = 90°$. Then $\tan i = \tan r = \dfrac{BK}{KC}$; since $BK = \sin r = \sin i$ and $KC = FG = \sin \rho$,

then $\tan i = \dfrac{\sin i}{\sin \rho} = n.$

Therefore the reflected and refracted rays are polarized at right angles to one another when the tangent of the angle of incidence is equal to the index of refraction. This angle is called the *polarization*

angle. It is assumed that the plane in which the vibrations of the reflected ray take place stands at right angles to the plane of reflection,
also called the plane of polarization; then the plane of vibration of the
refracted ray is the same as the plane of reflection.

b. Double Refraction in Anisotropic Media.

If one imagines a luminous movement to take place from any
point within an isotropic medium, then this will advance (be transmitted) in all directions with the same velocity, and the wave-surface
at any moment will be the surface of a sphere whose radius is proportional to the time which has elapsed since the beginning of the movement. But if the luminous movement starts from a point within an
anisotropic medium, in which the velocity of transmission varies with
the direction, it will advance in different directions with different
velocities; and the wave-surface can no longer be a sphere, but will be
a warped surface, whose form and position stands in the closest connection with the molecular structure of the anisotropic medium.

If now a ray of ordinary light from air, an isotropic medium, falls
on an anisotropic medium and penetrates it, since there are in the anisotropic medium different elasticities corresponding to all the possible
azimuths in which the vibrations of the ray of ordinary light take
place, then, for perpendicular incidence, these vibrations will be reduced to the two azimuths, which correspond to the directions of
greatest and least elasticity lying at right angles to the direction of
transmission of the ray. There arise therefore out of the incident ray
two rays, which in this particular case are transmitted in approximately
the same direction, with oscillations perpendicular to one another and
with different velocities, since their vibrations correspond to different
elasticities. Both rays are polarized because their vibrations in each
case lie in one azimuth, and they are polarized at right angles to one
another, because the directions of greatest and least elasticity cannot be
other than at right angles to one another within the plane normal to the
direction of the transmission of the incident ray. If both these rays
emerge into air through a surface parallel to that through which they
entered, no deflection will take place; but if the face of exit is inclined to that of entrance, the two rays which reach the exit face with
different velocities pass into the air at different angles.

If the incident ray coming through air strikes the surface of an
anisotropic medium not perpendicularly, but obliquely, the two rays
obeying the laws of elasticity in the anisotropic medium will traverse

it in different directions. The incident ray will therefore be separated into two different rays deflected or refracted to different degrees, and for this reason anisotropic media are also called *doubly-refracting*.

All crystalline bodies, not belonging to the isometric system, are anisotropic or doubly-refracting media, and therefore possess the common property of generally separating an incident ray into two, which traverse these bodies with different velocities, in different directions, and with planes of vibration or of polarization at right angles to one another. The characteristics and phenomena connected with the distribution of the elasticity of the ether vary according to whether a crystal belongs to a system with a principal axis (tetragonal and hexagonal) or to one without a principal axis (orthorhombic, monoclinic, triclinic).

Double Refraction in Crystals with a Principal Axis..

If one assumes—and this assumption explains the dioptric phenomena of uniaxial crystals—that the distribution of the particles of ether, as well as those of the mass, is symmetrical with respect to the principal axis, it follows that at right angles to this axis the disposition and consequent elasticity of the ether must be the same in all directions, and that in a direction parallel to the principal axis the elasticity must have a maximum difference from this; moreover, the elasticity in a direction which makes an angle $\phi < 90°$ with the principal axis must be intermediate between the first two, its amount depending on the angle ϕ, and it must be the same for all directions which have the same inclination to the principal axis.

From this it follows that if the square root of the elasticity of the ether at right angles to the principal axis and parallel to it be represented by two bisecting lines normal to one another, and a circle be described with radius equal to half the square root of the elasticity at right angles to the principal axis and an ellipse be formed about the two bisecting lines,—that is, one whose diameters correspond to the square root of the greatest and least elasticity,—and this be rotated about the diameter corresponding to the square root of the elasticity parallel to the principal axis, the resulting ellipsoid of rotation will represent the distribution of the elasticity of the ether in the crystal. This ellipsoid of rotation is also called the *ellipsoid of elasticity*.

Both the velocity and direction of vibration of the rays produced by the double refraction in a uniaxial crystal are obtained by passing a plane through the centre of the ellipsoid at right angles to the incident ray. The vibrations of both rays take place parallel to the great est and least diameters of this cross-section, and the lengths of the two diameters express the velocity of the rays.

Now the section through the ellipsoid at right angles to the principal axis (or axis of rotation) is a circle, and in this all the diameters are equal. Rays, then, which enter the crystal parallel to the principal axis suffer neither refraction nor polarization, but the light traverses the crystal in this case just as it would through an isotropic medium. The principal crystallographic axis is consequently a direction of simple refraction, and is for this reason called the *optic axis*. Thus it optically and morphologically a singular axis.

Every other section through the ellipsoid would be an ellipse which would be the more elongated the greater the angle made by the incident ray and the principal axis, and which would approach a circle as this angle diminished. In all these ellipses one diameter (the equatorial) remains the same, and is equal to the diameter of the circular section. That one of the two refracted rays which vibrates parallel to this diameter advances with a velocity which is independent of the direction of transmission, and is the same in all directions; it behaves like a ray in an isotropic medium, except for its polarization, and has a constant index of refraction. It is called the *ordinary ray (O)*; the plane passing through the incident ray and the principal axis (optic axis) is called the *principal optic section (optische Hauptschnitt)*, then the vibrations of the ordinary ray lie perpendicular to the principal optic section.

The second diameter of every possible section naturally lies in the principal optic section, its length is dependent on the inclination of the section to the principal axis, and is therefore variable. The ray vibrating parallel to this diameter—that is, in the principal optic section—will therefore traverse the crystal with a velocity varying with the angle of incidence, consequently it has no constant coefficient of refraction. This latter, in fact, must vary the more with the direction since it is inversely proportional to the velocity of transmission; it approaches the value of the index of refraction of the ordinary ray as the angle between the incident ray and the principal axis approaches zero, and reaches a maximum when this angle becomes 90°. This ray which vibrates in the principal optic plane is called the *extraordinary ray (E)*. When the index of refraction of the extraordinary ray is spoken of, it is understood to be the index of refraction for incidence perpendicular to the principal axis, and is designated by the letter ε while the index of refraction of the ordinary ray is ω. Naturally and ε are dependent on the wave-length, and therefore change with the color of the light, as n does for isotropic media.

As the principal crystallographic axis may be longer or shorter than one of the secondary axes, so the elasticity in the direction of the primary axis may be greater or smaller than at right angles to it. If the primary axis is the direction of greatest elasticity, the crystal is said to be *optically negative* or repulsive, and the extraordinary ray is less strongly refracted than the ordinary ray ($\omega > \epsilon$) and advances with greater velocity. *Optically positive* or attractive crystals are those for which the reverse relation holds; for these, then, $\omega < \epsilon$.

The optical character of crystals with a principal axis (tetragonal and hexagonal) which distinguishes them from those of the isometric system and from amorphous bodies is their double refraction, and that which distinguishes them from crystals of the remaining systems is the presence of a single optic axis coincident with the primary axis. Tetragonal and hexagonal crystals are collectively called *optically uniaxial crystals*. The optic axis is characterized by the fact that all rays propagated parallel to it traverse the crystal with the same velocity; that all rays vibrating at right angles to it advance with equal velocities in all directions; and finally, that every plane passing through the optic axis is a plane of symmetry for the ellipsoid of elasticity.

Double Refraction in Crystals without a Primary Axis.

The phenomena connected with the transmission of light through a crystal of the orthorhombic, monoclinic, or triclinic systems show that the distribution of the elasticity of the ether is not symmetrical with respect to a point, as in an isotropic medium; nor is it symmetrical with respect to a line, as in uniaxial doubly refracting media. It is, however, symmetrical to three planes. If the elasticity of the ether perpendicular to one of these planes is the greatest in the crystal, then there must be within this plane a direction which is parallel to the smallest elasticity of ether within the crystal; and, moreover, at right angles to this direction of least elasticity, there must be a direction in the same plane which corresponds to an intermediate elasticity. The distribution of the elasticities of the ether within a crystal not having a primary axis may be referred to three directions at right angles to one another, which are called the three axes of elasticity, and are distinguished as the axis of greatest elasticity (a), axis of mean elasticity (b), and axis of least elasticity (c). The form of the wave-surface of light (surface of elasticity) transmitted in crystals which are without a primary axis is derived in the following manner:

Let the length of the lines a, b, and c (Fig. 6) be proportional to the square root of the axes of greatest, mean, and least elasticity. Suppose that from the point of intersection, o, a luminous movement advances in all directions, and let us follow this movement in the plane bc. In the direction oc two rays will advance with different velocities, of which

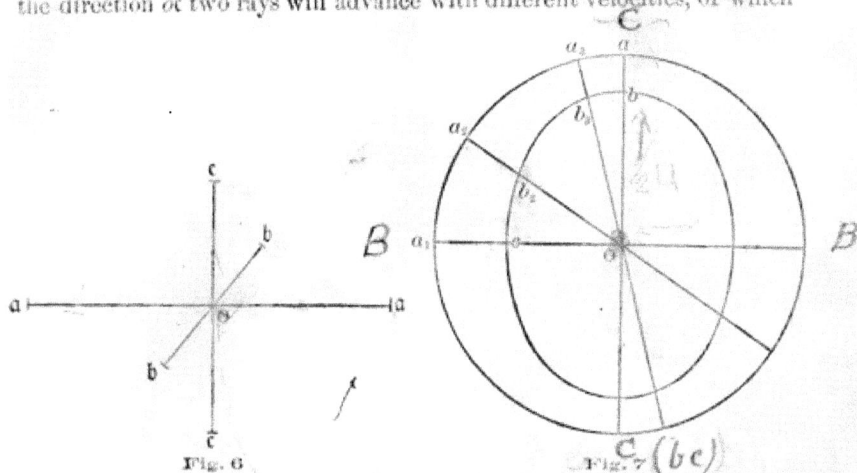

Fig. 6   Fig. 7 (bc)

one vibrating parallel to a will reach a in a unit of time, if (Fig. 7) $oa = \frac{1}{2}a$, while the second ray swinging parallel to b will reach b in a unit of time, if $ob = \frac{1}{2}b$. In the same manner, in the direction ob two rays will advance, of which one swinging parallel to a will traverse the distance $oa_1 = oa = \frac{1}{2}a$, in a unit of time, while the second swinging parallel to c will traverse $oc = \frac{1}{2}c$. In every other direction within the plane bc (Fig. 6) two rays will advance, one of which always swinging parallel to a, will traverse a distance oa_2, oa_3, etc., $= oa = \frac{1}{2}a$; while the second swinging parallel to an elasticity lying between b and c (and perpendicular to a) will traverse a distance equal to ob_2, ob_3, etc., if about o (Fig. 7) an ellipse be described with the half-axes $ob = \frac{1}{2}b$ and $oc = \frac{1}{2}c$.

Following in the same manner the movement of the light in the plane ab, we see that in the direction ob (Fig. 6) two rays proceed, one of which swinging parallel to a will traverse a distance $oa = \frac{1}{2}a$ (Fig. 8) in a unit of time, while the other swinging parallel to c will trav-
... ion oa (Fig.
... ce $oc_1 = \frac{1}{2}c$
... other direc-
... aneously, of
... nces oc_2, oc_3,

while the other will advance with a velocity $> \mathfrak{b} < \mathfrak{a}$, and will therefore pass over the distances $ob_{,,}$ $ob_{,,}$ etc., when $o\mathfrak{b} = \frac{1}{2}\mathfrak{b}$ and $oa = \frac{1}{2}\mathfrak{a}$ form the half-diameters of the ellipse.

Finally, for the movement in the plane $\mathfrak{a}\mathfrak{c}$ (Fig. 6) we find that from the point o in the direction oa, two rays are transmitted, one of which vibrating parallel $\mathfrak{b}$ traverses ob (Fig. 9) in a unit of time, while the second, vibrating parallel $\mathfrak{c}$, traverses oc. In the direction $o\mathfrak{c}$, the ray vibrating parallel $\mathfrak{b}$ will traverse ob, and that swinging parallel a will traverse oa. For the movement in every direction in the plane $\mathfrak{a}\mathfrak{c}$ we shall obtain the corresponding velocities if we describe about o a circle whose radius $o\mathfrak{b} = \frac{1}{2}\mathfrak{b}$, and an ellipse with the half-diameters $oa = \frac{1}{2}\mathfrak{a}$ and $oc = \frac{1}{2}\mathfrak{c}$, and draw radii in the direction in question. Since

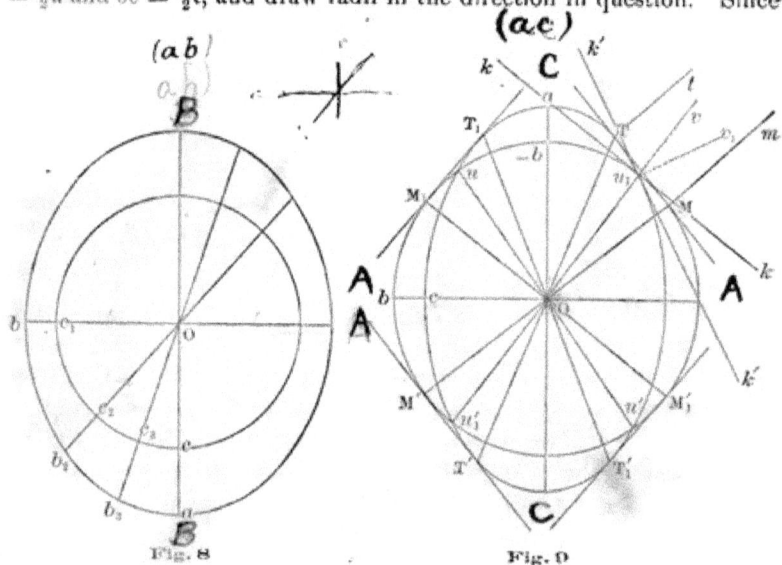

Fig. 8   Fig. 9

the diameter of the circle equals the square root of the mean elasticity, and the diameters of the ellipse equal the square root of the greatest and of the least elasticities, the circle and ellipse must intersect in four points.

The two rays traversing the crystal in the direction ou, have the same velocities, but different wave-surfaces (kk and $k'k'$), and therefore upon exit from the crystal will be differently refracted. On the other hand, the rays along oM and oT have the same wave-surface, when TM is tangent to both curves; they will therefore advance in the direction Tt, Mm as a single wave. The same is true of all the rays lying in the surface of a cone whose angle is ToM, for a plane through

TM is tangent to the surface of the ellipsoid at the exit of all these rays, and its contact with it is a circle whose diameter is *TM*. Therefore all the rays from *o* to the circumference of this circle have the same wave-surface, and will upon their exit advance as a hollow cylinder of rays. And since all rays traversing the crystal in the direction *Mm* or *Tt* possess but one wave-surface, then on emerging from the crystal they will not experience any double refraction. The direction normal to the plane *TM* being one in which rays traverse the crystal, and emerge without being doubly refracted, is called an *optic axis*. Therefore the directions *oM* and *oM,* are the optic axes, and such crystals are called *biaxial*. Moreover, a plane-wave coming from an isotropic medium in a direction normal to the tangential plane *TM* must produce in the biaxial medium a cone of rays which will emerge again as a cylinder of rays; the optic axes then are also called axes of the *inner conical refraction*.

The two wave-surfaces which emerge from the crystal at *u,* have different directions *u,v*, *u,v,*. which are normal to the tangent planes *kk* and *k'k'* ; they diverge, and, together with all those whose directions are normal to all the tangents to the surface of the ellipsoid at the point *u,,* give rise to a hollow cone of rays analogous to conical internal refraction. This phenomenon is known as *conical external refraction*.

This characteristic, as well as the fact that the optic axes of a biaxial medium are not axes of symmetry of the ellipsoid of elasticity, distinguish them essentially from the optic axis of uniaxial media. The movement of the light for every plane which does not pass through two axes of elasticity of the triaxial ellipsoid can be followed out in the same manner, and it will be seen that for every movement of light outward from the centre there will result two rays, advancing with different velocities and polarized at right angles to one another. Inversely, every ray entering an anisotropic biaxial medium with perpendicular incidence will be divided into two rays, which are polarized at right angles to one another, and which, with the exception of those parallel to an optic axis, proceed with different velocities. For oblique incidence the two rays produced by double refraction will advance with different velocities and in different directions. The directions of vibration of the two parts of a doubly refracted ray are the axes of the ellipse cut from the ellipsoid of elasticity by a central plane at right angles to the direction of the incident ray

Comparing the two parts of a doubly refracted ray in an anistropic biaxial medium with those in an anisotropic uniaxial medium, we see

that none of the first-mentioned rays have a constant velocity of transmission, and consequently that none have a constant index of refraction ; and since these values for both rays change with the direction,
they are both extraordinary rays.   Nevertheless one is called the
ordinary and the other the extraordinary ray, from analogy with those
of uniaxial media.   Three principal indices of refraction are distinguished in biaxial media: α is the index of refraction of rays advanc-
ing at right angles to $\mathfrak{a}$ and vibrating parallel to $\mathfrak{a}$; β is the index of
those advancing perpendicular to $\mathfrak{b}$ and vibrating parallel to $\mathfrak{b}$; and γ
the index of rays advancing perpendicular to $\mathfrak{c}$ and vibrating parallel
to $\mathfrak{c}$.   Since the refraction is inversely proportional to the square root
of the elasticity, we have

$$\alpha = C\frac{1}{\mathfrak{a}};$$

$$\beta = C\frac{1}{\mathfrak{b}};$$

$$\gamma = C\frac{1}{\mathfrak{c}}.$$

These indices naturally change with the wave-length of the light.
Fig. 9 shows that the plane of the optic axes in a biaxial medium must
always lie in the plane of the axes of greatest and least elasticity of
this medium, and that the angle between the optic axes must be
bisected by these axes.   These axes of elasticity are therefore generally
called the bisectrices; the one bisecting the acute angle of the optic
axes is called the *first* or *acute bisectrix*, and that bisecting the obtuse
optical angle is the *second* or *obtuse bisectrix*.   The axis of mean
elasticity stands at right angles to the plane of the optic axes, and is
called the optical normal.   The angle which the optic axes make with
one another, and consequently the angle each makes with a bisectrix, is
dependent on the relative values of $\mathfrak{a}$, $\mathfrak{b}$, and $\mathfrak{c}$, and of α, β, and γ.   If
the angle between one optic axis and the axis of least elasticity is called
V, then

$$\cos V = \sqrt{\frac{\mathfrak{b}^2 - \mathfrak{c}^2}{\mathfrak{a}^2 - \mathfrak{c}^2}} = \sqrt{\frac{\frac{1}{\beta^2} - \frac{1}{\gamma^2}}{\frac{1}{\alpha^2} - \frac{1}{\gamma^2}}}.$$

Now since the value of α, β, and γ changes with the wave-length of
the light, the angle between the optic axes and the bisectrix must

change with the wave-length also .This is known as the *dispersion of the optic axes*, $V_\rho \gtrless V_v$; that is, the angle between an optic axis and the bisectrix for red light is greater or less than that for blue light.

In every elliptical section through the triaxial ellipsoid of biaxial media, which is not at right angles to the plane of the optic axes, the projection of the optic axes $m\,m$ and $m_1\,m_1$ (Fig. 10) must be symmetrical to the diameter of the ellipse, which therefore represents their bisectrix. Now since the axes of the ellipse are the directions of vibration of the two parts of a doubly refracted ray which advances perpendicular to the plane of the ellipse, we may lay down the rule that the directions of vibration of both rays bisect the angles

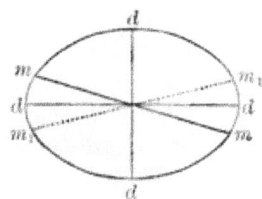

Fig. 10

between the optic axes. Therefore the direction of vibration of one part of a doubly refracted ray which strikes perpendicular to the face of a crystal is found by passing a plane through the ray (the normal to the crystal face) and the first bisectrix; the vibrations lie in this plane at right angles to the ray. The vibrations of the second part of the ray must be at right angles to those of the first part. If the plane through the incident ray and the bisectrix is called the *principal optic section*, then one ray vibrates at right angles to this plane and is called the ordinary ray, although it does not behave like the ordinary ray of a uniaxial medium; the ray vibrating in the principal section is called the extraordinary ray. If the elliptical section is at right angles to the plane of the optic axes, then these with the bisectrix and the principal optic section all fall together.

As two of the three axes of elasticity of a biaxial medium approach equality, the angle between the optic axes diminishes: it will be $= o$, and both optic axes will coincide with one another and with one axis of elasticity as soon as the difference between the other two axes of elasticity $= o$. This theoretical transition of biaxial media into uniaxial can take place by $\mathfrak{b}$ equalling $\mathfrak{c}$ or by $\mathfrak{a}$ equalling $\mathfrak{b}$. In the first case there arises an optically negative uniaxial crystal, and in the second a positive one; therefore in optically biaxial crystals those are considered as negative in which the axis of greatest elasticity is the acute bisectrix, and those as positive in which the axis of least elasticity is the acute bisectrix.

Optical Characteristics of the Three Crystal Systems without a Primary Axis.

Just as the crystal systems without a primary axis are distinguished from isotropic media by their double refraction, and from crystals with a primary axis by their having two optic axes, so they are distinguished from one another by the orientation of their ellipsoids of elasticity with respect to their crystallographic constants, and the consequent dispersion of their axes.

In the *orthorhombic* system the three axes of elasticity (a, b, c) coincide with the crystallographic axes of symmetry (a, b, c) because of the correspondence between the morphological and physical symmetry; any one of the first coinciding with any one of the second, without there being any connection whatever between the relative lengths of either group of axes. Such a connection is excluded by the fact that the choice of the vertical axis and of the fundamental form is arbitrary. But since every axis of elasticity coincides with a crystallographic axis of symmetry, a proper dispersion of the axes of elasticity (bisectrices) is rendered impossible. However, this does not prevent in one and the same crystal, as for instance in brookite, the bisectrices of the optic axes for light of different wave-lengths from coinciding with different crystallographic axes of symmetry. The plane of the optic axes always lies in one of the pinacoids, and light of different wave-lengths is dispersed symmetrically with respect to both bisectrices. Fig. 11 is the optical scheme for an orthorhombic crystal (∞P, oP) with optically negative character, whose axes lie in the macrodiagonal (principal) section with the vertical axis as the first or acute bisectrix. The dispersion is $\rho > v$.

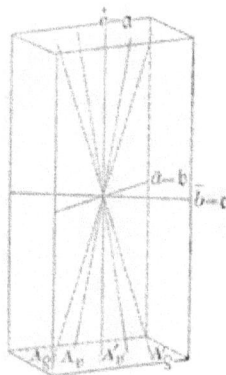

Fig. 11

In the *monoclinic* system only one of the so-called crystal axes, the orthodiagonal b, is an actual axis, that is, the normal to a plane of symmetry. This must therefore always coincide with one of the axes of elasticity, which naturally suffers no dispersion, and is an axis of elasticity for light of all wave-lengths. The two other axes must lie in the clinopinacoid, because they are at right angles to b, and since they correspond to no morphological axes of symmetry, they must generally suffer a small dispersion, so that they have different positions for rays of different colors. The plane of the optic axes must either lie in the plane of symmetry or at right angles to it, for in every case b is an axis of elasticity. According to the optical value of b, two groups of crystals are distinguished.

(1) $b = \mathfrak{b}$; the orthodiagonal is the axis of mean elasticity; it is so for all colors. The axes of greatest and least elasticity (bisectrices) for different colors lie dispersed in the plane of symmetry, in which also the optic axes for different colors are dispersed symmetrically with respect to their corresponding bisectrices. There is no common bisectrix for all wave-lengths. This kind of dispersion is called *inclined dispersion.* Fig. 12 presents the scheme of an optically positive crystal with inclined dispersion, in which $\mathfrak{c}_\rho$ (the positive bisectrix for red rays) has a greater inclination with respect to the vertical axis than $\mathfrak{c}_\nu$ (the positive bisectrix for blue rays). The inclined dispersion presents the most widely spread form of optical orientation of monoclinic crystals; such crystals are said to have a *symmetrical position* of the axes.

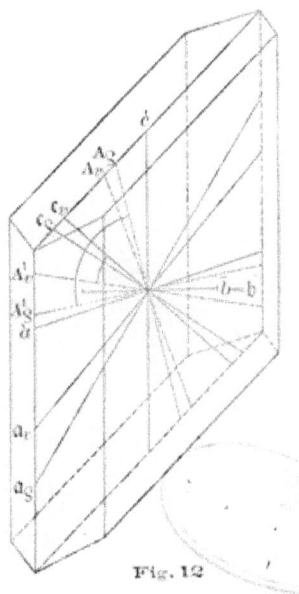

Fig. 12

(2) $b = \mathfrak{a}$ or $\mathfrak{c}$; the orthodiagonal is the axis of greatest or least elasticity, and is therefore one of the two bisectrices. Then the plane of the optic axes must lie at right angles to the plane of symmetry; these crystals have *normal symmetrical position* of the axes. They are divided into two groups, according as the orthodiagonal is the obtuse or the acute bisectrix. If the orthodiagonal is the obtuse bisectrix, there cannot be any dispersion of this bisectrix and the planes of the optic axes, for all colors must pass through the axis b. The dispersion is confined to the acute bisectrix and the axis of mean elasticity. Looked at in a direction at right angles to b, the planes of the optic axes for different wave-lengths lie horizontally over one another, for which reason this form of dispersion is called the *horizontal dispersion.* The scheme for this is shown in Fig. 13.

If the orthodiagonal is the acute bisectrix, only those axes of elasticity, the second bisectrix and the normal, which lie in the plane of symmetry can be dispersed, and the planes of the optic axes perpendicular to the symmetry plane must cross one another. Fig. 13 shows this kind of dispersion, which is *crossed dispersion,* if looked at in the direction of *b*. Horizontal and crossed dispersion constantly occur together, and it depends only on the size of the optic angle whether one or the other kind of dispersion is ascribed to a substance.

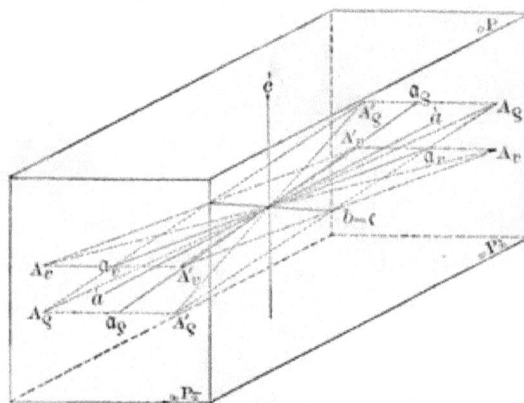

Fig. 13

Since in the *triclinic* or *asymmetric system* the so-called crystal axes are arbitrarily chosen co-ordinates, there can no longer exist between the axes of elasticity and the crystal axes any definite relationship. In general these directions do not fall together. For this reason there occurs a dispersion of all the axes of elasticity, and the ellipsoids of elasticity for light of different wave-lengths have no axes in common. This gives rise to the simultaneous occurrence of several axial dispersions.

Influence of Temperature and Pressure on Double Refraction.

As in isotropic media the index of refraction changes with the temperature and pressure, so in anisotropic substances there is a dependence of the optical constants on pressure and heat, which shows itself partly in a change in the absolute size of the index of refraction of a particular axis of elasticity, and consequently in the relative size of the two or three principal indices of refraction peculiar to an anisotropic substance, and partly in a change of position of the optical constants. So long as the changes of temperature in all parts of an anisotropic medium are the same, and its molecular construction and chemical composition remain the same, all the variations of the optical ellipsoid of elasticity occur in such a manner that this possesses at all temperatures the degree of symmetry corresponding to the crystal form of the medium. Consequently the ellipsoid of rotation of a uniaxial substance remains an ellipsoid of rotation for all temperatures, and never passes into a sphere for all kinds of light at any one time, nor becomes the triaxial ellipsoid of biaxial media.

In the same way the triaxial ellipsoid of a biaxial body remains such for all temperatures, or may become a rotation ellipsoid for each kind of light only at different temperatures, never for all kinds of light at one and the same temperature. Moreover, the axes of this triaxial ellipsoid are constant in their position so long as they coincide with crystallographic axes of symmetry. Therefore, in an orthorhombic body, the optical variations due to heating must be confined to the relative value of the three axes of elasticity, and consequently to the size of the angle and position of the plane of the optic axes.

With monoclinic crystals because of the variations in the relative value of the three principal coefficients of elasticity which are often considerable, and the consequent angle of the optic axes, there occurs not only a transition of the optic axial plane from normal symmetrical into symmetrical position or the reverse, but the triaxial ellipsoid of elasticity may be revolved about the symmetry axis common to itself and the crystal, and thus a change in the position of two axes of elasticity take place.

In the triclinic system the only limitation to the optical variations produced by heating is, theoretically, that for every temperature the elasticity of the ether must be expressed by a triaxial ellipsoid. The size and position of the three axes is, theoretically, wholly variable. In actual fact, for the few triclinic substances which have been investigated in this direction a great constancy in the optical relations for variations of temperature has been found.

A uniform pressure acting on all sides of a body must produce optical effects which would be subjected to the same regular variations.

In a great number of the cases investigated the crystal system of the substance and the uniformity of the variations in the optical ellipsoid of elasticity produced by heating remain the same.

But there is a considerable number of so-called "mimetic crystals" in which the outward crystal form appears to stand in more or less striking contradiction to their physical and especially to their optical behavior. These substances are characterized almost without exception by a very complicated twinning structure. Because of this apparent contradiction they are also called *optically anomalous* crystals. To these belong many garnets, alums, senarmontite, boracite, perofskite, analcite, leucite, tridymite, etc.

According to whether the outward form or the optical behavior of

these substances is considered to have the greater weight in determin
ing their crystal system, the apparent contradiction is explained either
as the result of strains which have disturbed the normal physical conditions belonging to the present crystal form, or by supposing that
many small individuals of a lower crystallonomic symmetry have been
combined by twinning to a compound individual of apparently greater
crystallonomic symmetry. The latter view is specially strengthened
by the fact that without exception the physical symmetry of such
mimetic structure is of a lower order than the crystallonomic, while
there appears to be no grounds *a priori* why strains of themselves
should not convert a less symmetrical physical condition into a more
symmetrical one.

The numerous studies of many investigators on these pseudosym-
metrical or mimetic forms have shown that a great number of these
apparent anomalies may be made to disappear upon heating. This is
explained by the fact that such mimetic substances are dimorphous,
and assumed a form through the physical conditions accompanying
their genesis which is not the position of equilibrium of their molecular structure, conformable with the subsequent physical conditions
in which they now exist. There arises therefore, with the changed conditions of existence, a molecular alteration within the outward crystal
form originally assumed, and which is more or less permanent, by
which the crystal endeavors to approach as near as possible to a condition of equilibrium corresponding to the altered conditions. Whether
this is actually attained,—that is, whether the symmetry of a mimetic
body indicated by optical investigation is actually the one which corresponds to present existing conditions of pressure and temperature,
or whether it is only occasioned by certain strains which may arise
through the exertions of a new molecular state of equilibrium within
an unyielding, rigid, outer form,—is not always easy to determine in
any given case.

For example, if we see plates of tridymite, which from their goniometric behavior are hexagonal, resolved optically into parts which show
the phenomena of triclinic penetration twins, and if we find that at
sufficiently elevated temperatures these plates show the normal optical
phenomena of uniaxial crystals flattened parallel to the base, the conclusion is certainly correct that we have in tridymite a holohedral
hexagonal form of silica, and that this form under certain conditions
of high temperature presents the normal form of silica.   But it
would be incorrect to conclude that there is a triclinic form of silica
capable of being formed under ordinary temperature and simple

atmospheric pressure. Much rather may the apparent twinning as well as the apparent triclinic optical behavior be explained by an abnormal condition of strain, which arises in the tridymite plate from the fact that a molecular alteration, possibly to the quartz form or to some unknown modification, is attempted, but is not attained because the rigidity of the outer form prevents it. Such a strain would act in the same way as an irregular lateral pressure or a many-sided unequal pressure.

Lateral pressure and irregular heating change the optical elasticity in an abnormal manner, and produce a contradiction between the crystallographic form and the optical behavior. Isotropic, that is, amorphous and isometric, bodies become anisotropic through lateral pressure or unequal heating, and there occurs a distribution of the optical elasticity, which expresses itself sometimes in an ellipsoid of rotation, sometimes in a triaxial ellipsoid. They thus become uniaxial or biaxial; and Brewster has shown that the occurrence of one or the other alteration is determined essentially by the form of the isotropic body. In the same way he found that optically uniaxial crystals which are compressed at right angles to their optic axis become biaxial; and Moigno and Pfaff showed that with positive crystals the plane of the optic axes stands parallel to the direction of pressure, and with negative crystals at right angles to it. This behavior is explained by the fact that pressure increases the elasticity, and the plane of the optic axes must lie in the plane of the axes of greatest and least elasticity. Since in positive crystals $c = \mathfrak{c}$, then the original elasticity which is the greatest in all directions perpendicular to $\mathfrak{c}$ becomes still greater in the direction of pressure, and at right angles to this it remains unaltered; the plane of the optic axes therefore passes through the primary axis and the direction of pressure. It is the reverse when $c = a$. H. Bücking found that a small pressure is sufficient to bring about a biaxial condition, but that the pressure must be considerably greater to increase the axial angle afterwards. The differences of elasticity due to pressure, therefore, are not directly proportional to the pressure. He also investigated the effect of pressure on biaxial sanidine, and found that a pressure parallel to the axis of mean elasticity diminishes the angle of the optic axes when they lie at right angles to the clinopinacoid, and increases it when they lie in the clinopinacoid; that is, it acts like a uniform heating (increase of temperature). W. Klein showed that a lateral heating perpendicular to the primary axis converts a uniaxial crystal into a biaxial one, and in such a manner that the elasticity becomes smaller in the direction of the application of the heat. Upon the lateral heating of plates of biaxial crystals cut at right angles to the bisectrix a deformation of the ellipsoid of elasticity takes place, until the heating becomes uniform ; when the alterations are those shown in uniformly heated plates.

c. *Investigation of Minerals in Parallel Polarized Light.*

Ordinary light in its passage through doubly refracting media in any direction except that of an optic axis is always separated into two rays, which are polarized at right angles to one another, and are generally transmitted with different velocities and in different directions. It differs essentially from a ray of polarized light in that the latter is not separated into two, if its plane of vibration is parallel to or perpendicular to the principal optical plane of the doubly refracting medium through which it passes. In such cases the polarized ray only suffers a change of velocity. But if the plane of vibration of the polarized light makes any other angle than 0° or 90° with the principal section of the anistropic medium, it is separated into two rays perpendicular to one another, which are generally transmitted with different velocities in different directions, just as in the case of ordinary light. The ray of polarized light becomes depolarized or rather repolarized.

In consequence of the fact that polarized light is not separated into parts when its plane of vibration is parallel or perpendicular to the principal plane of the medium traversed, and because of the interference of the separated rays in all other positions when the vibrations are reduced to one plane, doubly refracting media exhibit certain differences from singly refracting ones, and give rise to interference phenomena when investigated in polarized light, which lead not only to the distinction of isotropic and anistropic media, but also to the determination of the position of the axes of elasticity and of the optic axes. Now since the position of the axes of elasticity, as already pointed out, stands in the closest relation to the crystal structure of the media, so the optical investigation makes possible a determination of the crystal system with the same or even greater sharpness than the goniometric investigation does. For establishing the position of the axes of elasticity and the distinguishing of isotropic and anisotropic media, the investigation in parallel polarized light is to be preferred ; convergent polarized light is used for determining the optic axes and their inclination.

Polarizing Instruments.—Every instrument by which refracting media may be investigated in polarized light is called a polarizing in-

strument : it always consists of two parts. The first part transfor[ms]
ordinary light into polarized light, and is called the *polarizer;* t[he]
second part tests or analyzes the polarized light either by itself or af[ter]
its passage through the medium under investigation : it is called t[he]
analyzer. To transform ordinary light into polarized light, it m[ust]
either be reflected at the Brewster angle from a non-metallic mirror,
be allowed to pass through a doubly refracting medium in any dir[ec]
tion but that of an optic axis. One of the two polarized rays tl[us]
produced must then be eliminated.

The simplest polarizing instrument, the *tourmaline tongs,* consi[sts]
of two brown or dark-green tourmaline plates cut parallel to the pr[in]
cipal section, and set in frames which may be rotated in the end ri[ng]
of elastic wires bent into the shape of shears. The mineral under
vestigation is held between the tourmaline plates. If a ray of ordin[ary]
light falls on the first tourmaline plate it will be divided into two ra[ys]
according to the laws governing the movement of light in uniaxial [me]
dia : these rays will advance parallel to each other for perpendicular [in]
cidence, but with different velocities, one vibrating parallel to the [op]
tic axis (E), the other vibrating at right angles to it (O). Now si[nce]
tourmaline possesses the property of extinguishing the vibration[s at]
right angles to its optic axis,—that is, of absorbing the ordinary r[ay]
when the plate is sufficiently thick,—there emerges from it only a [ray]
vibrating parallel to *c.* The tourmaline plate is thus a polarizer ; [or]
dinary light upon entering it is transformed through double refract[ion]
and absorption into polarized light, whose plane of vibration is kno[wn.]
If the second tourmaline plate, which is to serve as an analyzer, is [so]
placed that its optic axis is parallel to that of the polarizer, then [the]
extraordinary ray which comes from the polarizer, and whose pl[ane]
of vibration lies parallel to the principal section of the analyzer, [will]
experience no separation, but will also pass through the second tour[ma]
line plate as an extraordinary ray with unchanged direction of vib[ra]
tion. If one looks through both plates in this position (with [the]
principal sections parallel) there is a uniform green or brown fiel[d of]
view, as though there were but a single tourmaline plate.

If the analyzer is rotated until its principal section stands at ri[ght]
angles to that of the polarizer, the extraordinary ray, which emer[ges]
from the latter, will still be undivided, as its plane of vibration [now]
stands at right angles to the principal section of the analyzer ; it w[ill]
enter the latter with unaltered direction of vibration. The ray t[hat]
traverses the analyzer with vibrations at right angles to the optic a[xis,]
that is, as an ordinary ray. It will therefore be absorbed, and will

pass through it. If one looks through the tourmaline plates in this position (with the principal sections crossed at right angles) the field of view will be dark. In every other position of the polarizer with respect to the analyzer, the ray emerging from the former, its plane of vibration being no longer parallel or perpendicular to the principal section of the analyzer, will be separated in the same manner as a ray of ordinary light—that is, into an ordinary ray, which vibrates at right angles to the optic axis of the analyzer and is absorbed; and an extraordinary one, which vibrates parallel to the optic axis and passes through. The component of the light coming from the polarizer which forms the extraordinary ray, that is, the intensity of the light emerging from the analyzer, must naturally be dependent on the intensity of the incident ray and the inclination of the principal optic sections of the polarizer and analyzer to one another. It is proportional to the cosine of this inclination. Let ab (Fig. 14) be the principal section of the polarizer, ef that of the analyzer, x the angle included between them. Let $mg = I$ represent the amplitude of vibration of the ray emerging from the polarizer; then, if gh is perpendicular to ef, this ray will be separated in the analyzer into an ordinary ray vibrating at right angles to ef with the amplitude hg, which is absorbed, and into an extraordinary ray vibrating parallel to ef with the amplitude $hm = I_{,}$, which passes through the analyzer. Then

$$hm = mg \cdot \cos x;$$
$$I_{,} = I \cdot \cos x.$$

Fig. 14

If $x = 0°$, that is, if the principal sections of the polarizer and analyzer are parallel, $I_{,} = I$; for $x = 90°$, that is, when the principal sections are crossed at right angles, $I_{,} = 0$. The intensity of light is proportional to the square of the amplitude of vibration.

The deep color of tourmaline renders it unfit for use in microscopical investigations, and it is generally replaced by the *nicol prism*. Such a nicol prism is made from a natural cleavage piece of calcite which is three times as long as thick. The upper and lower faces of this rhombohedron, which make angles of 71° and 109° with the edges ... in the principal section, are replaced by others whose inclination

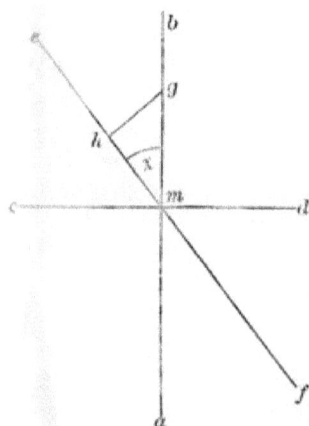

to these edges is 68° and 112°; the rhombohedron is then sawn across at right angles to the principal section and to these newly cut faces, and the faces of the section after being thoroughly polished are cemented together in their original position by Canada balsam. The cross-section of such a nicol prism in the principal section is shown in Fig. 15. It is blackened on the outside, and fastened with a cork in a metal tube.

If now a ray of light, mn, parallel to the long edge of the prism falls upon the end face of the same, then it will be separated within the prism into an ordinary ray, no, with an index of refraction of 1.658, and into an extraordinary ray with a considerably smaller index of refraction. The index of refraction of the Canada balsam is 1.536 ; from which the critical angle for the transition of the ordinary ray is found to be 67° 53'. Now since the angle of incidence of the light is $90° - 68° = 22°$, then the angle of refraction in calcite is 13° 4' and the angle of incidence on the layer of balsam is 76° 56', and the ordinary ray must therefore experience a total reflection in the direction oo_1. The extraordinary ray traverses the layer of balsam and the second half of the prism, and emerges at q in the direction qe parallel to mn. Its plane of vibration lies parallel to the short or inclined diagonal of the end faces of the nicol prism, which has a rhombic form. Two nicol prisms act in exactly the same manner as two tourmaline plates ; the extraordinary ray which emerges from the first nicol will experience no separation in the second prism, which serves as an analyzer, if its principal section is parallel or at right angles to that of the first. In a parallel position the ray traverses the analyzer as an extraordinary ray, and suffers no total reflection from the layer of balsam. The field of view is completely clear. In a crossed position the extraordinary ray coming from the polarizer is converted into an ordinary ray in the analyzer, and experiences total reflection from the layer of balsam. The field of view is dark. For an inclined position of the principal sections of both nicols to one another we must have $I_1 = I \cdot \cos x$, where x is the inclination of the principal sections to one another, and the illumination of the field of view is $I^2 \cdot \cos^2 x$.

The original construction of the nicols has been modified in various ways, resulting in the shortening of the prism, the strengthening of the transmitted light, together with the more complete polarization even of the inclined incident rays.

Fig. 15

In order to apply the microscope to investigations in polarized light, a nicol prism is inserted in the path of the light between the mirror and the object to serve as a polarizer, and a second as analyzer is placed between the object and the eye of the observer, either within the tube or above the ocular lens. A microscope thus furnished with nicols is called a *polarizing microscope*. The insertion of the nicols, however, only accomplishes the desired results when the instrument satisfies the following conditions: (1) The object under investigation must be capable of rotation in its own plane about the optical axis of the instrument while the nicols remain crossed; (2) The angle between any two positions of the object must be measurable with requisite accuracy; (3) The principal sections of the nicols must have a known position, which may be restored after being displaced.

The general construction of a polarizing microscope may be learned from the description of that manufactured by Nachet et fils of Paris: it differs from others in having the ocular wholly independent of the objective (Fig. 16). The tube is cut across, and the lower part *B*, bearing the objective, is united to the rotating stage of the microscope. Thus the objective follows the movement of the stage during its rotation, and consequently every point of the thin section which has been brought to the intersection point of the cross wires remains in the centre of the field of view for every position of the stage; it cannot rotate other than concentrically. The rough adjustment of the objective is effected by the rack-and-pinion movement above *L*, the fine adjustment by the micrometer-screw *L*. The head of the latter is divided into 100 parts, its position is read with a vernier to the tenth of a part; the height of the thread (the pitch) of the screw is 0.25 mm.

The objective is not screwed on, but is held in place by a spring in a very convenient and solid manner. The upper part of the tube is held firmly by the outer (bent) metal column, and is also raised and lowered by means of a rack-and-pinion movement. At the upper end is an opening through which the cross wires in the ocular may be illuminated by means of a mirror *M*; this is sometimes desirable when the nicols are crossed, and the field of view is very dark. At the lower end of the ocular tube, in front, is a second larger opening into which the analyzer *A* can be moved. When this is not in use the opening may be closed by means of a sleeve. At the extreme end of the ocular tube at *U* is a slit in which may be inserted a quartz plate for observation with the sensitive tint, or Bertrand's lenses for magnifying interference figures.

The stage of the microscope consists of a circular plate, which can

be rotated either by the hand or by the screw E, which works when it
is pushed forward, and can be thrown out of gear by being pulled
back. The rotating plate carries a vernier P', which moves upon the

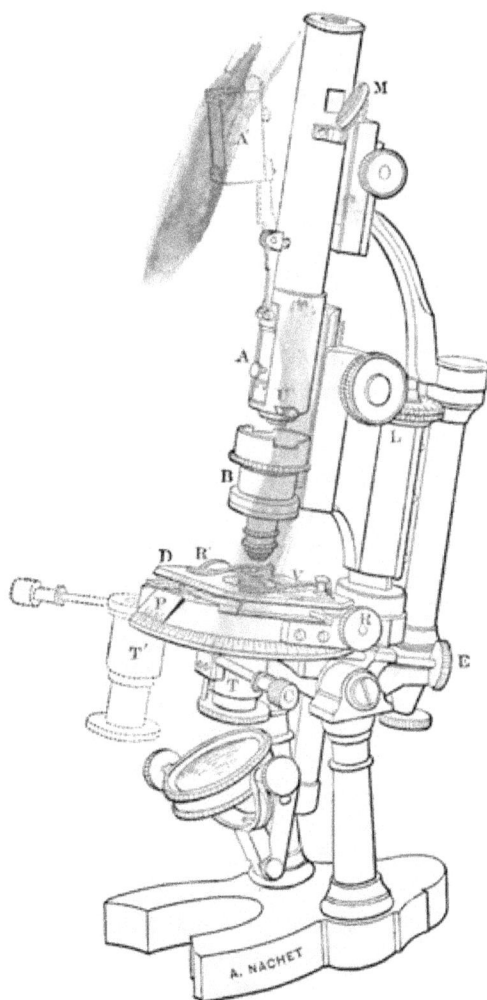

Fig. 16

circular scale on the rim of the lower stationary plate of the stage, and
reads to the tenth of a degree. The object does not lie directly on the
rotating plate, but on the mechanical stage D, which is moved by means
of the screws R and R' working at right angles to one another. Thus

the thin section is not moved by the hand, but mechanically ; and every point of it can be brought into the centre of the field without any spot of the thin section escaping observation. The movement of the object by the screws R R' can be read off on linear scales, on which the mechanical stage D glides. The thin section rests against the small ledge V, and is held by two weak springs. Beneath the object-table is the tube T, which can be raised and lowered by means of the rack-and-pinion c, and when lowered may be pushed aside to T' for the purpose of changing the apparatus for illumination. In the tube are placed the polarizer, and the condensing lenses used for investigations with different magnifying powers and in parallel or convergent light.

Isotropic Mineral Plates in Parallel Polarized Light.

If a thin plate of an isotropic mineral (amorphous or crystallizing in the isometric system) be placed in the path of a polarized ray between the polarizer and analyzer the ray of light will experience no alteration of its plane of vibration, no matter in what direction the plate was cut from the mineral, nor in what position it lies between the polarizer and analyzer. Since the elasticity of the ether is the same in all directions through such a mineral, its rotation about any axis whatever will effect no change of the plane of vibration of the polarized light. If the mineral is also colorless, it will not influence the color or the brightness of the field of view, except for the small absorption which a ray of light experiences in passing through any medium ; if it is colored, the field of view will show a color somewhat different from that of the mineral. But this color does not change in any manner with the position of the plate. Moreover, the direction of the ray will not be changed if the plate has parallel faces, and is set at right angles to the direction of the ray ; if the latter is not the case, the ray within the plate will be deflected from its course, but on emerging from the plate will advance parallel to its direction at incidence. If the faces of the plate are inclined to one another, the direction of the ray after leaving the plate will differ from that at incidence in proportion to the inclination of the two faces. Assuming that the principal sections of the analyzer and polarizer are in crossed position, then the consequent darkness of the field of view will not be disturbed by the insertion of an isotropic plate. *This property of remaining dark in every position between crossed nicols, and for a rotation of 360° in its own plane, is the most important characteristic of an isotropic plate in contra-distinction to an anisotropic one.*

Plates of amorphous or isometric minerals often show the phenomena of becoming partially or completely light between crossed nicols. Such anomalies are the results of internal strains produced either by inclusions of gases or fluids which exert a pressure on their surrounding walls, or by solid bodies which in contracting exert a tensile strain on the adjacent parts of the inclosing mineral ; or they depend on conditions connected with the genesis of the mineral, that is, with its molecular structure. Such phenomena are distinguished from regular double refraction by the fact that the appearance is not generally alike in all parts of the plate, but differs from place to place. Such double refraction is called an optical anomaly.

Thin Plates of Doubly Refracting Minerals in Parallel Polarized Light.

If a transparent plate of a doubly refracting mineral, which is not cut at right angles to an optic axis, is placed between the polarizer and analyzer when their principal sections make any angle whatever with one another, it generally gives rise to phenomena of chromatic interference. Beginning with the simplest case, suppose that the plate has parallel faces and is everywhere of the same thickness, that the rays fall at right angles to it, and consequently traverse equal thicknesses at all points ; that the light is homogeneous, and that the principal sections of the polarizer and analyzer make an angle $>0° < 90°$ with one another. Upon this supposition, a ray (Fig. 17) which strikes the plate at a is separated into two rays which traverse the plate in like directions, but with vibrations at right angles to one another and with different velocities. Upon egress at the point b, they

Fig. 17

pass into air again without deflection and advance parallel to each other ; but since the velocity is different for each ray within the plate, then at b one ray must have advanced a certain number of wave-lengths ahead of the other. The rays are therefore in different phases of vibration, and retain this difference of phase on their way through the air. One of the rays, the extraordinary, vibrates parallel to the principal optic section of the plate, the ordinary ray vibrates at right angles to this principal optic plane, and these directions of vibration do not change on their passage into air.

Let Fig. 18 lie in the plane of the doubly refracting plate, and let the projection of the principal section of the polarizer on this plane be $PP_{,}$, that of the principal section of the analyzer be $AA_{,}$, ϕ be the angle between them, IIII, be the principal section of the plate, and ρ the angle this makes with PP_1, and $OM = i$ be the amplitude of vibration of the ray coming from the polarizer which vibrates parallel to PP_1. Since the plane of vibration of the ray neither coincides with the principal section of the plate nor is at right angles to

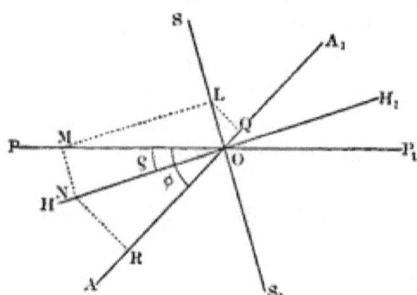

Fig. 18

it, then, according to the parallelogram of forces, it will be divided into two rays, an ordinary ray vibrating at right angles to IIII_1, with the intensity $OL = \sin \rho$, and an extraordinary ray vibrating parallel to IIII_1, with the intensity $ON = \cos \rho$, i being taken as unity.

If the velocity of the ordinary ray within the plate is c_o and that of the extraordinary ray c_e, and the thickness of the plate d, then $\dfrac{d}{c_o} = o$

and $\dfrac{d}{c_e} = e$ are the times in which O and E traverse the plate. The vibrations of a particle of ether about its point of equilibrium follow the laws for the motion of a pendulum. The velocity of vibration is $= 0$ at the moment of its greatest distance from the position of equilibrium; it increases with its approach to this, and reaches its maximum the moment when this point is passed. This maximum is proportional to the amplitude of vibration (the intensity). If t denotes the time which has elapsed since the particle of ether was at the greatest distance from its point of equilibrium, T the duration of an oscillation (that is, the time which the particle of ether takes to travel from one position of maximum elongation to the other and back again), then the velocity of vibration of a particle of ether in the path of the ordinary ray at its entrance into the plate $= \sin \rho \sin 2\pi \dfrac{t}{T}$, and that

of the extraordinary ray $= \cos \rho \sin 2\pi \dfrac{t}{T}$. Upon their egress from the plate the velocity of vibrations of O and E, since they advance more slowly in the plate than in air, are respectively

$$v_o = \sin \rho \sin 2\pi\left(\frac{t-o}{T}\right),$$

$$v_e = \cos \rho \sin 2\pi\left(\frac{t-e}{T}\right).$$

If the homogeneous light used has in air the wave-length λ, the time of its vibration T, and the velocity of transmission V, then $T = \dfrac{\lambda}{V}$ and the above equations become

$$v_o = \sin \rho \sin 2\pi\left(\frac{t}{T} - \frac{o V}{\lambda}\right),$$

$$v_e = \cos \rho \sin 2\pi\left(\frac{t}{T} - \frac{e V}{\lambda}\right).$$

Upon its passage into the analyzer the ordinary ray furnishes one component vibrating at right angles to the principal axis $AA_{\prime}$, having the intensity LQ, which is removed through total reflection in the analyzer, and a component vibrating in the principal section of the analyzer, with the intensity

$$OQ = OL \sin (\phi - \rho) = \sin \rho \sin (\phi - \rho).$$

In the same manner the extraordinary ray upon its entrance into the analyzer separates into one component which disappears through total reflection and vibrates at right angles to $AA_{\prime}$, with the intensity NR, and one vibrating parallel to $AA_{\prime}$, with the intensity

$$OR = ON \cos (\phi - \rho) = \cos \rho \cos (\phi - \rho).$$

Since both rays traversing the analyzer are extraordinary, they have the same velocity of transmission, and their intensities, being derived from the ordinary and extraordinary ray of the plate, are respectively

$$I_{r_o} = \sin \rho \sin (\phi - \rho) \sin 2\pi\left(\frac{t}{T} - \frac{o V}{\lambda}\right),$$

$$I_{r_e} = \cos \rho \cos (\phi - \rho) \sin 2\pi\left(\frac{t}{T} - \frac{e V}{\lambda}\right).$$

In the doubly refracting plate the rays did not interfere, since their planes of vibration stood at right angles to one another; in the analyzer they have the same plane of vibration and must therefore form an interference ray, whose intensity must be equal to the sum of the intensities of the rays producing it. Since OQ and OR stand in opposite sense to one another, we have for the interference ray

$$I = I_{\rho_0} - I_{\tau_0} = \cos \rho \cos (\phi - \rho) \sin 2\pi\left(\frac{t}{T} - \frac{eV}{\lambda}\right)$$

$$- \sin \rho \sin (\phi - \rho) \sin 2\pi\left(\frac{t}{T} - \frac{oV}{\lambda}\right)$$

which may be reduced to the form

$$I' = \cos^2 \phi + \sin 2\rho \sin 2 (\phi - \rho) \sin^2 \pi\frac{(o-e)\,V}{\lambda}. \quad . \quad . \, (I)$$

This expression shows that in the general case the intensity of the interference ray emerging from the analyzer is composed of two factors, one of which, $\cos^2 \phi$, is independent of the wave-length and only varies with the inclination of the principal sections of the polarizer and analyzer. For the relation which is almost exclusively used in practice, namely, the crossed position of the polarizer and analyzer, $\phi = 90°$; therefore $\cos^2 \phi = 0$ and the equation becomes

$$I' = \sin 2\rho \sin 2(\phi - \rho) \sin^2 \pi\frac{(o - e)\,V}{\lambda}$$

or
$$I' = \sin^2 2\rho \sin^2 \pi\frac{(o - e)\,V}{\lambda}.$$

The intensity of light between crossed nicols shown by a doubly refracting plate which is not cut at right angles to an optic axis is primarily dependent on the quantity $\sin^2 2\rho$, that is, on the inclination of the principal optic section of the plate with reference to the principal sections of the polarizer and analyzer. The value of I' is a minimum, and becomes zero when $\sin^2 2\rho = 0$, that is, every time that the principal optic section of the plate coincides with the principal section of the polarizer ($\rho = 0$) or with that of the analyzer ($\rho = 90°$).

Upon rotating the plate 360° in its plane, or for a complete rotation of the stage of the microscope, this coincidence occurs four times, from which is derived the rule that *doubly refracting plates become dark four times during a complete rotation between crossed nicols, the positions of darkness occurring every 90° from one another.*

A maximum of brightness ($I' = $ max) must occur when $\sin^2 2\rho = 1$, that is, when the principal section of the plate is inclined 45° to the principal sections of the polarizer and analyzer. Thus if a doubly refracting plate is set at a position of darkness in homogeneous light, then by rotating it an illumination will set in which will increase with the rotation until it reaches 45°, beyond which it will diminish, becoming $= 0$ when the angle of rotation $= 90°$. For every position of darkness the principal optic section of the plate or a plane at right angles to it is parallel to the principal section of the polarizer, and this observation furnishes a means not only of distinguishing anisotropic from isotropic plates, but also of determining the position of the axes of greatest and least elasticity in the plate.

The brightness of the plate is further dependent on $\sin^2 \pi \frac{(o-e)\, V}{\lambda}$, that is, on the color (wave-length) of the light used; on $o-e$, that is, on the difference of phase of the two rays traversing the plate, consequently on the difference between the axes of greatest and least elasticity in the plate, and its orientation in the crystal; and on its thickness, since $o = \frac{d}{c_o}$ and $e = \frac{d}{c_e}$. The quantity $\sin^2 \pi \frac{(o-e)V}{\lambda}$ becomes $= 0$ when $\frac{(o-e)\,V}{\lambda}$ is a whole number, that is, when one ray precedes the other by a number of whole wave-lengths. On the other hand, $\sin^2 \pi \frac{(o-e)\,V}{\lambda}$ is a maximum when $\frac{(o-e)\,V}{\lambda} = \frac{2n+1}{2}$, that is, when one ray precedes the other by an uneven number of half wavelengths. Therefore a plate in homogeneous light is dark in every position between crossed nicols when the difference of phase between the two rays is measured by whole wave-lengths, and it has a maximum brightness for every position in which this difference of phase is measured by unequal half wave-lengths.

If in equation (I) we suppose $\phi > 0 < 90°$ and then increase it by 90°, the expression becomes

$$I^2 = \sin^2 \phi - \sin 2\rho \sin 2(\phi - \rho) \sin^2 \pi \frac{(o-e)\,V}{\lambda}. \quad . \quad (II)$$

This expression added to equation (I) reduces the second member to 1, from which it follows that a rotation of the analyzer 90° to the polarizer reverses the phenomena. What was dark between crossed nicols must be light with parallel nicols.

If the observations are made in white light, the phenomena may also be explained by equation (I), if it is remembered that white light is composed of innumerable kinds of homogeneous light of different wave-lengths.

The first part of the expression in equation (I) is independent of the wave-lengths. In the second part of the expression,

$$\sin 2\rho \sin 2(\phi - \rho) \sin^2 \pi \frac{(o - e) V}{\lambda},$$

$\sin 2\rho \sin 2(\phi - \rho)$ is influenced by the fact that the principal sections of the plate do not generally fall together for different kinds of light. But the differences due to dispersion are for the most part so small that they may be neglected. The part of the expression chiefly affected is $\sin^2 \pi \frac{(o - e) V}{\lambda}$. And indeed all those rays must disappear from the white light for which $\frac{(o - e) V}{\lambda} = n$, when n signifies any whole number, while all those rays will contribute to the illumination of the plate for which $\frac{(o - e) V}{\lambda}$ is a fraction. The plate will therefore appear colored in every case, and the color will be composed of those kinds of light for which $\frac{(o - e) V}{\lambda} = \frac{2n + 1}{2}$, or approach nearest to this value.

The quantities $\sin 2\rho$ and $\sin 2(\phi - \rho)$ do not influence the color in any way, but only the intensity of the color. Therefore the color shown by a plate in polarized light does not change in kind during a rotation, but only in intensity. If we again assume the case which occurs almost exclusively in practice, namely, that the nicols are crossed, we have, in equation (I),

$$\cos^2 \phi = 0 \quad \text{and} \quad I^2 = \sin 2\rho \sin 2(\phi - \rho) \sin^2 \pi \frac{(o - e) V}{\lambda} \Sigma i_\lambda,$$

when Σi_λ expresses the sum of the endless number of expressions which correspond to all values of λ.

The discussion of this equation pursued in the same manner as for that of homogeneous light leads to the rule that *doubly refracting plates, not cut at right angles to an optic axis, generally show an interference color in parallel polarized white light, which is dependent on their thickness; on the position of the plates in the crystal, and on the relative size of the axes of elasticity; or on the indices of refraction of the substance. The intensity of this color depends on the inclination of the principal section of the plates to the principal sections of the polarizer and analyzer; it reaches a minimum four times in a complete rotation (the plate is dark) when this inclination is $0°$ and $90°$;*

it appears at a maximum four times when the inclination is 45°. For the parallel position of the principal sections of the polarizer and analyzer the complementary phenomena appear: in what is the dark position between crossed nicols the plate is white between parallel nicols; in all other positions the colors are complementary to what they were between crossed nicols.

These interference colors of doubly refracting plates in polarized light belong to the category of Newton's colors (of thin plates), and such a doubly refracting plate will show the same interference colors as an isotropic plate of the thickness d, if $d = (o - e) V$. These interference colors belong to the most characteristic phenomena of microscopical investigation, and for a known thickness and orientation of the plate directly indicate the value of $(o - e)$, which is among the constants of every substance. It is evident that for a constant thickness and the same substance the interference color will be higher, as there is a greater difference between the two axes of elasticity to which the vibrations of the rays are parallel. Therefore optically uniaxial bodies, other things being equal, must give the highest interference colors in sections parallel to the optic axis, and optically biaxial bodies in sections at right angles to the axis of mean elasticity. The interference colors must diminish as the plate is cut more nearly perpendicular to an optic axis, and the colored interference ceases when the light traverses the plate exactly parallel to an optic axis.

Newton determined the order of succession of the interference colors shown by thin plates of increasing thickness and arranged them in a color-scale bearing his name. It will be seen from the accompanying table that certain tones of color recur periodically; the colors which lie between two analogous tones are called an " order." A knowledge of this color scale greatly facilitates the estimation of the amount of double refraction peculiar to a particular mineral, and is absolutely necessary in the determination of the optical characters of a mineral cross-section.

Behavior of Doubly Refracting Plates cut at Right Angles to an Optic Axis in Polarized Light.

In every direction at right angles to an optic axis in a doubly refracting mineral the elasticity of the ether is the same, and for all rays travelling exactly parallel to this axis the mineral should behave like an isotropic medium. If a section at right angles to an optic axis be examined between crossed or parallel nicols in parallel, homogeneous light, then in the first case one would expect it to remain dark for a

NEWTON'S COLOR-SCALE ACCORDING TO QUINCKE.[*]

No.	Millionths of Millimeters.	Interference Color between Crossed Nicols.	Interference Color between Parallel Nicols.	
1	0	Black.	Bright white.	First Order.
2	40	Iron-gray	White.	
3	97	Lavender-gray	Yellowish white.	
4	158	Grayish blue.	Brownish white.	
5	218	Clearer gray.	Brownish yellow.	
6	234	Greenish white.	Brown.	
7	259	Almost pure white.	Light red.	
8	267	Yellowish white.	Carmine-red.	
9	275	Pale straw-yellow.	Dark reddish brown.	
10	281	Straw-yellow.	Deep violet.	
11	306	Light yellow.	Indigo.	
12	332	Bright yellow.	Blue.	
13	430	Brownish yellow.	Gray-blue.	
14	505	Reddish orange.	Bluish green.	
15	536	Red.	Pale green.	
16	551	Deep red.	Yellowish green.	
17	565	Purple.	Lighter green.	Second Order.
18	575	Violet.	Greenish yellow.	
19	589	Indigo.	Golden yellow.	
20	664	Blue (sky-blue).	Orange.	
21	728	Greenish blue.	Brownish orange.	
22	747	Green.	Light carmine-red.	
23	826	Lighter green.	Purplish red.	
24	843	Yellowish green.	Violet-purple.	
25	866	Greenish yellow.	Violet.	
26	910	Pure yellow.	Indigo.	
27	948	Orange.	Dark blue.	
28	998	Bright orange-red.	Greenish blue.	
29	1101	Dark violet-red.	Green.	
30	1128	Light bluish violet.	Yellowish green.	Third Order.
31	1151	Indigo.	Impure yellow.	
32	1258	Greenish blue.	Flesh-colored.	
33	1334	Sea-green.	Brownish red.	
34	1376	Brilliant green.	Violet.	
35	1426	Greenish yellow	Grayish blue.	
36	1495	Flesh-colored.	Sea-green.	
37	1534	Carmine-red.	Green.	
38	1621	Dull purple.	Dull sea-green.	
39	1652	Violet-gray.	Yellowish green.	Fourth Order.
40	1682	Grayish blue.	Greenish yellow	
41	1711	Dull sea-green.	Yellowish gray.	
42	1744	Bluish green.	Lilac.	
43	1811	Light green.	Carmine.	
44	1927	Light greenish gray.	Grayish red.	
45	2007	Whitish gray.	Bluish gray.	

[*] Ueber Newton'sche Farbenringe und totale Reflexion des Lichtes bei Metallen. Pogg. Ann. 1866. CXXIX. 177.

complete rotation in its own plane, and in the second case to remain illuminated. The actual appearance, however, is different for an optically uniaxial and biaxial substance. In tetragonal and hexagonal crystals the principal axis is also the optic axis, that is, the direction of single refraction for light of every color; and a basal section of such minerals for perpendicular incidence in parallel light acts the same for every color and every color combination, consequently for white light it acts like an isotropic body. The distinction of such a basal section of a uniaxial mineral from a section of an isotropic substance is made by investigation in polarized light which is not parallel.

In the orthorhombic, monoclinic, and triclinic systems the optic axes no longer coincide with the axes of symmetry; they therefore suffer a dispersion, and, strictly speaking, there can no longer be any section which shall be perpendicular to an optic axis for two different colors at the same time. Consequently it is not to be expected that such a section would behave like an isotropic plate, but rather that in every position between two nicols, making any angle whatever with one another, such a plate in parallel white light would be illuminated by a color approaching the lowest tints of Newton's color scale. But even with the use of homogeneous light thin sections of a biaxial mineral cut at right angles to an optic axis are not dark, but light, and they are light in every position during a complete rotation in their own plane. This apparent anomaly as recently shown by E. Kalkowsky (Z. X. 1884, ix. 486–497) is the necessary consequence of the fact that the optic axis of biaxial bodies are axes of internal conical refraction. A ray of light falling parallel to an optic axis on a biaxial plate cut at right angles to this axis is divided within the same into an infinite number of rays which lie on the surface of a cone and are polarized in all directions. They thus emerge as a cylinder of rays in which each ray vibrates in a different azimuth from the rest; between crossed nicols, then, such a plate must show the same illumination in all positions. This phenomenon distinguishes sections in this direction when they are sufficiently thin from those of any other direction.

Behavior of Several Doubly Refracting Plates lying upon one another in Polarized Light.

If two doubly refracting plates overlie one another, the resultant phenomena in polarized light depend on the inclination of the principal sections of the polarizer and analyzer to one another and to the principal sections of the plates, as may be seen by a further application of the methods employed in discussing Fig. 18. By using white light there will be an interference color whose height is dependent on the sum of the thicknesses of both plates and the sum of the differences of phase attained by the rays in both plates, and whose intensity is determined by the inclination of the principal sections of both plates to one another and to those of the nicols.

If the principal sections of the two plates are perpendicular or parallel to one another, then the system of the two plates with respect to the four occurrences of the extinction of light between crossed nicols will act just as a single plate. The interference color, which appears when the principal sections of the plates are other than at 90° or 0° to those of the nicols, will rise in comparison with the interference colors of each single plate if with the parallel position of their principal sections to one another equivalent axes of elasticity fall together; or with the crossed position if unequivalent axes of elasticity fall together. On the other hand, it will be lowered when the opposite conditions exist. If one therefore knows the optical character of one plate, then by observing the rising or sinking of the interference color upon the insertion of a second plate in parallel or right-angled position of the principal sections the relative value of the axes of elasticity in this second plate may be determined.

Plates of Anisotropic Twinned Crystals in Polarized Light.

In sections of twinned crystals the parts belonging to each individual must in general behave differently in polarized light, since their axes of elasticity are differently oriented with respect to the principal section of the nicols. The position of darkness for each lamella will naturally be reached when its axes of greatest and least elasticity coincide with the principal sections of the polarizer and analyzer. The application of the rules previously given for the behavior of doubly refracting lamellæ in polarized light shows that for certain positions between crossed nicols the lamellæ belonging to a twin must appear equally bright, and for sufficiently thin lamellæ must also be of the same color. This happens when the principal sections of each half of the twin are equally inclined on opposite sides of the principal sections of the nicols.

If the section through a system of twinned lamellæ is not perpendicular to their composition-plane, then there must be strips between each two adjacent lamellæ which consist of wedges of both lamellæ overlapping one another. The behavior of these strips between crossed nicols will be understood by considering their action as twins and also as superimposed plates.

Stauroscopic Methods for determining the Direction of Extinction in Doubly Refracting Plates.

Since the determination of the direction of the extinction of light in a doubly refracting plate furnishes criteria for the recognition of the position of the axes of elasticity in the mineral with respect to the crystal axes, and consequently for the discovery of the crystal system, it is one of the most important determinative expedients. Now the eye is relatively insensible to small variations in the brightness of light, and it is evident that the readings of the positions of greatest darkness of doubly refracting plates when using white light may differ considerably. It is more correctly effected by using monochromatic light, but this is not convenient, and numerous methods have been sought which would furnish greater exactness in the adjustment to the maximum of darkness without using monochromatic light. A calcite plate cut at right angles to the optic axis was employed by Kobell (Pogg. Ann. 1855, xcv. p. 320). Placed between the object and the analyzer, it shows an interference figure consisting of a dark cross and a number of concentric isochromatic rings when the principal sections of the object and of the nicols coincide, the interference figure being distorted when they are not coincident. Such an instrument is called a stauroscope. Brezina improved the sensitiveness of this method by substituting for the single calcite plate a system of two plates cut nearly at right angles to the axis. If the calcite plate is placed between the ocular of a microscope and the nicol above it, it becomes a stauroscope.

A more exact method for detecting the direction of extinction in doubly refracting plates than the use of maximum darkness is the use of a particular color. This is most conveniently accomplished by inserting between the crossed nicols a quartz plate cut parallel to the axis, and of such a thickness that it will show a violet interference color (18 of Newton's color-scale); if the axis of the quartz plate be set at 45° to the principal sections of the nicols, the whole field will be equally colored violet. If a doubly refracting plate be placed so as to cover part of the field only, it will appear of a different color from the violet, because the difference of phase for the rays emerging from the plate is added to that derived from the quartz. If now the plate is rotated till its axis of greatest and least elasticity coincide with the principal sections of the nicols, a dissection of the light by the plate will no longer take place, and the plate will be colored **the** same as the **quartz** plate. This adjustment to the color of the quartz plate is extremely sensitive for colorless or very slightly colored **minerals.**

E. Bertrand inserted in the ocular of the microscope a quartz plate composed of **two** pairs of **right- and** left-handed quartzes of **the** same thickness **which are cut perpendicular to the** axis and cemented together so **that** each pair occupies opposite quadrants (Fig. 19). **This** plate is set in **the ocular so** that the **lines of contact between the** four **parts,** which appear **as two** dark **lines at right** angles **to one** another, shall be exactly parallel **to** the principal sections **of the nicols.** When the nicols are crossed all four **quartz quadrants** present the same tint of color. **Upon introducing** a doubly refracting plate on the stage **of the microscope,** the opposite sectors of the plate **are** similarly colored and the adjacent ones dissimilarly colored. They all **become alike** when the principal sections of the plate **are** made parallel **to those of** the nicols. The Bertrand ocular undoubtedly furnishes **the most** exact stauroscopic determination, and is in the most **convenient form.**

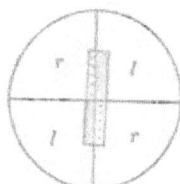

Determination *of the* **Relative** *Value of Both* **Axes** *of* **Elasticity** *in a* **Doubly Refracting** *Plate.*

In the microscopical determination of minerals it is frequently **necessary to** determine which of **the directions of** extinction in a plate corresponds **to** the axis of greatest elasticity, and which to that of least elasticity. This problem, called the determination of the optical character, is solved by means of a plate of known character.

When the position of **the axes of** elasticity in the plate is determined the plate is rotated so that its principal sections make angles of 45° with those of the crossed nicols ; **the** interference color is thus at its maximum. **A** thin **mica** plate **is** then placed either in the lower end of the tube of the microscope or between the ocular and upper analyzer in such a position that its previously determined axes of elasticity are parallel **to those of the** plate under investigation. The difference **of** phase of the rays will **then** be increased when equivalent **axes of** elasticity **in the two** plates fall together, and will be diminished when unequivalent axes of elasticity cover one another. In the first case the mica plate acts like a thickening of the plate, and the interference color must rise in the scale ; in the second case it acts like a thinning of the plate, and the interference color must fall. Since in the mica plate the axis of smallest elasticity lies parallel to the plane of its optic axes, which is easily determined, the observation leads directly to the desired result. The same result may be obtained by using a quartz plate cut parallel to its axis.

For strongly doubly refracting plates (for example the microscopic zircons of rocks) it is well to use a quartz wedge to determine the relative values of the axes of elasticity. Such a quartz wedge is cut so that one of its faces is exactly parallel to the principal axis (optic axis, axis of least elasticity), while the other face makes a very small angle with it. The long side of the wedge gives the direction of the principal axis. If the wedge is pushed between crossed nicols so that its principal axis is inclined 45° to the principal sections of the nicols, then the whole series of Newton's colors from iron-gray of the first order through the second or third order appears in a succession of bands if one moves the wedge forward toward its thin edge ; when moved in the opposite direction the succession is reversed. If at the same time the plate to be investigated lies with its principal section inclined at 45° to those of the nicols, the color of the quartz wedge will be changed in the place where it covers the plate, and the new color will be that of a thicker part of the quartz wedge when the axis of smallest elasticity of the plate lies parallel to the axis of the wedge. On the other hand, the new color will correspond to that of a thinner part of the wedge when the axis of greatest elasticity of the plate is parallel to the axis of the wedge.

Such a quartz wedge also serves to determine the order of the interference color of a doubly refracting plate. Suppose the plate shows red, and that its principal section is turned 45° to those of the nicols. If the quartz wedge is pushed between the ocular and upper nicol with its thin edge forward, so that its axis of smallest elasticity is parallel to the axis of greatest elasticity in the plate, then the interference color must descend in the scale as thicker parts of the quartz wedge come to lie over the plate. The plate then shows one after another the Newton colors in descending order, until the acceleration of one of the rays in the plate exactly corresponds to the retardation of the same in the quartz wedge. At this instant it is the same as though the plate were crossed by an exactly similar plate of the same substance. The plate must appear gray or black according to the strength of its dispersion of color. If during this operation the original color of the plate (red) recurred n times, then the original color of the plate must have been of the $n + 1$ order.

Determination of the Index of Refraction in Doubly Refracting Plates.

Owing to the extreme thinness of the sections used in microscopical investigation, and to the generally small double refraction of the rock-making minerals, the same methods may be used for determining the coefficient of refraction which were given for isotropic media in thin plates, with no greater error than the conditions of the case necessarily impose. Consequently the plate to be investigated must first be brought into its position of darkness between crossed nicols, and after the analyzer has been removed, the polarizer being retained, the index of refraction for the ray vibrating parallel to the principal section is determined according to the method given on page 28, the position of the plate of course remaining unchanged. If the plate is then rotated 90° in its plane, the index of refraction for the second ray may be found in the same manner. The values found are only the principal indices of refraction of the mineral, when the plate from which they were derived has been cut at right angles to an axis of elasticity The best signal which can be used is a microscopic photograph on glass of a newspaper clipping with various-sized type, or a system of crossed lines. This may be fastened with wax to the lower end of the polarizer, and the reduced image which is projected above the condensing lens of the polarizer used to focus on. The following table gives the indices of refraction of the most important rock-making minerals, arranged in descending order.

Rutile	2.759	Diopside	1.680	Brucite	1.567
Anatase	2.524	Axinite	1.680	Dipyre	1.554
Cassiterite	2.029	Olivine	1.678	Muscovite	1.551
Zircon	1.987	Bronzite	1.668	Quartz	1.551
Titanite	1.910	Sillimanite	1.660	Cordierite	1.542
Pyrope	1.812	Glaucophane	1.644	Nepheline	1.540
Aegerine	1.808	Andalusite	1.638	Albite	1.532
Almadine	1.766	Apatite	1.637	Sanidine	1.524
Corundum	1.764	Anthophyllite	1.636	Cancrinite	1.515
Grossular	1.761	Tourmaline	1.635	Leucite	1.508
Epidote	1.756	Actinolite	1.629	Haüyne	1.499
Staurolite	1.753	Melilite	1.629	Sodalite	1.488
Vesuvianite	1.726	Tremolite	1.623	Analcite	1.488
Disthene	1.724	Dolomite	1.622	Natrolite	1.480
Spinel	1.717	Topaz	1.620	Opal	1.455
Zoisite	1.695	Calcite	1.601	Fluorite	1.435
Coccolite	1.690	Meionite	1.578	Tridymite	1.428
Hypersthene	1.685	Chlorite	1.577		

d. Investigation of Minerals in Convergent Polarized Light.

As observation in parallel polarized light furnishes the means of determining the position of the axes of elasticity in a doubly refracting plate, and of establishing by a number of such determinations on plates from different positions in a crystal the orientation of the axes of elasticity with respect to the crystal axes, and consequently the crystal system, so observation in convergent polarized light serves to distinguish uniaxial crystals from biaxial, to determine the dispersion and the optical character, and finally makes it possible to decide whether a plate which appears isotropic in parallel polarized light belongs to an isotropic substance or to an anisotropic one that has been cut at right angles to its optic axis.

Means of Observation.—The Nörremberg polarization instrument is commonly used for macroscopical investigations of minerals in convergent polarized light. But for microscopical investigation this is not applicable, therefore the microscope has been arranged for observations in convergent light.

If on the same metal tube which holds the polarizer a strong condensing lens or system of lenses be screwed, and this be pushed as close as possible to the object by raising the polarizer, the glasses on either side of the object being as thin as possible, and if a strong objective lens be brought as close as possible to the object, then the plate under investigation will be in exactly the same condition as in the Nörremberg apparatus, and the strongly divergent bundle of rays which traverse the plate will be united by the objective to form an image. If the ocular be removed and the analyzer be in place, the diminished image will appear at a somewhat greater distance than before. It will be extremely sharp and clear, but very small. This method was first proposed by A. v. Lassaulx.

In order to obtain a larger image and retain the cross-wires which are situated in the ocular, E. Bertrand introduced into the tube of the microscope above the focus of the objective a weak condensing lens, which unites with the ocular of the microscope to form a new microscope for observing the image projected by the objective system of lenses. The field of view is larger the stronger the system of lenses above the polarizer and in the objective; and in order to prevent total reflection on the layers of air between the plate and systems of lenses with strong convergence, it is well to introduce between them a strongly refracting fluid, such as almond-oil, etc., which cannot injure the lenses or their metal frames.

Interference Phenomena of Uniaxial Plates cut Perpendicular to the Axis in Convergent Polarized Light.

It has already been pointed out that a plate of a uniaxial mineral cut parallel to a basal plane behaves exactly the same in parallel polarized light as a plate of an isotropic mineral. Both remain dark during a complete rotation in their plane between crossed nicols, and are bright between parallel nicols. If the parallel light is replaced by convergent there will be no change of this phenomena with isotropic plates, for in no case can there be a separation or change of the planes of vibration of the rays coming from the polarizer. On the other hand, basal sections of uniaxial crystals show polarization phenomena in convergent light which they do not exhibit in parallel light. We will confine ourselves to an explanation of the phenomena which occur between crossed and parallel nicols in convergent light, these being the only positions of practical importance.

Let A (Fig. 20) be the cross-section of a basal plate of a tetra-

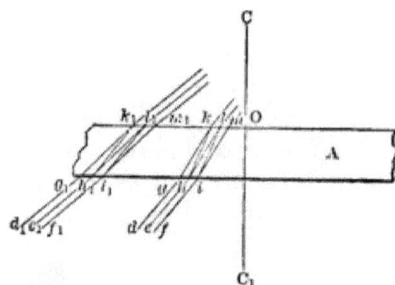

Fig. 20

gonal or hexagonal mineral which is between crossed nicols and is traversed by strongly converging rays of homogeneous light. Let cc_1 be the principal crystallographic axis, and consequently the optic axis. Then all rays from the polarizer entering the plate at this point with perpendicular incidence, and therefore passing through it parallel to the optic axis, will experience no alteration whatever ; when they reach the analyzer they enter it as ordinary rays, and are totally reflected from the balsam film ; the middle of the plate will then appear dark. A ray, fi, on the contrary, upon entering the plate will be separated into two rays, im and il, of which one vibrates in the principal section, $fice_1$, the other at right angles to it. In the same way the ray eh parallel to fi will be divided into the rays hl and hk, and so on. Thus there emerge, as at l, from every point on the plate two rays in parallel directions, an ordinary and an extraordinary ; one of each of these pairs of rays has traversed the plate with a different velocity from the other, and moreover their paths in the plate il and hl are of different lengths. They are thus at l in different phases of vibration, and have planes of vibration at right angles to one another. Reaching the analyzer, each of two such rays splits up into an ordinary and an extraordinary component. The ordinary ones are totally reflected from the balsam film ; the extraordinary come to an interference through the difference of phase which they acquired in the plate, since they are now reduced to the same plane of vibration. The intensity of these interference rays will be expressed approximately by the formula on page 58,

$$I^2 = \sin 2\rho \sin 2(\phi - \rho) \sin^2 \pi \frac{(o - e) V}{\lambda}.$$

In this, as previously shown, $\frac{(o - e) V}{\lambda}$ expresses the difference of phase dependent upon the difference of the axes of elasticity in the plate, or of their reciprocals ; upon the thickness of the plate, and upon the wave-lengths. If this difference of phase is given by $(o-e) V = \lambda$, that is, one wave-length, then $I^2 = 0$ and the point l must appear dark between crossed nicols ; while between l and the focus of the axis it is light. Now since it is evident that for a plate of uniform thickness the difference of phase must be the same for all rays which emerge at the same distance, lo, from the axis of the plate, and have the same inclination to it, so there must appear a continuous row of dark points at a distance lo from the locus of the axis, that is, a dark circle with the radius lo.

At a somewhat greater distance than lo the difference of the phase of the rays which emerge with greater inclination to cc_1 will be $> \lambda$, because the difference $\alpha - \gamma$ increases with the distance from cc_1 and the difference in the paths of the rays within the plate also increases. The plate will be light in such places, and the maximum of illumination will lie at that distance from o, for which the difference of phase of the rays is $\frac{3}{2} \lambda$, for then

$$I^2 = \sin 2 \rho \sin 2(\phi - \rho).$$

Naturally the same must apply to all points equidistant from o, and there must be outside of the dark ring whose radius is lo a bright one with a radius greater than lo.

For still greater distances from o the difference of phase will be $\frac{(o - e) V}{\lambda} > \frac{3}{2} < 2$, the brightness diminishes and reaches a minimum when $\dfrac{(o - e)\,V}{\lambda} = 2$, and at this distance there will be another dark ring.

Proceeding in the same manner, it is evident that a basal plate of a uniaxial mineral in homogeneous light between crossed nicols must show a dark centre, and a succession of light and dark rings. The distance of the dark rings apart depends on λ, consequently it is different for red, yellow, and other lights ; it depends on $o - e$, that is, on the strength of the double refraction in the plate, or on the difference between its indices of refraction ω and ϵ, and on the thickness of the plate, with which indeed both the length of the path of the rays and their difference of phase increase. The diameter of the rings is proportional to the wave-lengths, and inversely proportional to the thickness of the plate and the strength of the double refraction. The distance of the rings from one another decreases with the distance from the centre of the field. The number of visible rings is naturally inversely proportional to their diameters.

Since for the dark rings of such a plate $I^2 = 0$, the darkness at all points of such a ring is absolute. But for the light rings

$$I^2 = \sin 2\rho \sin 2(\phi - \rho) \; ;$$

their brightness therefore is not the same at all points, but is dependent on ρ, that is, on the angle which the principal section of the plate makes with the principal sections of the nicols. Now, from a previous definition, the principal section in a uniaxial crystal is the plane passing through a ray and the optic axis. Consequently for the rays emerging at m, n, r, m_1, n_1, r_1, (Fig. 21,) of the light ring HR, mo, no, or, etc., are the principal planes, and $< mop$, nop, rop, etc., correspond to the angle ρ of the formula, and $< moa$, noa, roa, etc., to the angle $\phi - \rho$. I^2 is evidently a maximum when $\rho = \phi - \rho = 45°$, and a minimum when $\rho = 0$, $\phi - \rho = 90°$, $\rho = 90°$, $\phi - \rho = 0°$. The light ring has therefore a maximum of intensity at n and n_1, and at the corresponding points of both the other quadrants, if pp_1 and aa_1 are the principal sections of the nicols and $nop = 45° = n_1op_1$. On the other hand, the intensity of the light ring is 0 at a, a_1, p, and p_1. The same is true for all the other light rings, and the whole figure of alternating concentric dark and light rings is therefore traversed by a dark cross,

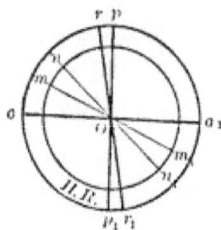

Fig. 21

whose arms are parallel to the principal sections of the polarizer and analyzer (Fig. 22). Since all radial directions about the optic axis are alike, a rotation of the plate in its plane does not alter the phenomena, the cross and rings remaining fixed.

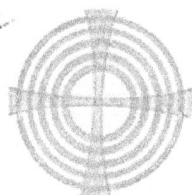 

Fig. 22   Fig. 23

Between parallel nicols the appearance is reversed, those parts which were dark before being light, and the light parts being dark (Fig. 23).

If white light be used instead of homogeneous light, then in the place where a black ring occurs for red light will be a light ring for yellow, green, and other kinds of light. Since the color depends on λ, and the value $\dfrac{(o-e)\,V}{\lambda}$ can only be a minimum for one color, therefore with white light between crossed nicols there will be a series of colored rings, or isochromatic circles. The dark cross parallel to the principal sections of the nicols must also be present for white light, because it is determined by the factor $\sin 2\rho \sin 2(\phi - \rho)$, which is independent of the wave-length. With parallel nicols a white cross will replace the dark one, and the colors of the isochromatic circles will be the complementary ones to those which appear between crossed nicols.

If the observed plate is cut exactly parallel to the basal plane of the crystal, the locus of its optic axis will coincide exactly with the optic axis of the microscope; it will lie at the intersection of the cross-wires, which will bisect the arms of the cross of the interference figure, and this will not alter its position upon the rotation of the stage of the microscope. But if the section is not parallel to the base, the optic axis of the plate will be inclined to the axis of the microscope, and the interference figure will be eccentric in the field of view. During a revolution of the section the centre of the interference figure will describe a circle about the point of intersection of the cross-wires, whose radius is proportional to the inclination of the plate to its optic axis.

The inclination may be so great that only a small peripheral part of the interference figure can be seen at one time. Such uniaxial in-

terference figures are distinguished from those of biaxial bodies by the
fact that for every position during a complete rotation of the plate the
arms of the cross move parallel to themselves and to the cross-wires.

*Uniaxial plates cut parallel to the axis observed in convergent po-
larized light* exhibit in homogeneous light alternating dark and light
curves, and in white light colored curves of hyperbolic form whose
distinctness increases with the thinness of the plate. They are of no
practical importance for mineral diagnosis, but must not be confounded
with isochromatic curves of biaxial crystals, as may easily happen upon
a superficial inspection.

Plates of Biaxial Crystals cut Perpendicular to an Axis in Convergent Polarized Light.

Since in all biaxial crystals the optic axes have different
positions for different colors (are dispersed), strictly speaking a
mineral plate can only be normal to an optic axis for light of a particular wave-length. But the dispersion of the optic axes in most
cases is so small that it may be neglected. Let us assume that the
angle between the optic axis is great (60° − 90°), for otherwise the
phenomena in plates perpendicular to an axis would be identical with
those in plates but slightly inclined to a bisectrix. Suppose the light

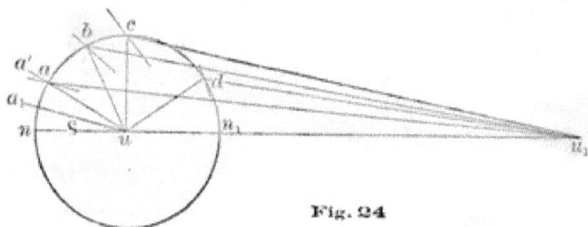

Fig. 24

employed is homogeneous, and that the nicols are crossed. In Fig. 24
let u be the point of emergence of the optic axis to which the plate is
perpendicular, u_1 the point of emergence of the second axis, nn, the
projection of the principal section of the polarizer, and at right angles
to it that of the principal section of the analyzer. Those rays coming
from the polarizer which traverse the plate parallel to the optic axis
do not alter their plane of vibration, and consequently are totally reflected in the analyzer without decomposition. The plate is dark at
u. For the same reason all rays from the polarizer emerging along
the line nu, experience no decomposition in the plate, because its
principal plane, nu n_1u_1, is parallel to their plane of vibration, consequently for them sin $2\rho = 0$. There must be a dark bar in the inter-

ference figure to which the projection on it of the plane of the optic axes in the plate is parallel. Rays from the polarizer emerging at any other point of the plate must be separated into an ordinary ray vibrating at right angles to the principal section, and an extraordinary one vibrating parallel to the principal section; and, for the same reason as that given for uniaxial plates at right angles to the axis, there will emerge at every point of the plate two rays with parallel direction and perpendicular planes of vibration.

The principal section for the rays emerging at a is found by connecting a with u and $u_{,}$, and, bisecting the angle $uau_{,}$, the extraordinary ray emerging at a vibrates parallel to the line bisecting the angle $uau_{,}$, the ordinary ray vibrates at right angles to it. If $a_{,}u$ is drawn parallel to the line bisecting the angle $uau_{,}$, then the angle $nua_{,} = \rho$. Both of the rays emerging at a will be separated in the analyzer into ordinary and extraordinary components; the ordinary components will be totally reflected on the balsam film, and the extraordinary components will produce an interference ray of the intensity

$$ I^{2} = \sin^{2} \rho \, \sin 2(\phi - \rho) \, \sin^{2} \pi \, \frac{(o - e) V}{\lambda}. $$

If $\dfrac{(o - e) V}{\lambda}$ is a whole number, that is, if one ray has gained upon the other in the plate by a number of whole wave-lengths, then $I^{2} = 0$ and the plate is dark at a. All points for which the difference of phase of the two rays is the same number of whole wave-lengths, as at a, may be united with the point a to form a dark curve. But since in biaxial crystals the differences of elasticity are not the same for all directions which have the same inclination to an optic axis, but are different and are of such a kind that the difference from 90° to 90° reaches a maximum and a minimum, then this dark curve cannot be a circle, but must be an ellipse. The eccentricity of this ellipse, however, is very small, because the three axes of elasticity differ but little from one another. If we consider a point a' at such a distance from u in the direction ua that the difference of phase of the two rays will be $\dfrac{(o - e) V}{\lambda} = \dfrac{2n + 1}{2}$, then the point a' will appear light, and there must be moreover an endless number of points for which $\dfrac{(o - e) V}{\lambda}$ has the same value, and which therefore unite with a' to form a light elliptical curve, concentric with the dark one through a.

Continuing in the same manner, it is evident that plates which are cut at right angles to an optic axis of an orthorhombic, monoclinic, or triclinic crystal when observed between crossed nicols in convergent polarized light must show concentric, light, and dark curves of nearly circular form, which are traversed by a dark bar parallel to the projection of the plane of the optic axes of the plate.

The number of rings is dependent on the difference $o - e$, the thickness of the plate, and the wave-length of the light employed. The form of the interference figure is represented in Fig. 25.

Fig. 25

If white light be used in place of homogeneous light, the dark bar will remain unchanged, since its presence is independent of that part of the formula containing λ, while for the same reasons that were given in discussing the interference figure of a uniaxial crystal there must appear isochromatic curves in the place of the dark and light ones. That a dark bar should appear, and not a dark cross as in the case of uniaxial interference curves, is due to the fact that for no other rays than those emerging along the line nn, are the principal sections parallel or perpendicular to that of the analyzer, as Fig. 24 shows for the rays emerging at b,c,d.

If a plate cut at right angles to an axis of a biaxial mineral be rotated between crossed nicols from the position in which it has just been discussed, the bar will bend into the arm of a slightly curved hyperbola whose convexity is turned toward the second axis, and it will straighten itself to a bar again when the plane of the optic axis lies parallel to the principal section of the analyzer. Upon further rotation it assumes the hyperbolic curve, and after a rotation of 180° assumes the original position. If the plate was not cut exactly at right angles to an optic axis, the centre of the interference figure will lie eccentrically in the field of view, and during a rotation of the plate in its plane it will describe a circle about the point of intersection of the cross-wires. The dark bar always bisects the field of view when the plane of the optic axes is parallel or at right angles to the principal plane of the polarizer; in all other positions it shows distinctly the form of an hyperbola whose pole coincides with the locus of the optic axis within the field of view, and whose convexity is turned toward the pole of the other axis.

Plates of biaxial minerals cut perpendicular to the acute bisectrix, examined in convergent polarized light between crossed nicols with homogeneous light, when the angle between the optic axes is not too great, exhibit light and dark curves, which may be easily derived from what has been said of the origin of such curves in plates cut perpendicular to an optic axis. If the plane of the optic axes in the plate lie parallel to the principal section of the polarizer or analyzer, then the loci of both the axes lying at equal distances from the centre of the field of view will be dark, and each axial point will be surrounded by closed, light, and dark curves, part of which enclose only one axis, and part both axes together, and belong to a group of lemniscates. The innermost axial rings have an oval form, consequently there is one curve which has the shape of an ∞, and whose point of intersection lies in the centre of the interference figure.

For very weak double refraction or very slight thickness of the plate, thus for a small value for $\dfrac{(o-e)\,V}{\lambda}$, and consequent great distance apart of points for which $\dfrac{(o-e)\,V}{\lambda} = n$, the inner, separated curves are wanting from about the axial points and the first dark lemniscate encloses both axes.

The interference figure in the position given is traversed by a small and sharp dark bar which connects the two axial points, and also by a broad dark stripe increasing rapidly in width outward which stands perpendicular to this direction (Fig. 26). Both dark bars arise from the fact that the rays from the polarizer have not suffered any

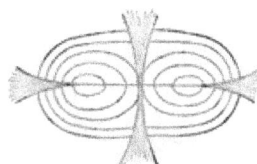

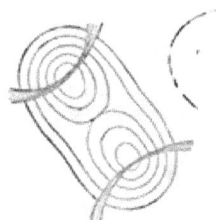

Fig. 26   Fig. 27

decomposition in the plate along these lines because their plane of vibration stands parallel or perpendicular to the principal section of the plate; they enter the analyzer as ordinary rays and become totally reflected from the balsam film.

If the observation is made with white light, the lemniscates instead of being light and dark will be variously colored, and isochromatic. The dark bars, as before, will remain unchanged.

If the plate is rotated between crossed nicols from the position in

which the plane of the optic axes is parallel to the principal section of
the polarizer or analyzer (parallel position), the isochromatic curves
about the axes remain unchanged except that they rotate with the axes
in the field.   But the dark cross opens to equilateral hyperbolas whose
poles each lie at the locus of an optic axis, and after a rotation of 45°
(diagonal position) the interference figure has the appearance shown
in Fig. 27.

If the principal sections of the nicols are parallel, the complementary phenomena are presented, as has been explained for plates of uniaxial substances.   The distance apart of the loci of the optic axes is
naturally a measure of the size of the angle between the optic axes,
and is therefore independent of the thickness of the plate.

Dispersion of the Axes.—The interference figures of biaxial
plates cut perpendicular to the acute bisectrix serve to show the dispersions of the optical constants, which are different for each of the
crystallographic systems without a principal axis, and therefore furnish
conclusive evidence as to the crystal system of the plate under investigation.   It can be taken as a rule that the grade of symmetry of an
interference figure is the same as that of the crystal face on which it
lies, and that the same planes are planes of symmetry for the crystal
faces and for the interference figure.

In the *orthorhombic system* the bisectrices cannot experience any
dispersion, but the optic axes may, under condition that their dispersion, $\rho > v$ or $\rho < v$, must be symmetrical with respect to the bisec-

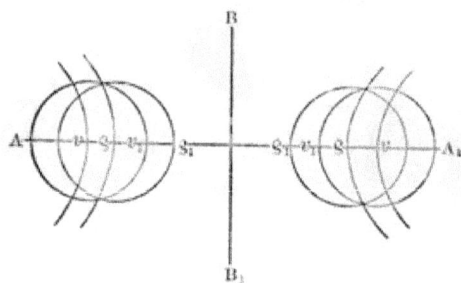

Fig. 28

trices.   Therefore if a plate cut perpendicular to the bisectrix of an
orthorhombic mineral be observed between crossed nicols in red and
then in blue light, the loci of the axes and also the dark rings will not
coincide in the two cases, but according as the angle of the red or of
the blue axes is greater they will have different positions.   In Fig. 28
it is assumed that the angle of the red axes is the smaller, and that at ρ

will lie the dark axial spots and the poles of the hyperbolas in the diagonal position, and ρ_1 will be the first dark ring in red light. In blue light the poles of the dark hyperbolas may be assumed to lie at v, and the first ring at v_1. Between ρ_1 and v_1 will lie the first dark rings, and between ρ and v the dark hyperbolas, for orange, yellow, and green. Then in white light there will appear on the first ring at ρ_1 a combination color in which red is excluded; at v_1 one in which blue is excluded. This first colored ring, therefore, will appear red at the spot lying nearest the middle point or within, that is, nearest the pole of the hyperbola; the second, which is without or farthest from the pole, will be blue. The colors in the outer rings blend into one another because so many light rings are superimposed, but within they are more distinct. The inner red, therefore, is very distinctly seen, while the outer blue is less so. In that part of the first ring which is farthest from the centre of the interference figure the order is naturally reversed, blue being within and red without. The first is distinct, the second indistinct or blended. It is the same with the other rings, but only the innermost ring is used for observation because the phenomenon in it is clearest.

On the dark hyperbolas which occur between v and ρ, Fig. 28, red is extinguished on the convex side, and this must therefore appear edged with blue. On the other hand, blue is extinguished on the concave side, and this must be edged with red. From this is derived the

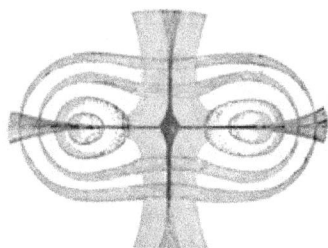

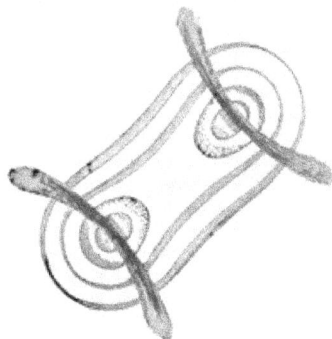

Fig. 29   Fig. 30

rule that the axial angle is smallest for the color which appears within that part of the first ring which is nearest the centre of the figure, and which in the diagonal position borders the concave side of the hyperbola. On the other hand, that color appears on the convex side of the hyperbola and in that part of the innermost ring farthest from the centre of the figure for which the axial angle is the greatest. Figs. 29

and 30 show orthorhombic interference figures in parallel and diagonal position with the dispersion $\rho<v$ (the red being indicated by stippling, the blue by lining). They show as in Fig. 28 that the figures are symmetrical with respect to the axial plane $AA_{,}$, and to one, $BB_{,}$, normal to it. The interference figure lies on a pinacoid of the orthorhombic system, and like this is therefore bisymmetric.

In the *monoclinic system* the relative size of the axial angle may be recognized by the same phenomena as in the orthorhombic interference figures. But there occurs in this a particular disposition of the colors which is determined by the dispersion of the axes of elasticity

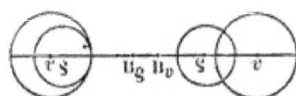

Fig. 31

lying in the plane of symmetry, and which is added to the dispersion of the optic axes. If the orthodiagonal is the axis of mean elasticity and the plane of the optic axis, therefore, lies in the plane of symmetry, then both bisectrices for the different colors are dispersed and one cannot properly speak of a section at right angles to the acute bisectrix, but only of one perpendicular to one of the acute bisectrices. For this dispersion (the inclined) the dark axial spots between crossed nicols in red light would appear at ρ (Fig. 31) equally distant from the locus of

Fig. 32

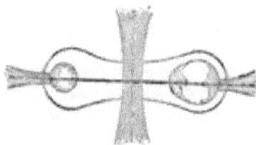

Fig. 34

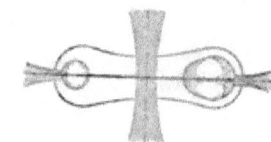

Fig. 33

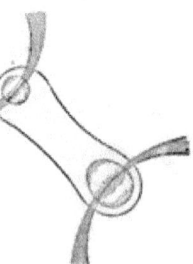

Fig. 35

the bisectrix B_ρ, and the first dark rings in place of the small circles about ρ. For the case $\rho<v$ the dark axial spots in blue light would lie at v equally distant from B_v, and the first dark rings in place of the larger circles about v. The axial figures are thus displaced with

respect to one another, and on using white light the disposition of the colors within the inner rings and on the edges of the hyperbolas is symmetrical to the plane of the bisectrices, but not to one at right angles to it. If the dispersion of the bisectrices is not large, then the colors lie symmetrical with respect to the dark bar at right angles to the plane of the optic axes as their order of succession goes, but they have different intensities on either side, and the inner rings are more elliptical on one side than on the other, as is shown in Figs. 32 and 33, which have been derived from the scheme for gypsum, Fig. 31. With more strongly inclined dispersion, as in diopside, the inner rings in analogous places and the hyperbolas on the same sides are colored differently; that is, on one side they are red within and blue without, and on the other blue within and red without, as shown in Figs. 34 and 35. In other cases both of these forms of appearance are combined.

For horizontal dispersion, that is, when b is the obtuse bisectrix

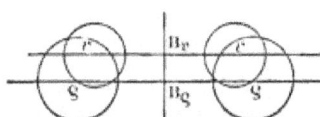

Fig. 36

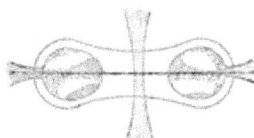

Fig. 37

with the normal symmetrical position of the optic axes, the axial figures with crossed nicols for red and blue color will have the position with respect to one another indicated in Fig. 36, and there can be no section which will be at right angles to all acute bisectrices at once.

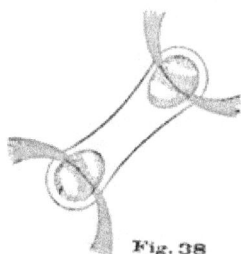

Fig. 38

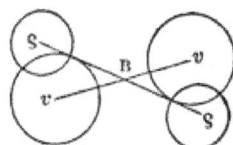

Fig. 39

The obtuse bisectrix is the same for all colors; since it coincides with the axis of symmetry, the acute bisectrices and the axes of mean elasticity are dispersed. In white light the superposition of the axial figures for the different wave-lengths must produce a disposition of the colors which, when the dispersion is $\rho > v$, is shown in Fig. 37 for the parallel position, and in Fig. 38 for the diagonal position. The disposition of the colors is now symmetrical with respect to a plane normal to all of the axial planes, for the different colors, which corresponds to the crystallographic plane of symmetry. Finally, if the axis of symmetry b is the acute bisectrix, the dispersion is crossed, for the axial planes for red and blue light will show a displacement with respect to one another, as shown in Fig. 39. The acute bisectrix is then the same for all colors; the obtuse bisectrix and axis of mean elasticity are dispersed. In white light the partially overlapping axial figures would produce a disposition of colors, for $\rho > v$, which is shown in Fig. 40 for the parallel position, and in Fig. 41 for the diagonal posi-

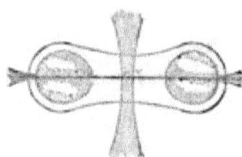

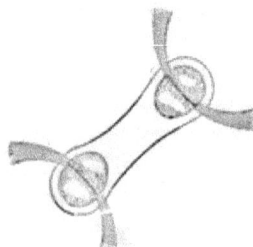

Fig. 40   Fig. 41

tion. There is no longer any plane with respect to which the interference figure is symmetrical, for it lies within the plane of symmetry. But the disposition of the colors is symmetrical with respect to the locus of the acute bisectrix, since this is the crystallographical axis of symmetry.

In *triclinic crystals* all the axes of elasticity are dispersed; the dispersions shown by the interference figure in convergent light are therefore different for different colors. The disposition of the colors in the axial figures is consequently entirely unsymmetrical.

Determination of the Optical Character of Doubly Refracting Plates in Convergent Light.

The commonest methods for determining the optical character of doubly refracting minerals in convergent light are based on the fact that the isochromatic curves of an interference figure, since they unite all the points of a plate at which the emerging rays suffer the same retardation, must experience an alteration if the difference of phase of the rays is increased or diminished. With uniaxial crystals this is accomplished most conveniently by the insertion of a mica plate of such thickness that the two component rays produced in it by

double refraction emerge with a phasal difference of $\frac{1}{4}$ of a wave-
length (Viertelundulationsglimmerblättchen). Since mica is optically
negative, and the plane of the optic axes stands very nearly perpendicular to the cleavage plane, this latter plane contains the axes of smallest
and of mean elasticity; the former coincides with the line connecting
the loci of the optic axes, and the latter is at right angles to it. The
direction of the plane of the optic axes, and consequently of the axis
c, is marked on the mica plate or on the frame holding it. If one ob-
serves the interference figure of a uniaxial crystal after the mica plate
has been so placed between the object and the analyzer that its axial
plane is inclined 45° to the principal sections of the crossed nicols,
the dark cross will have resolved itself into two dark spots, d and $d_{,}$,
in Figs. 42 and 43, and the isochromatic curves in two opposite quad-

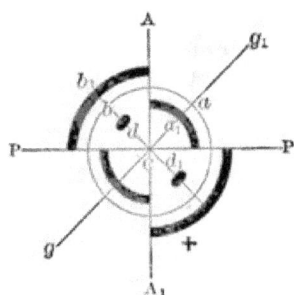

Fig. 42

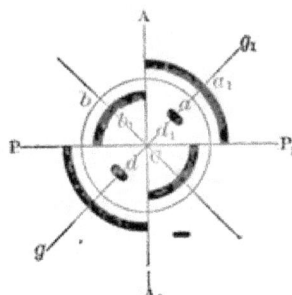

Fig. 43

rants will have contracted, and in the other two will have expanded.
The crystal under investigation is optically positive (Fig. 42) when the
line joining the dark spots is at right angles to the axial plane of the
mica plate (forms with it the sign $+$); the crystal, on the other hand,
is negative (Fig 43) when this connecting line is parallel to the axial
plane of the mica plate ($-$). This is explained by the following consideration: Let $PP_{,}$ and $AA_{,}$ (Fig. 42) be the principal sections of
the polarizer and analyzer, $gg_{,}$ the axial plane of the mica plate, and in
homogeneous light let aba be the first dark ring, that is, the location
of all points of the plate for which $\dfrac{(o-e)\,V}{\lambda} = 1$. The plate is assumed to be optically positive, $c = a$. Of the rays emerging at the
point a of the first ring, the extraordinary will vibrate in the principal
section ac, and will be retarded a wave-length behind the ray vibrating
at right angles to the principal section. In passing through the mica

plate the extraordinary ray will be again retarded $\frac{1}{4}\lambda$ behind the ordinary ray, since it vibrates parallel to the axis of least elasticity in the mica. The phasal difference of both rays when they enter the analyzer is therefore $\frac{3}{4}\lambda$ and the point a can no longer be dark; the dark ring has therefore moved, and will be found at a point a_1 where the phasal difference of the rays emerging from the plate is $\frac{3}{4}\lambda$, which will become λ on passing through the mica plate. The same takes place in the opposite quadrant PcA_1.

For the rays emerging at the point b of the first dark ring the principal section is bc. Here also the extraordinary ray is retarded a wave-length behind the ordinary. But upon their entrance in the mica plate the extraordinary ray vibrates parallel to the axis of greater elasticity. The extraordinary ray is therefore accelerated with respect to the ordinary one, and by $\frac{1}{4}$ of a wave-length. Upon entering the analyzer the phasal difference will be only $\frac{3}{4}\lambda$ and this spot will not appear dark. The first dark ring will be moved to a point b_1, where the phasal difference of the rays emerging from the plate is $\frac{5}{4}\lambda$. The first dark ring is thus divided into four arcs of 90°. In those quadrants through which the axial plane (gg_1) of the mica plate passes there occurs a contraction, and in the other two quadrants a dilation. The same change affects the other rings of the figure.

The reverse phenomenon of Fig. 43, which is found for negative uniaxial crystals, is explained in a similar manner.

In plates of biaxial crystals, also, cut perpendicular to the bisectrix the character of the double refraction may be determined by means of the mica plate. The plate for investigation is placed in parallel position between crossed nicols, and the mica plate is inserted so that its axial plane bisects the angle between the principal sections of the nicols. If the plate in question is optically positive, its bisectrix being the axis of least elasticity, then the axial rings are contracted in the quadrants through which the axial plane of the mica plate passes, and are widened in both the others.

With negative character of the plate under investigation the axial rings are dilated in the quadrants through which the axial plane of the mica plate passes, and contracted in both the others.

The phenomenon of the dilation and contraction of the axial rings effected by the mica plate is not always easily recognized. It is then better to employ a quartz wedge such as already described for determining the optical character of doubly refracting plates through the change in their interference color. In such a quartz wedge, since quartz is optically positive the axis of greater elasticity lies parallel to the thin edge of the wedge ; the axis of least elasticity lies parallel to the long edge.

If a plate cut perpendicular to the bisectrix is placed in the diagonal position between crossed nicols, there will be a certain number of axial rings in the interference figure which surround each axis separately, and around both of these groups will be the lemniscates. For optically positive crystals the bisectrix perpendicular to the plate is c; the line connecting the hyperbolas $= a$, that at right angles to this $= b$. If the quartz wedge is inserted horizontally anywhere between the plate and the analyzer, so that the long edge (c) will advance parallel to the line connecting the hyperbolas, then the ray which in the plate vibrates parallel to a and is accelerated will vibrate parallel to c in the quartz wedge and will be retarded. The ray vibrating parallel to b in the plate and there retarded will in the quartz wedge vibrate parallel to a and be accelerated ; therefore the two rays emerging at any point of the first dark circle whose phasal difference in the plate is λ will have a smaller difference ; the same is true of all points of the second dark ring, and so on : the rings must therefore widen. The farther forward the quartz wedge is pushed the wider the rings must become. They move from the axial spots toward the centre of the interference figure, and finally open into lemniscates. It is as though the plate became thinner and thinner.

If the long edge of the quartz wedge be moved at right angles to the line connecting the hyperbolas, then a in the quartz coincides with a in the plate, and c in the quartz coincides with c in the plate. The effect is as though the thickness of the plate were increased, and with the advance of the quartz wedge the axial rings will approach the axial spot. The movement of the interference figure will be found to be from without inward.

For negative crystals the phenomena are reversed.

c. Color of Minerals.

The color of minerals in reflected light can only be used as a criterion when it is an inherent color, and especially when the minerals will not yield transparent plates, which is particularly the case with the ores, many of which in the series of oxide ores, as magnetite, ilmenite, and hematite, are among the most widely distributed constituents of rocks, while others in the group of the sulphide ores, as pyrite and pyrrhotite, are frequently accessory constituents. To the idiochromatic transparent minerals which are widely distributed in rocks belong certain oxides, as rutile, and especially silicates with heavy metallic bases, as the micas, pyroxenes, amphiboles, garnets, tourmaline, etc. Color can seldom be used for determining minerals in consequence of the great variety of the colors which are due to the stage of oxidation of the metals (iron and manganese), and to their relative proportions in combination with isomorphous molecules of other elements. In the case of stained minerals, which are of themselves colorless and only derive their color from foreign substances of inorganic nature, the color has naturally no determinative importance. The microscopical investigation of such allochromatic substances shows either that their pigment is present in well-defined and recognizable minute plates, needles, or grains, in the form of inclusions, and then very often irregularly disseminated through the whole mass, or that the coloring matter cannot be recognized as separate from the colored mass. When a pigment is present in the latter form it is called a dilute one, and it is a peculiarity of such dilutely stained bodies that their color disappears more or less completely when they are cut sufficiently thin. It is generally assumed—and the phenomenon of pleochroism in allochromatic, dilutely stained bodies necessitates this assumption—that in this case the pigment is dispersed through the intermolecular spaces of the substance. Chemical investigation has shown that extremely small amounts of dilute pigments can often produce a very intense coloration.

If the color of a body in incident light arises from the fact that not all of the rays of incident white light but only those of certain wave-lengths are reflected, while those of other wave-lengths are absorbed, then the colors of this body in transmitted light are determined by the absorption of certain rays. It is known that luminous waves are always weakened by their passage through transparent media; that they are indeed completely extinguished when the thickness of the layer traversed is sufficiently great. Now, since the elasticity of the luminiferous ether in isotropic media is the same in all directions, it must be assumed that the weakening of a luminous wave traversing them will be independent of its direction. Their degree of transparency therefore must only depend on a coefficient of absorption peculiar to the substance, and on its thickness, if the rays of all wave-lengths are equally absorbed. If the absorption is especially confined to rays of certain wave-lengths, its color must still be independent of the direction.

It is different with anisotropic media, since in them the elasticity of the ether and therefore the velocity of the light changes with the direction, and it may be assumed that the absorption of the light would also be different in different directions, and that for like absorption of rays of all wave-lengths the transparency of equally thick plates may be different if the plates have been cut in different directions. And if the absorption of the rays for all wave-lengths is not the same, rays of different wave-lengths may be absorbed more in one direction than in another. Then plates of such isotropic media which have been cut in different directions will show different colors in transmitted light. This phenomenon of color-absorption of doubly refracting bodies, which changes with the direction, is called *pleochroism*.

Pleochroism of Uniaxial Minerals.—If a uniaxial crystal is looked at in such a way that the rays of light strike at right angles to its base, only ordinary rays will reach the eye, and the color shown by the crystal in this direction (basal color) is determined exclusively by the absorption which the ordinary ray experiences. If these rays are investigated in any way by a nicol, the color will always remain the same in whatever way the nicol may be turned about its axis. There is no double refraction in the direction of the principal axis. If now the crystal (tourmaline, beryl, vesuvianite, etc.) be viewed in a direction inclined to the principal axis, the color will be changed, and will differ more from the basal color as the inclination of the ray to the principal axis is greater. The maximum difference in color must be seen when the crystal is viewed at right angles to the principal axis. This phenomenon is explained by the fact that for an inclined position of the principal axis to the direction of the rays a double refraction of the rays takes place; with the ordinary ray is associated an extraordinary ray whose velocity and absorption differ the more from those of the ordinary ray the nearer the principal axis lies to the direction of the rays. Consequently the colors shown by a uniaxial crystal in any other direction than that of its principal axis are determined by the combination of the ordinary and extraordinary rays, each of which is absorbed differently.

It is customary to consider only the extreme cases, that is, when the principal axis is parallel and perpendicular to the direction of the rays, and to say that uniaxial crystals possess dichroism; which is not strictly correct, since the color changes steadily with the direction.

The absorption belonging to each of the rays traversing a doubly refracting plate may be observed by means of the polarizing microscope. If the polarizer be in place and the analyzer be removed, then, by rotating the stage of the microscope until the principal section of

the plate be brought first parallel and then at right angles to the
principal section of the polarizer, the plate in the first case will be
traversed by an extraordinary ray only, and in the other by an ordinary
ray, and the particular absorption of each can be tested. If the polarizer be removed and the analyzer retained, it will be found that a portion of the light reflected from the mirror is polarized, which interferes
with the observation.

Pleochroism of Biaxial Minerals.—Plates of biaxial minerals cut
perpendicular to an optic axis usually show no pleochroism, but all
other plates of such minerals show a color which changes with the
direction of the plate, if there is any appreciable color-absorption present. This color is composed of the colors of both of the rays which
traverse the plate. If with Haidinger we call these *facial colors*
(Flächenfarben), then, as in the case of uniaxial plates which are not
cut perpendicular to the principal axis, these colors may be separated
by a nicol prism into axial colors, that is, into the colors of the individual rays which traverse the plate. For example, let Fig. 44 repre-

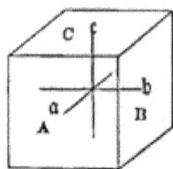

Fig. 44

sent a cube cut from an orthorhombic crystal in such
a manner that each face is perpendicular to an axis
of elasticity; then if we observe perpendicularly incident light through the faces A, B, and C, we shall
have three facial colors with a maximum difference between them. The facial color C is composed of rays
vibrating parallel to a and b. In the same way the
facial colors A and B are composed of rays vibrating
parallel to b and c, and a and c, respectively; therefore three facial
colors and three axial colors are distinguished in biaxial crystals.

It was formerly assumed that the directions of strongest color-
absorption in biaxial crystals, which may be termed the axes of absorption, were coincident with the axes of elasticity. H. Laspeyres has
shown that this is only true so long as the axes of elasticity coincide
with the crystallographic axes of symmetry; consequently it is true for
all three axes of elasticity in the orthorhombic system, and in the
monoclinic system for the axis coinciding with the axis of symmetry,
b; on the other hand, it is not necessarily true for the two axes of elasticity lying in the plane of symmetry of monoclinic crystals, nor for all
the axes of elasticity of triclinic crystals. Indeed H. Laspeyres found
that in manganese-epidote a dispersion of the axes of absorption takes
place in the plane of symmetry independent of the dispersion of the
axes of elasticity a and c.

Pleochroic Halos.—Many minerals, as andalusite, cordierite, muscovite, biotite, diopside, etc., show a peculiar phenomenon, namely, that particular spots in them possess a marked pleochroism, especially in the immediate vicinity of microscopic inclusions, so that, in one of the positions of **extinction**, after the analyzer has been removed there are strongly colored halos around the microscopic inclusions, which halos disappear more or less completely after a rotation of the plate through 90°. In cordierite* and andalusite these halos are bright yellow, and arise from a local aggregation of an organic pigment; they disappear after the mineral has been heated to redness. A. Michel-Lévy † and H. Gylling‡ found that very high heating did not destroy these halos in certain micas, and the former concluded that in this case the phenomenon did not arise from an organic pigment, but might be occasioned by a greater local percentage of ferruginous molecules. Against this explanation is the fact that the phenomenon is confined to the immediate vicinity of an inclusion, and also that the halos are always oval, while one would expect a crystallographic boundary if the phenomenon were confined to an isomorphous shell. Moreover it is well known that micaceous and fibrous substances lose water with great difficulty, even upon very strong and continued heating at redness. §

Colorless minerals may sometimes be rendered pleochroic by artificial coloration. Bořický observed that many minerals occurring in rocks (olivine, bronzite, cordierite) which in their natural state show no pleochroism, or only a weak one, become distinctly pleochroic, and even strongly so, upon being heated to redness. He obtained the best results when the substance (in thin section) was exposed on platinum foil to a bright red heat for 1.5 to 2 minutes. Not infrequently the olivine in rocks of the melaphyre and basalt series is colored red by a more or less advanced separation of Fe_2O_3. This is accompanied quite often, but not always, by distinct pleochroism.

---

* *H. Rosenbusch*, Die Steiger Schiefer und ihre Contactzone an den Granititen von Barr-Andlau und Hohwald. Strassburg, 1877. p. 221.

† Sur les noyaux à polychroisme intense du mica noir. C. R. 1882. xciv. 1196.

‡ Några ord om Rutil och Zirkon med särskild hänsyn till deras sammanväxning med Glimmer. G. F. F. 1882. vi. 167.

§ E. Cohen has recently demonstrated that the pleochroic halos in the biotite of certain granite-porphyries and gneisses are also due to organic pigments, but that it requires a higher temperature to destroy them than is necessary to dissipate those in muscovite and cordierite. (N. J. B. 1888. B. I. 165.)

Aggregates.

Literature.

E. BERTRAND, Du type crystallin auquel on doit rapporter le Rhabdophane, d'après
les propriétés optiques que présentent les corps crystallisés affectant la forme
sphérolithique. Bull. Soc. min. Fr. 1880. III. 58–62 and 1881. IV. 60–61.
— De l'application du microscope à l'étude de la minéralogie. ibidem 1880. III.
93–96.
— Sur les propriétés optiques des corps crystallisés présentant la forme sphérolithique. C. R. 1882. XCIV. 542.
D. BREWSTER, On circular crystals. Trans. Roy. Soc. 1853. XX. part 4. 607–623.
E. MALLARD, Sur quelques phénomènes de polarisation chromatique.   Bull. Soc.
min. Fr. 1881. IV. 66–71.
A. MICHEL-LÉVY, Des différentes formes de sphérolithes dans les roches éruptives
in Mémoire sur la variolite de la Durance. Bull. Soc. géol. Fr. (3). V.
257–266.
— Sur la nature des sphérolithes faisant partie intégrale des roches éruptives. C. R.
1882. XCIV. 465.
H. ROSENBUSCH, Einige Mittheilungen über Zusammensetzung und Structur grani-
tischer Gesteine. Z. D. G. G. 1876. XXVIII. 369–390.

For reasons already given this subject has been transferred from
the chapter on the morphological characters of rock-making minerals
to this place. The term *aggregates*, as here used, includes only those
aggregations which are homogeneous or cannot be shown to be heterogeneous. They may consist of amorphous or of crystalline substances. In general the texture of amorphous aggregates can only be
detected microscopically when they have been rendered doubly refracting from mechanical causes. They then behave like those crystalline
aggregates which do not belong to the isometric system. The characteristic of aggregates lies in the fact that the arrangement of the
more or less regularly bounded individuals, which are crowded together as an aggregation, is neither parallel nor symmetrical. This
irregular crystallographic arrangement causes the optical orientation to
vary with each individual grain of the aggregate. In such an aggregation, when viewed between crossed nicols, the extinction for all the
individuals can never be in the same azimuth; they will show different
colors or different degrees of light and shade, which will depend upon
their thickness, the position of the thin section with respect to their
axes, and the inclination of their principal plane to those of the nicols.
This optical appearance of aggregates between crossed nicols is
called *aggregate-polarization* (Pl. VIII. Figs. 4 and 5), in distinction

from the optical behavior of crystals, which is uniform throughout their whole extent. The boundaries of the individuals forming an aggregate, which in ordinary light often are scarcely noticeable, are very marked between crossed nicols, and show the manner of arrangement or the texture of the aggregate.

Spherical aggregates, so common in the mineral kingdom, deserve special notice. Of these the spherulites (*Sphærocrystals*) already mentioned are a particular case. They consist sometimes of a singly refracting amorphous substance; at others of a crystalline mass arranged in concentric shells, or in radial fibres; sometimes both forms of arrangement occur together, so that the spheres consist of concentric shells which in turn are made up of individuals set at right angles to the shells. More rarely there are spherical aggregates in which both radial and concentric arrangement is wanting.

Radial and concentric aggregates occur with the most different minerals, as for instance calcite and other carbonates, quartz, chlorite, dellesite, feldspar, etc.

If one considers a spherical aggregate of an amorphous substance that is built up of concentric shells each of which exerts a pressure on all those within it, then the density of the sphere will increase toward its centre. Such a sphere may be considered as composed of radial cylinders in which the elasticity in the direction of the axis of the cylinder is greater than at right angles to it, that is, as a radially fibrous aggregate of optically uniaxial, negative crystals.

A central cross-section through such a sphere, or through one made up of orthorhombic crystals, when viewed in parallel polarized light between crossed nicols is divided into four light quadrants separated by a dark cross, whose arms are at right angles to one another and parallel to the principal planes of the nicols. On rotating the section through a complete circle the actual position of the cross does not change with respect to the planes of the nicols, though it appears to rotate in opposite direction to the rotation of the section. The lightest part of the quadrants is along the radii inclined 45° to the principal planes of the nicols, from which it diminishes gradually on both sides to complete extinction along the radii parallel and perpendicular to these principal planes (Pl. IX. Figs. 1 and 2). The cross shades gradually into the light quadrants. If the sphere consists of an amorphous substance, and its double refraction is the result of centripetal condensation, then the color of the quadrants will diminish from the centre outward, which is not the case with a proper spherulite. Such amorphous spheres may therefore show colored rings in parallel polarized light under certain conditions. If the analyzer be rotated until it comes into parallel position with the polarizer, the dark cross will gradually open until a white one at last replaces it, when the light-colored quadrants will appear in their complementary colors.

Sections which have not been cut exactly through the centre of such spherical aggregates show the same phenomena in a less precise form.

If the individuals of a radial aggregate are not grouped about a point, but along a line or plane, there arise distorted spherulites, which Zirkel has called *axiolites*, whose dark cross between crossed nicols can only be closed in four definite positions, which are at right angles to one another, while on rotating the section the cross must be open or be resolved into two hyperbolas in every other position.

Radial aggregates of monoclinic or asymmetric crystals might present the same phenomenon except that the arms of the cross would not in general lie parallel to the principal plane of the nicols, but would be inclined to it at an angle depending on the position of their principal optical plane with reference to the direction of the rays of crystals. But the cross would have four arms at right angles to one another only for the case in which all the needles within the section plane had the same crystallographic and optical orientation with respect to that plane; for if the needles were variously rotated about their axes, they would give rise to a many-armed cross whose arms would be irregular in size and position. Homogeneous spherical aggregates of such crystals are only known at present for certain triclinic feldspars (oligoclase in variolites).

Spherical aggregates of amorphous substances not subjected to strain appear dark in all positions between crossed nicols, while spherical aggregates of crystallized substances in which all the individuals lie parallel to one another must behave like simple crystals, being dark in four positions of rotation at right angles to one another, and light-colored in all other positions. Such spherical aggregates may be intergrown with more or less amorphous material without the phenomena changing.

Spherical aggregates composed of granular individuals, whose dimensions may be greater at the centre, or the periphery, are called *granospherites* (Pl. IX. Fig. 3) in contradistinction to radial and shelly spherulites.

III. Chemical Properties.

Literature.

H. Behrens, Mikrochemische Methoden zur Mineralanalyse. Verslag. en Mededeel. der Kon. Akad. van Wetensch. (2). XVII. Amsterdam. 1881.

E. Bořicky, Elemente einer neuen chemisch-mikroskopischen Mineral- und Gesteinsanalyse. Prag. 1877.

— Beiträge zur chemisch-mikroskopischen Mineralanalyse. N. J. B. 1879. 564.

R. Bréon, Séparation des minéraux microscopiques lourds. Bull. Soc. min. 1880. III. 46.

W. C. Brögger, Om en ny konstruktion af et isolationsapparat for petrografiske undersögelser. G. F. i St. Forhdl. VII. No. 91. 417. 1884.

E. Cohen, Ueber eine einfache Methode, das specifische Gewicht einer Kaliumquecksilberjodidlösung zu bestimmen. N. J. B. 1883. II. 87.

C. Doelter, Ueber die Einwirkung des Elektromagneten auf verschiedene Mineralien und seine Anwendung behufs mechanischer Trennung derselben. S.W.A. 1882. LXXXV. I. 47.

— Die Vulkane der Capverden und ihre Produkte. Graz. 1882.

F. Fouqué, Nouveaux procédés d'analyse médiate des roches et leur application aux laves de la dernière éruption de Santorin. Mém. prés. par divers savants à l'Acad. des sc. 1874. XXII. No. 11.

— Étude microscopique et analyse médiate d'une ponce du Vésuve. C. R. 1874. 12. Oct. 869.

P. Gisevius, Beiträge zur Methode der Bestimmung des specifischen Gewichts von Mineralien und der mechanischen Trennung von Mineralgemengen. Inaug.-Diss. Bonn. 1883.

V. Goldschmidt, Ueber Verwendbarkeit einer Kaliumquecksilberjodidlösung bei mineralogischen und petrographischen Untersuchungen. N. J. B. B.-B. 1881. I. 179.

K. Haushofer, Ueber die mikroskopischen Formen einiger bei der Analyse vorkommender Verbindungen. Z. X. 1880. IV. 42.

— Beiträge zur mikroskopischen Analyse. S. M. A. 1883. III. 436.

— Mikroskopische Reactionen. S. M. A. 1884. IV. 590.

D. Klein, Sur une solution de densité 3.28 propre à l'analyse immédiate des roches. C. R. 1881. XCIII. 318.

— Sur la séparation mécanique par voie humide des minéraux de densité inférieure à 3.6. B. S. M. 1881. IV. 149.

P. Mann, Untersuchungen über die chemische Zusammensetzung einiger Augite aus Phonolithen und verwandten Gesteinen. N. J. B. 1884. II. 172.

A. Michel-Lévy et L. Bourgeois, Sur les formes cristallines de la zirkone et sur les déductions à en tirer pour la détermination qualitative du zirkon. C. R. 1882. 20 Mars; — B. S. M. 1882. V. 136.

K. Oebbeke, Beiträge zur Petrographie der Philippinen und der Palau-Inseln. N. J. B. B.-B. 1881. I. 456.

C. Rohrbach, Ueber die Verwendbarkeit einer Bariumquecksilberjodidlösung zu petrographischen Zwecken. N. J. B. 1883. II. 186.

— Ueber eine neue Flüssigkeit von hohem specifischem Gewicht, hohem Brechungsexponenten und grosser Dispersion. Ann. Chem. Pharm. N. F. 1883. XX. 169.

A. STRENG, Ueber einige mikroskopisch-chemische Reaktionen. N. J. B. 1885.
    I. 21.

J. THOULET, Séparation mécanique des éléments minéralogiques des roches.
    B. S. M. 1879. II. 17.

— Sur un nouveau procédé pour prendre la densité des minéraux en fragments très-
    petits. B. S. M. 1879. 189.

— Triage mécanique des éléments minéraux contenus dans les roches. B. S. M.
    1880. III. 100.

— Contributions à l'étude des propriétés physiques et chimiques des minéraux micro-
    scopiques. Inaug.-Diss. Paris. 1880.

M. WEISKY, Die Mineralspecies nach den für das specifische Gewicht derselben
    angenommenen und gefundenen Werthen. Breslau. 1868.

L. VAN WERVEKE, Ueber Regeneration der Quecksilberjodidlösung und über einen
    einfachen Apparat zur mech. Trennung mittelst dieser Lösung. N. J. B. 1883.
    II. 86.

A chemical investigation of the mineral constituents of a rock may
be undertaken not only for the purpose of confirming an optical diagnosis, but may be necessary in many cases in order to determine the
particular species within a family, or to take the place of an optical
determination, when this is insufficient, as for opaque or isometric
minerals. From the nature of microscopical investigations it often
happens that the chemical methods used in mineral analysis are not
serviceable. The small quantities to which it is necessary to apply the
reagents require an unusual sharpness in the reactions; the impossibility of distinguishing colorless and amorphous precipitates microscopically determines the use of only those reactions which furnish
characteristically distinct coloration or easily recognized crystallizations. In general those methods are to be preferred which furnish
crystallizations that are independent of the relative proportion of the
substances taking part in the reaction, and also of the physical conditions under which the experiment takes place.

The reactions given in the following pages may all be carried on
with easily-devised apparatus and under the ordinary microscope. The
chemical tests may be made on the thin sections themselves, or on the
minerals which have been isolated from the rock mechanically. In the
first case uncertainties might often arise as to which constituent took
part in the reaction; but this uncertainty may often be entirely obvi-
ated. On the other hand, there are particular reactions which can
scarcely be carried on except on a thin section, and so the testing of
the isolated powder must be supplemented by that of the section.

Chemical Investigation of Thin Sections.

Thin sections which are intended primarily for chemical investigation should be left uncovered, and it is better not to polish their upper surface, as the surface exposed to the reagents will then be greater and the chemical action more energetic. If only a part of the section is to be tested, this is separated from the rest by a thread of viscous balsam in the form of a ring, which prevents the drop of the reagent from spreading; the latter may be applied through a capillary pipette. If an already covered section is to be investigated, the glass cover is removed by a knife-edge, and the balsam washed off with a brush dipped in alcohol or ether. If only a part of the section is to be tested, the glass covering this part may be carefully cut across with a diamond, the glass cover removed as before, and the section cleaned in the same way. If the portion of the section to be investigated is very small, and the reagent should not touch any other part, then the glass cover may be entirely removed and replaced by one in which a fine funnel-shaped hole has been bored. When the opening is properly adjusted over the right spot, the balsam is heated and the cover made fast. The balsam in the opening is removed with alcohol. The reagent is then confined to the spot beneath the opening. Glass covers may be prepared beforehand for such purposes by covering them with wax, and after a small circle, 0.5–1 mm. in diameter, has been cleaned in the proper place, subjecting them to hydrofluoric acid until they are eaten through. If the acids to be used on the section would attack glass, perforated platinum foil may be used instead. The treatment of thin sections with weaker or stronger acid serves to detect or remove easily soluble constituents, to distinguish gelatinizing silica from non-gelatinizing, or, finally, to produce etched figures on minerals.

To the more or less easily soluble minerals which are widely distributed in rocks belong the carbonates, phosphates, and many iron ores. Upon the solution of the carbonates, of which calcite is soluble in acetic acid, others in cold hydrochloric acid, and still others only in hot acid, there occurs an effervescence through the escape of carbonic-acid gas which will not elude observation except for very small amounts of the carbonate. But if the particles of carbonates are very small and isolated in the section, the development of carbonic acid may be easily overlooked. In such cases it will be well to cover the section with water and a glass cover, and to place the drop of acid so that it may diffuse slowly in the water over the section. With a low power the formation of bubbles of carbonic acid may be observed wherever the carbonates exist, since the glass cover prevents the bursting of the bubbles.   If the section must be warmed during the operation, it may be laid on a perforated copper plate, the aperture of which is over the diaphragm of the stage of the microscope, and the necessary temperature may be obtained by heating two long tongue-shaped projections by means of an alcohol lamp or a gas jet.   The bases with which the carbonic acid was combined are found in the solution covering the section.   This may be taken up in a capillary tube and transferred to a clean object-glass, and the bases determined by the ordinary methods of analysis, or by those to be given later on.   The capillary tubes should be kept in large numbers and thrown away after being used once, because of the difficulty of cleaning them.   In many cases it is well to carry on the reaction within the capillary tube itself by admitting the solution to be investigated at one end and the reagent at the other, and letting them act on one another within the tube.

Gas may be generated upon the solution of many sulphides; the gas in this case being hydrogen sulphide.   If this gas has been generated under a glass cover, it may be detected by the coloration of a strip of filter-paper moistened with lead-water, which is dipped into the solution covering the section.

Certain phosphates and the oxides of iron and manganese dissolve in mineral acids without the evolution of gas; the acid mostly employed is hydrochloric.   The principal phosphate met with in rocks is apatite, which is widely distributed.   If apatite is present in the section, the phosphoric acid may be detected by an addition of ammonium molybdate.   If the test is applied directly to the apatite, it is better not to treat the section with hydrochloric acid, but to use a drop of ammonium molybdate which is dissolved in nitric acid.   After the action has been completed the solution is put upon a clean object-glass, and there forms, sometimes after a slight warming, a great amount of very small crystals, mostly resembling rhombic dodecahedrons, which are greenish in transmitted light and yellow by incident light; they occur sometimes singly, sometimes united in more or less regular groups (Pl. XIII. Fig. 5).   If the rock contains silicates which are easily attacked by acids, the phosphoric acid cannot be determined in the manner just given, since soluble silica gives a similar reaction with ammonium molybdate.   In this case the solution obtained by diluted nitric acid must be evaporated on the object-glass; and after sufficient heating, by which the silica passes into the insoluble state, it is again brought into solution and the reagent applied.

Among the iron oxides limonite is the most readily soluble in hydrochloric acid, then magnetite; and hematite and ilmenite with the most difficulty. They all dissolve more slowly in thin section than in powder because of the smaller surface attacked. Chromic iron is insoluble or nearly so: therefore a thin section is rarely treated with hydrochloric acid for the purpose of distinguishing these iron ores; more frequently it is necessary to remove them by acids in order to observe minerals or structural relations which they conceal. This is often necessary with porphyritic rocks and clay slates or phyllites. To test for the presence of native iron in a thin section, it is covered with a solution of copper sulphate from which there is deposited on the metallic iron, if present, a coating of metallic copper. In order to avoid confusion with a coating of rust, A. von Lasaulx recommended the use of the solution of cadmium borotungstate employed for the mechanical separation of minerals, this becomes deep violet-blue through reduction in the vicinity of metallic iron. Zinc and copper have the same action, and therefore should not be present.

The treatment of thin sections with acids is to be specially recommended for proving the presence of gelatinizing silica. The method of procedure is governed by the object in view. If it is only a question of the presence of such silicates as belong to the family of olivine, nepheline, zeolite, the more basic feldspars, chlorite, or serpentine, then the carefully cleaned section is covered with a thin coating of the acid employed. If the layer of fluid on the section is too thick, the resulting gelatine spreads itself over the whole section, and gives those portions which have not gelatinized the appearance of having been attacked. When the acid has acted sufficiently after being warmed, it is removed by rinsing with water, and, if necessary, with the addition of a drop of ammonia to neutralize the last trace of acid. The action should last only long enough to form a very thin film of gelatinous silica over the substances attacked, through which the polarizing phenomena of the minerals may be observed. So it is better to repeat a test several times than to permit it to work too strongly the first time. In order to render the transparent gelatinous silica more apparent, the section is covered with a drop of water to which a dilute solution of fuchsine in water has been added, and is allowed to stand for some time. In this way the gelatinous film is saturated with the pigment; the section is then rinsed thoroughly with water, when the color disappears from all places except those in which a gelatinization has taken place. If the acid has not attacked the mineral sufficiently, the fuchsine is destroyed by a drop of acid and the experiment repeated. Any other coloring material, which will be absorbed by the gelatinous silica, may be used. In many cases it is well to cover the partially gelatinized section with the solution of a salt, as an iron salt, and after this has penetrated sufficiently into the gelatine, to add a reagent which will produce a coloring or precipitate in the imbibed solution of salt (ferrocyanide of potassium or ammonia). This method is especially recommended for permanent preparations.

If one wishes to determine the bases which were present in the gelatinized substances, the acids are allowed to act longer and more strongly. The solution is then removed with a capillary pipette, is evaporated on an object-glass in order to render the dissolved silica insoluble, is again dissolved in acidulated water, and tested by methods to be described later on.

Finally, if it is desired simply to remove the gelatinized substances (in the case of zeolitic decomposition of feldspar rocks, of chloritic and serpentinous alteration of pyroxenes and amphiboles, etc.), then they are destroyed as completely as possible, and the section is thoroughly rinsed with a strong jet of water, in order to wash off the gelatinous silica, which often adheres stubbornly. Minerals are sometimes discovered in this way whose presence would scarcely be suspected on observing the sections before they were attacked.

Etched figures have been of much less service in the microscopical investigation of mineral aggregates and rocks than in the physical researches in crystallography, because of the uncertain position of the sections of minerals composing aggregates, and the essential dependence of the symmetry of the etched figures on that of the crystal face on which they have been produced. Nevertheless they may often be employed to advantage for determining the presence of twinning. to prove the law of twinning derived from the apparent form, or to furnish evidence of the parallel growth of closely related minerals which belong to different systems. Etched figures furnish definite conclusions, especially in the study of minerals of the pyroxene and amphibole families, when other methods leave one in doubt. They may also be advantageously employed in the investigation of the orientation of the optical ellipsoid of elasticity in minerals of the mica, chloritoid, and chlorite series, when their outward form is wanting. Finally, etched figures, in certain cases, give criteria for the determination of substances which otherwise are distinguishable with difficulty, as, for example, quartz and cordierite, when there are no sections which permit a positive optical determination.

Etched figures are obtained by the use of various acids, or of caustic alkali, according to the substance under investigation, which also determines the conditions under which the acids are allowed to act. The action should be as gentle as possible to produce sharp‐ figures whose form and symmetry may be plainly recognized. After the corrosion of the reagent the etched substance is thoroughly freed from compounds, which may have resulted from the reaction, by wash‐ ing in water or acid, and the thin section should be examined in a weakly refracting medium (water). If it is placed in a strongly refract‐ ing medium, the etched figures may be completely overlooked unless strongly divergent light is transmitted through the section by sinking the condensing lens. In every case the objective is focused on the surface of the section.

The forms of the etched figures differ on one and the same face of any mineral, according to the corrosive agent employed; the degree of their symmetry alone appears independent of the latter and of its con‐ centration. The sharpest etched figures are produced on crystal faces and cleavage planes; they are only moderately precise and clear on artificially made faces (ground faces) when these are well polished.

Heating thin sections to a red heat serves to reveal hydrous min‐ erals and carbonaceous substances, or to produce colorations which are characteristic of certain compounds. Most hydrous minerals, such as zeolites and chlorites, become clouded through the high heating of thin sections containing them. For this purpose the section is re‐ moved from the object-glass, carefully cleaned of balsam by means of alcohol or ether, and brought into the flame on thin platinum foil. Colorless hydrous minerals simply become clouded; colored ones change their color; chloritic substances upon sufficient heating become rust-brown or black. Carbonaceous particles scattered through a sec‐ tion may be distinguished from those of iron oxide by heating to red‐ ness, by which process the carbonaceous matter is consumed. As these two sometimes occur mechanically combined, it is well in free‐ ing a section of such impurities to alternate the processes of treating with acid and of heating to redness. The combustion of carbonaceous substances varies greatly; in many cases graphite is not consumed even by continued and strong heating.

Colorless silicates containing protoxide of iron are colored red and reddish brown on being heated to redness. C. W. C. Fuchs first observed this property in olivine. Pyroxene and hornblende, when colorless or only faintly colored, act in the same manner. Olivine sometimes becomes pleochroic; hornblende always so, and often extra‐ ordinarily strong. With the latter mineral the colors and pleochroism are the same as those in the hornblende of rock inclusions in lavas and volcanic ejectamenta. The phenomenon may be referred to an extremely fine distribution of sesquioxide of iron freed from combination. H. Vogelsang showed that minerals of the haüyne group may become blue upon being heated, if they did not already possess this color.

Colorless aluminous minerals are colored blue if the section is moistened with very dilute cobalt solution on platinum-foil, is very strongly heated, and then digested with dilute hydrochloric acid. To increase the temperature sufficiently it is covered with a platinum cover. Often the reaction only takes place after repeated heating.

Microchemical Investigation of Loose Grains.—Preparation of the Material for Observation.

In order to investigate the constituents of a mineral aggregate or of a rock in a pure condition it is necessary to separate them from the mixture. The separation of a mixture into its mineral components is seldom effected by the successive application of a single method. Generally several methods must be used in connection with one another, which are based partly upon the different specific gravities of the constituents, partly on their different susceptibility to chemical reagents, and partly on their behavior towards stronger or weaker magnets. For all these kinds of separation it is necessary to bring the mixture into the form of a powder, and to give the powder not only such dimensions that each grain or the greater number of them shall be homogeneous—that is, shall consist only of one kind of mineral—but the size of the grains must be as uniform as possible. The coarseness of a powder in a given case depends on the grain of the mixture, for the grain of the powder should be as large as possible, since the separation is easier and more successful the larger the grain of the powder to be separated. The finer the powder is, the slower and more difficult will be the mechanical separation, but the quicker and easier the chemical. It is very desirable that the grains should not lose their crystal form, if they possess any, and that in the absence of crystal form their boundary should be made up of cleavage faces. This object is best accomplished by reducing the material in a metal mortar, by striking it with the pestle and avoiding the rubbing and grinding of the powder as far as possible. When the proper-sized grain has been approximately reached, the powder is separated into portions of like-sized grain by means of a series of fine wire sieves with meshes of about 1 to 0.2 sq. mm. In place of the wire sieves, a series of sieves may be made by covering one end of a number of wooden or tin cylinders with different grades of bolting-cloth, held in place by tightly fitting rings, which project far enough below the bottom of one sieve to fit over the top of the next, and so form a closed set of boxes which can be shaken together. The different grades of powder within these boxes are examined microscopically to see which furnishes the requisite homogeneity of the single grains ; the whole powder is then reduced to this size of grain, and is put in a large vessel and washed free of the fine mineral dust which remains suspended in the water. For chemical separation this is not necessary.

The order in which the separations by specific gravity, by magnets, or by chemical action are to be applied to a powder will depend on the problem presented in each particular case.

Separation according to Specific Gravity.—An actual separation of a mixed powder according to the specific gravity of its constituents is only obtained by the use of fluids which are heavier than the powder, so that it floats on the fluid, and which can be diluted by the addition of lighter fluids and made specifically lighter. The fluids most generally employed are the so called Thoulet's and Klein's solutions.

Thoulet's solution was first proposed by E. Sonstadt and afterwards by Church, but became generally known through the researches of Thoulet, and was thoroughly investigated by V. Goldschmidt. It is a solution of potassium-mercuric iodide, whose maximum density, according to V. Goldschmidt, is 3.196. According to Goldschmidt's statement, the highest specific gravity is obtained when a mixture of mercuric iodide is dissolved in cold water with potassium iodide in the proportion $KI : HgI_2 = 1 : 1.24$, and this solution is evaporated on the water-bath until a crystalline coat forms on the surface, or until a crystal of tourmaline or fluorite floats on it (sp. gr. = 3.1). Upon cooling, the density of the solution rises to 3.196 through contraction. According to van Werveke's observations, an excess of KI does no harm. Upon filtering, the solution is perfectly transparent, and of a yellowish-green color. So long as the relation of the two salts, KI and HgI_2, is correct, the solution may be continually diluted as far as a sp. gr. 1.0, and by evaporation on a water-bath be brought back to a maximum 3.196. If the relation of the salts is changed, then with an excess of HgI_2 there separates out a yellow hydrous double salt in acicular crystals ; with an excess of KI this substance separates in cubes. The same separations take place when the solution stands a long time in dry air.

Through long usage, the solution loses its green color and becomes reddish brown from the separation of iodine.   One may avoid this decomposition, or bring the altered solution back to its original condition by adding metallic mercury during evaporation.   The free iodine then combines with the mercury to form mercurous iodide, which coats the metallic mercury with a fine grayish-green dust, and causes it to fall apart in small spherules upon being stirred up; these unite only with great difficulty.   Upon further evaporation HgI is changed to HgI_2 and Hg, and the mercuric iodide combines with the excess of potassium iodide.   The solution changes in the air through the giving off and taking up of water, and in this way its density is altered.   Separations in this solution, therefore, must be carried on with constant temperature or in closed vessels, and with the greatest possible dispatch. Its specific gravity is nearly constant when it is about 3.01–3.1; below this limit it increases by losing water, and above it it decreases by taking up water.   Consequently the concentrated solution may be exposed to the air without its altering noticeably.

The dilution of the concentrated solution to a particular specific gravity by the addition of water cannot be accomplished with certainty by the introduction of a measured amount of water, because of the contraction which takes place.   One must proceed, therefore, empirically, and place in the solution a piece of mineral of the required specific gravity as an indicator, and then proceed to add water very carefully, drop by drop, or, for a small difference between the initial and desired density, add a dilute solution until the indicator is suspended in the solution.

Since metallic iron decomposes the solution with the separation of mercury, all splinters from the mortar which might have gotten into the powder must be removed by a magnet or by acid, before putting the powder in the solution.

The solution discovered by Dr. Klein, and named after him, is that of a cadmium borotungstate, with the formula $2H_2O$, $2CdO$, B_2O_3, $9WoO_3 + 16aq$.   This salt dissolves at 22° C. in less than 10 times its weight of water; the light yellow-colored solution has a specific gravity of 3.28 at 15° C.   If a diluted solution of this salt is evaporated on the water-bath, the violet color which is frequently observed disappears as soon as the sp. gr. 2.7 is reached.   If the evaporation is continued until an augite crystal floats on the warm solution, crystals are formed upon its cooling, which, when dissolved in a little water, yield a solution in which olivine will float; by combining these two solutions one is obtained with sp. gr., 3.3–3.6.   The highest possible specific gravity,

3.6, is obtained by evaporating on a water-bath until olivine floats on the warm solution. Cadmium borotungstate is deposited in crystalline masses which consist of rhombic individuals. If these are cleaned by drawing off as much of the mother-liquor as possible, and are then heated in a tube in the water-bath, they melt at 75° C. in their water of crystallization, and form a somewhat mobile fluid, on which spinel floats. This concentration may also be reached by the evaporation of the solution on the water-bath. At a very high specific gravity the Klein solution is quite oily, and its applicability for the separation of powder is very limited, and only coarse powder can be separated by it. By evaporating the dilute solution until a crystalline coating is formed, and after its subsequent filtration, a cold solution is obtained with sp. gr., 3.36–3.365, which is generally serviceable. This solution, like Thoulet's, is miscible with water under all conditions without decomposition.

It has the advantage of higher specific gravity and of being innoxious, but its preparation is far less simple than that of Thoulet's solution. The solution is decomposed by metallic iron, zinc, and lead, as well as by carbonates. Consequently these substances must be removed from the powders with acids before they come in contact with the solution.

C. Rohrbach suggested the use of a solution of barium mercuric iodide, which with proper treatment reaches a sp. gr., 3.588, and is still quite mobile. The solution, however, cannot be diluted with water without being decomposed, which prevents its general application. It is only employed in cases where the specific gravity is above that of Thoulet's and Klein's solutions, and where a separation cannot be made by chemical or magnetic methods.

R. Brauns [*] has recently suggested the use of methyl iodide for separating minerals with high specific gravity. Methyl iodide, CH_2I_2, is a yellow fluid, strongly refracting and very mobile. It is easily miscible with benzole, but not with water or alcohol, and does not attack metallic substances. Its specific gravity, which at 16° C. is 3.3243, varies considerably with the temperature; thus at 10° C. it is 3.3375, and at 20° C. 3.3155, the variation being about 0.0022 for each degree.

A solution that has been diluted with benzole may be concentrated by evaporation on the water-bath, or, if there is only a small amount of benzole present, by evaporation in a draught. The concentrated solution does not change upon exposure to the air, which, together with its high index of refraction,—$n_{na} = 1.74092$ at 16° C.,

—renders it specially useful for determining indices of refraction by means of total reflection.

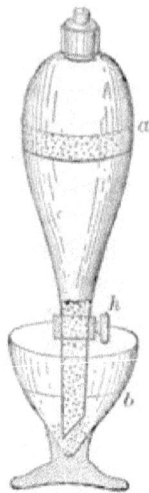

The vessels used for mechanical separations are the same for all of the solutions just mentioned. They have been made with a variety of forms; but the handiest, most solid, and most convenient form is that devised by T. Harada, represented in Fig. 45. A long, pear-shaped vessel of thick glass is closed at the upper end by a ground-glass stopper, and at the lower, narrower end by a glass cock. The solution and powder are introduced from above, the stopper inserted, and the mixture vigorously shaken up. The powder is then allowed to rise or sink, and as soon as a clear stratum of the fluid appears between the upper and lower portions, a small glass is placed under it so that the lower end of the apparatus rests firmly on the bottom of the glass; the cock h is then opened. A small part of the solution falls out, only a few drops, until the pressure of the air balances that of the column of fluid, and the separation of the heavier powder which falls into the glass proceeds automatically. One must avoid letting an air-bubble into the narrow part of the apparatus. When all the descending powder has passed the cock h, it is closed, and a layer of water is put over the solution in the glass; the apparatus is raised until the lower end reaches the layer of water. The water then rises up to the cock h, and allows all the powder beneath it to fall into the glass. The further dilution of the solution for a second separation of the powder is accomplished by adding a few drops from above, or better, by reversing the apparatus, and allowing the solution in the lower part of it, which has been diluted by the water, to enter through the opened cock h. The mixture is again thoroughly shaken and the operation repeated.

Harada's separating apparatus, together with all narrow and tube-shaped apparatus, has the disadvantage that the heavier powder in falling carries down with it mechanically a certain part of the lighter, floating powder, and in the same way the lighter powder holds up mechanically a part of the heavier; and also that in the space between the solution and the stopper a mixed powder remains sticking to the walls of the vessel in consequence of the shaking. C. W. Brögger sought to obviate this by modifying Harada's apparatus. He placed in the middle of the vessel a wider-bored cock (Fig. 46a), the aperture of which is the same as that of the vessel. Fig. 46*a* shows the apparatus after the first settling of the heavier powder S_1, with the middle cock

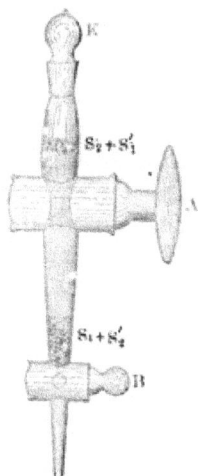

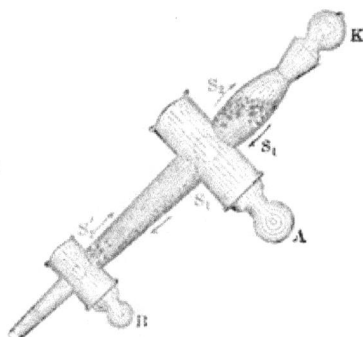

Fig. 46 a   Fig. 46 b   Fig. 46 c

A open. The powder S_1 over the lower cock B contains a part of the lighter powder S_2'; the lighter powder S_2 at the top contains a part of the heavier S_1'. If the cock A is shut, the apparatus shaken vigorously and inverted, then after some time there will be a separation of the powder in both parts of the apparatus, which is represented in Fig. 46*b*. Now if the apparatus is carefully turned in the position 46*c*, the heavy powder S_1 and S_1', and the lighter powder S_2 and S_2' will move in the directions indicated by the arrows without getting mixed. If this movement is continued until S_2' is directly under the cock A, and S_2 directly above it, then by carefully opening the cock the powder S_2' glides along the upper side of the apparatus into the upper part, while the powder S_1' descends along the lower side into the lower portion. The cock A is then closed and the separation completed, the heavier powder S_1 and S_1' being drawn out through the lower cock B.

Fig. 47

It is well in any case to repeat the separation several times, working with each of the portions of the powder separately. In treating large quantities, especially for the first separation, it is advisable to use an ordinary separating funnel, with a stopcock placed some little distance below the funnel proper (Fig. 47). The mixture is stirred with a glass rod.

The substances to be separated are first determined optically in order to know approximately the specific gravity required for the solution. The densities of the minerals most widely distributed in rocks are given in a table on page 110.

In order to bring the separating solution to a particular density, which lies between the specific gravity of the bodies to be separated, it is necessary to use the so-called indicators (or floats), or to employ a balance for the determination of the specific gravity of the fluids. As indicators, one may use mineral fragments whose specific gravity has been determined with the utmost exactness, and which may be held in readiness in great numbers in glasses. V. Goldschmidt has prepared such a scale, with convenient intervals, from which the following are selected as sufficient for ordinary purposes:

No.	Name.	Locality.	Sp. gr.		No.	Name.	Locality.	Sp. gr.
1.	sulphur	Girgenti	2.070		11.	quartz	Middleville	2.650
2.	hyalite	Waltsch	2.160		12.	labradorite	Labrador	2.689
3.	opal	Scheiba	2.212		13.	calcite	Rabenstein	2.715
4.	natrolite	Brevig	2.246		14.	dolomite	Muhrwinkel	2.733
5.	pitchstone	Meissen	2.284		15.	dolomite	Rauris	2.868
6.	obsidian	Lipari	2.362		16.	prehnite	Kilpatrick	2.916
7.	pearlite	Hungary	2.397		17.	aragonite	Bilin	2.933
8.	leucite	Vesuvius	2.465		18.	actinolite	Zillerthal	3.020
9.	adular	St. Gotthard	2.570		19.	andalusite	Bodenmais	3.125
10.	elaeolite	Brevig	2.617		20.	apatite	Ehrenfriedersdorf	3.180

A more convenient form of indicator has been devised by W. H. Hobbs: it consists of a small glass tube, closed at both ends, and partly filled with some heavy metal, which should be confined to the lower end, so that the glass float will maintain a vertical position in the heavy solution. In the upper end is inserted a loop of platinum wire, by which it may be readily lifted out of the solution. By varying the weight of metal within the glass tubes a series of indicators may be made, having any desired interval between them.

If the solution is to have exactly the specific gravity of one of the indicators in the scale, this is placed in the solution, and water or more concentrated solution is added until the indicator is suspended. When this remains in every place in the solution in which it is brought, the solution is adjusted, and has exactly the desired density. Should this lie between those of two indicators of the scale, both of these are placed in the solution, and its density brought to a state in which the heavier indicator sinks and the lighter rises. After the solution has been adjusted the mineral indicators are lifted out with a glass rod flattened and bent at one end.

The balance for determining the specific gravity of fluids (which is made by G. Westphal, of Celle, province of Hannover, and is called by his name) is better adapted for measuring the density of a particular solution than for bringing a fluid to a certain density. This measurement may be easily made even during a separation, if a separating funnel be used, and a small glass tube like Fig. 48 be let down into it. This tube is held up by the three arms which rest on the edge of the funnel, and is filled with solution through the bent tube *a* without the admission of the powder. The glass sinker of the balance is then sunk in the solution within the tube *r*.

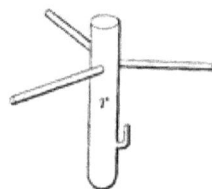

Fig. 48

Westphal's balance (Fig. 49) consists of a beam *f*, suspended on a fulcrum at *l*, which terminates behind in a pointer *s*, whose position can be read from a scale *a*. For a horizontal position of the bar the

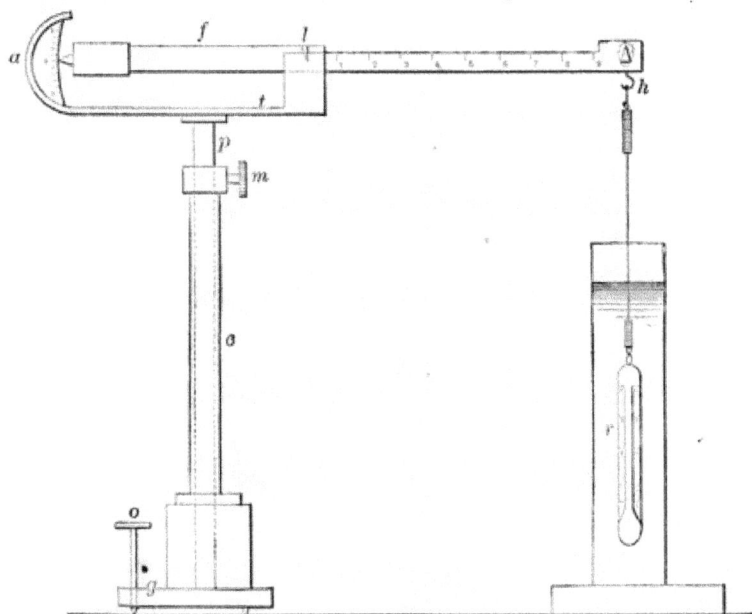

Fig. 49

pointer *s* must be at the zero-point of the scale. The bar *f*, from its point of suspension to the small hook *h*, is divided into 10 parts. The support *t* of the beam is firmly connected with the scale *a*, and together with the rod *p* sinks in the hollow cylinder *c*, being held at any height by the screw m. The hollow cylinder terminates in a massive foot resting on a circular base, which can be brought into a horizontal position by means of the screw o. From the hook h is hung the sinker r, which is immersed in the fluid. The weights, in the shape of riders, which counteract the buoyancy of the fluid, and restore the beam to its position of equilibrium, are hung on the hook h, to indicate whole numbers. The riders indicating the decimals are placed on the beam f, and their weight read from the scale on the beam. For example, if the bar swings about the zero-point of the scale s when two of the riders hang from the hook h, the rider for determining the first decimal stands at 8, and the smallest rider for determining the second decimal is half way between 4 and 5; then the specific gravity of the fluid is 2.845.

The complete success of a mechanical separation according to specific gravity, and especially the exact quantitative separation of a mixture, cannot be attained because of the following hindrances: (1) The impossibility of preparing a powder which shall consist of nothing but homogeneous grains; (2) the variations in the specific gravity of the constituents; (3) the change of specific gravity which minerals experience through weathering, decomposition, and alteration. For these reasons one must be satisfied with as close an approximation as possible to the homogeneity of the separated powder. Moreover, every separation is accompanied by a variable amount of intermediate products, that is, mixtures of several minerals grown together, together with more or less altered grains which are useless.

Laminated minerals, such as mica, often float much longer in a solution than they should, according to their specific gravity, and therefore render all subsequently separated portions impure. They may be removed by allowing the powder to glide down several times over paper, by which means the laminated minerals remain sticking to the paper; or by letting the powder fall in small portions on the slightly moistened sides of a funnel. The mica plates accumulate on the sides, while the grains of other minerals roll down into a vessel placed beneath.

Mechanical Separation of a Rock Powder by means of the Electro-magnet.

As magnetite can be easily extracted from a mixed mineral powder with an ordinary bar magnet, so all iron-bearing minerals may be separated from non-ferruginous minerals by means of an electro-magnet. It is not known with certainty what are the factors which influence the attraction of a mineral by an electro-magnet, since it is not always proportional to the percentage of iron, for many minerals rich in iron (chromite, biotite) are less strongly attracted than others much poorer in iron. Thus the attraction of a mineral can often be increased by heating it to redness, by which means the percentage of iron is not altered, but only the form of its occurrence is changed. Minerals of the amphibole, pyroxene, olivine, epidote, garnet, and similar series, may often be separated with an electro-magnet by regulating the magnetic moment of the electro-magnet. It may also be used to advantage in separating individuals rich in interpositions from those which do not contain them, when the mineral itself is free from iron. Thus, it is easy to separate the leucite, rich in inclusions of a rock (Capo di Bove), from those free from them ; the brownish-clouded plagioclases of gabbros also may be completely separated from the colorless ones.

The magnetic power of the electro-magnet is best regulated by using an electro-magnet in the shape of a horseshoe, on whose poles are screwed movable plates of soft iron formed as in Fig. 50. Their wedge-shaped projections are turned towards one another; the magnetic force increases very rapidly the closer the wedges are brought together, and diminishes the farther they are moved apart. The plates are screwed fast at

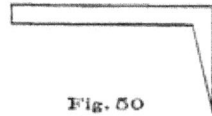

Fig. 50

the proper distance apart, the current allowed to circulate through the magnet, and the powder brought near the edge of the wedges on a piece of paper, or when necessary brought in contact with them. The paper with the powder is then withdrawn, and the powder which was attracted is allowed to fall on paper spread beneath it. The rapidly-repeated opening and closing of the circuit is most conveniently accomplished by introducing into one connecting wire a small cup of mercury, in which one end of the wire can be dipped and drawn out by the right hand, while the left hand is manipulating the powder.

Not infrequently it is necessary to separate single grains from a mixture by hand, when the other methods fail. It is then well to use a thick strip of glass in which a longitudinal groove has been cut. The powder is placed in this groove so that the grains may not lie too close together. The glass plate is slowly moved along under the microscope, and as soon as a grain of the mineral sought for is in the field, it is removed with a thin waxed thread, or with the sharpened end of a match slightly moistened.

The Separation of Minerals by Chemical Means.

The chemical methods which may be used for the separation of rock constituents are so manifold, and vary so for particular cases, that a general scheme for their application cannot be formulated. An experienced chemist will easily combine them himself, and an inexperienced one cannot employ them with success. Without dwelling upon the separation of the carbonates by weak acids, and of the well-known methods of partial silicate analysis, attention should be called to the far-reaching application of hydrofluoric acid, either alone or in combination with hydrochloric or sulphuric acid. If into a platinum dish containing pure concentrated hydrofluoric acid the powder of a rock is gradually introduced, not too rapidly lest a strong boiling be produced, but rapidly enough to produce a sufficient elevation of temperature, then the constituents of the rock will be attacked in a certain succession—first the glassy portions, then the feldspars and related substances, then the quartz, finally the constituents rich in magnesia and iron belonging to the pyroxene, amphibole, olivine, and kindred families. Consequently, if the process is interrupted at the proper time by suddenly adding water, it is possible after some practice to decompose certain substances and retain others unattacked. In this way the microlites of glasses, the feldspars of porphyritic ground-masses, or the older, more basic secretions may be isolated, often completely retaining their crystal form. During the action of the acid the powder is stirred with a platinum rod, which treatment is continued during the addition of the water in order to break up the lumps of gelatinous silica, and aid their removal by the water. As soon as it may be done without danger, the fingers should be used to rub the unattacked crystal powder against the sides of the dish in order to free it from the gelatinous film. Finally, the water is drained off and the crystal powder is carefully heated to redness, by which means the gelatinous silica still adhering is converted into pulverulent silica, which can be easily and completely washed away. By proper treatment the small crystals of the magnesia and iron silicates retain brilliant crystal faces.

Sauer,[*] Cossa,[†] and Cathrein [‡] have shown how a mixture of hydrofluoric and hydrochloric or sulphuric acids can be employed to isolate rutile from slates. The same method permits the separation of zircon, tourmaline, spinel, andalusite, disthene, etc., from other silicates.

---

[*] N. J. B. 1879, 571 ; 1880, I. 280.   [†] N. J. B. 1880, I. 162–164.   [‡] N. J. B. 1881, I. 172.

Determination of the Specific Gravity of the Isolated Powder.

If the specific gravity of a powder which has been obtained by separation was not already discovered during the separation, and if the picnometric determination cannot be made, it may be found by bringing a grain into suspension in a separating fluid and then ascertaining the density of the fluid by means of Westphal's balance. If one does not possess such a balance, then the adjusted fluid is poured from the powder into a calibrated and weighed flask, holding 20–25 ccm., filled exactly to the mark and weighed. The weight divided by the volume gives the specific gravity. Care should be taken in the first process that no air-bubbles are attached to the sinker of the balance, and in the second the determination should be repeated three times. In both cases the work should be done as quickly as possible because of the hygroscopic nature of the fluid. The first process is the shorter; the second is the more exact.

The specific gravity of bodies also whose density is considerably greater than that of the separating fluid can be determined by means of such fluids when their absolute weight is not too small (greater than 0.01 gr.). A small sphere of wax is weighted by a mineral grain enclosed in it, and its weight determined $= g$. To this is attached the grains whose specific gravity $= d'$ is to be measured, and whose absolute weight $= g'$ has been determined. The system is now placed in the separating fluid, which is adjusted, and its density $= D$ determined. The wax sphere is removed from the solution, the loosely attached grains are carefully removed, the sphere replaced in the fluid, and its specific gravity $= d$ determined. From this its volume, $v = \dfrac{g}{d}$, is known, and we have the equation

$$D = \frac{g + g'}{v + v'} = \frac{g + g'}{v + \dfrac{g'}{d'}};$$

therefore

$$d' = \frac{g'D}{g + g' - Dv}.$$

The following table presents the most important rock-making minerals arranged according to their specific gravities. The numbers given signify only average values, and may prove to be somewhat inexact in particular cases.

Cassiterite	6.84	Olivine	3.41	Anorthite	2.76
Hematite	5.30	Vesuvianite	3.40	Lazulite	2.75
Magnetite	5.20	Hypersthene	3.39	Talc	2.74
Ilmenite	4.75	Epidote	3.39	Meionite	2.73
Chromite	4.46	Zoisite	3.35	Beryl	2.72
Zircon	4.45	Diopside	3.30	Baslite	2.70
Rutile	4.25	Axinite	3.29	Dipyre	2.66
Melanite	4.15	Ottrelite	3.26	Quartz	2.65
Brookite	4.14	Sillimanite	3.23	Albite	2.63
Picotite	4.08	Hornblende	3.22	Elæolite	2.60
Perofskite	4.06	Andalusite	3.20	Sanidine	2.56
Corundum	3.95	Bronzite	3.19	Nepheline	2.55
Hercynite	3.94	Fluorite	3.18	Leucite	2.47
Anatase	3.90	Anthophyllite	3.17	Cancrinite	2.46
Pleonast	3.82	Apatite	3.16	Haüyne	2.45
Pyrope	3.75	Spodumene	3.14	Petalite	2.39
Staurolite	3.74	Glaucophane	3.10	Brucite	2.36
Allochroite	3.70	Actinolite	3.02	Gypsum	2.31
Spinel	3.60	Biotite	3.01	Sodalite	2.28
Disthene	3.60	Gehlenite	2.95	Natrolite	2.23
Topaz	3.56	Prehnite	2.94	Opal	2.21
Ouvarovite	3.51	Melilite	2.93	Analcite	2.19
Grossular	3.50	Dolomite	2.90	Hyalite	2.17
Acmite	3.49	Wollastonite	2.86	Chabasite	2.10
Titanite	3.48	Muscovite	2.85		
Arfvedsonite	3.45	Chlorite	2.78		

Determination of the Hardness of the Isolated Powder.

The grains of powder are pressed firmly into the smoothly filed end of a leaden stamp a few millimetres thick, which is used as a handle in carrying out experiments in scratching the faces of minerals of known hardness. Approximate determinations may be made by rubbing the powder between two object-glasses and observing whether these are scratched or not. In the latter case the grating and the easier or harder crushing of the grains between the glasses indicates the hardness.

Chemical Reactions.

These are the same whether carried on upon isolated powder or directly on the thin section. It is first necessary to bring the substance to be investigated into solution. For non-silicates this is done by the well-known methods. Silicates are decomposed by direct treatment with hydrofluosilicic acid or with hydrofluoric acid. It is always preferable to use the substance to be investigated in the form of isolated powder. The results of the hydrofluosilicic-acid method have been carefully worked out and described by Bořicky; those derived by using hydrofluoric and sulphuric acids, by Behrens and others.

Bořicky's method of treating the silicates with hydrofluosilicic acid is as follows: An object-glass is covered with a thin, even coat of Canada balsam; on this is placed one or more grains of the mineral about the size of a poppy-seed. These may be fastened to the balsam and thus kept in place by gently heating the object-glass. The grains are covered with a spherical drop of hydrofluosilicic acid, which should not be allowed to spread over the balsam. The balsam should not be cracked, as the acid would attack the glass. The hydrofluosilicic acid must be absolutely pure, and should leave no residue upon being evaporated. The mineral fragment to be investigated must be dissolved as completely as possible; for otherwise the crystallizations formed upon the drying up of the acid would give a false idea of the composition of the substance. The evaporation of the solution, which is naturally very slow, may be accelerated and the action of the reagent advanced by gently warming the object-glass over a spirit-lamp. Upon the drying of the solution there arise characteristic crystallizations in the form of fluosilicates of the univalent and bivalent elements which were present in the mineral investigated. The fluosilicate of aluminium is gelatinous. If the crystallization is incomplete in consequence of too rapid evaporation, it should be redissolved in water or in a drop of dilute hydrofluosilicic acid, transferred to a new object-glass, and allowed to crystallize anew. If the mineral was not completely dissolved, it is treated again with a fresh drop of hydrofluosilicic acid. Many silicates, especially mica, cannot be completely decomposed even by concentrated hydrofluosilicic acid; these are then decomposed by hydrofluoric acid in a small platinum dish, evaporated to dryness after the addition of an excess of hydrofluosilicic acid, taken up with distilled water, and the solution evaporated on an object-glass. In studying these extremely minute crystallization products it is best to improve the illumination by lowering the polarizer.

The reactions and crystallizations most frequently used in the diagnosis of rock-making minerals are the following:

Potassium.—Upon the drying of the hydrofluosilicic compound there form isotropic, colorless crystals of K_2SiFl_6, in cubes, octahedrons, or combinations of these forms with the rhombic dodecahedron (Pl. XII. Figs. 1 and 2). From a very acid solution and at a low temperature there sometimes arise anisotropic crystals of apparently orthorhombic form, especially when the solution contains a high percentage of sodium. If these are dissolved in hot water and again allowed to crystallize out, they assume the normal forms. In hydrochloric-acid or sulphuric-acid solutions after the liberation of the potassium with hydrofluoric acid there are formed with platinic chloride sharply developed yellow octahedrons of potassium platinic chloride (K_2PtCl_6), more rarely cubes or crystals rich in combinations.

Sodium.—From the hydrofluosilicic-acid solution there arise upon evaporation crystals of Na_2SiFl_6 in hexagonal combinations (Pl. XI. Figs. 4-6) $\infty P . oP$ ($10\bar{1}0$) (0001) or $\infty P . P$ ($10\bar{1}0$) ($10\bar{1}1$). They lie sometimes on the prism faces, sometimes on the base; they are colorless, very weakly doubly refracting, with negative character. The pyramid is quite obtuse ($10\bar{1}1$) : ($10\bar{1}1$) = 66° 6'. The crystals are generally longer the more calcium there is in the solution. This test is exceedingly sharp and certain even for very small amounts.

Calcium.—Upon the evaporation of the hydrofluosilicic-acid solution there form monoclinic crystals of $CaSiFl_6 + 2aq$, which assume a great variety of forms. Sometimes they are in pointed, thorn-like or ramified groups and single crystals, sometimes in rhomboid plates, most frequently in spindle-shaped individuals, with not very strong double refraction (Pl. XII. Figs. 4 and 5). It is very characteristic of all the forms that they seldom have straight-edged boundaries, but generally crooked ones. Upon the addition of dilute sulphuric acid they are decomposed, and there is deposited in their place long prismatic crystals of gypsum. Upon decomposition with hydrofluoric and sulphuric acids only a part of the calcium sulphate goes into solution, if the percentage of calcium in the silicate is large; with a smaller percentage of calcium and an excess of dilute sulphuric acid, all of the calcium goes into solution without residue, and there forms about the edge of the drop the characteristic prisms and plates of gypsum ∞P . $\infty P \infty . P$ (110) (010) ($11\bar{1}$), usually lying on (010) (Pl. XIII. Fig. 1). This is the most delicate and the surest test for Ca.

Magnesium.—From the hydrofluosilicic-acid solution there crystallizes $MgSiFl_6 + 6aq$ in rhombohedral crystals, which most frequently exhibit the combination $\infty P2 . R$, ($11\bar{2}0$) $\pi(10\bar{1}1)$. more rarely $R . \infty P2$ or $R . oR$. They are always very sharp-edged, and have plane faces; are strongly doubly refracting, with positive character; generally polarize in bright colors of the second order, and exhibit when in the proper position a very distinct interference figure. They are colorless (Pl. XII. Fig. 6). The formation of Struvite crystals ($NH_4MgPO_4 + 6aq$), with their coffin-like, hemimorphic forms, is a very characteristic reaction (Pl. XIII. Figs. 3 and 4). The crystals may be obtained most perfectly from a very dilute solution, to which ammonium chloride and ammonia is added to distinctly alkaline reaction. A grain of salt of phosphorus is placed at the edge of the solution, or a drop of sodium phosphate is added to it. The crystals separate slowly in the cold solution, rapidly in a warm one. But in the latter case there arise forms of growth very difficult to recognize; these also form at first in concentrated solutions, and it is only after the greater part of the salt has separated out that the characteristic crystals begin to form.

Iron.—Upon the evaporation of the hydrofluosilicic-acid solution crystals of $FeSiFl_6 + 6aq$ are deposited, which are completely isomorphous with those of the magnesium salt, and have the same optical characters. They may be distinguished from the latter by being moistened with potassium ferrocyanide or with ammonium sulphide: in the first case they become blue; in the second, black. The amorphous precipitate with potassium ferrocyanide or with ammonia is also very easily recognized.

Aluminium.—This separates out of a hydrofluosilicic-acid solution in a gelatinous state. From a sulphuric-acid solution after the addition of a slight amount of cæsium chloride or of cæsium sulphate, there are deposited isometric crystals of cæsium alum. The predominant forms of these are $O, O . \propto O \propto$; the less frequent form is $\propto O \propto$, which usually separates from neutral solutions. The crystals apparently never exhibit the optical anomalies so common to the alums. If the solution is too concentrated, there arise many branched forms of growth; these may be dissolved in water and again allowed to crystallize. Too great an excess of sulphuric acid retards the formation of crystals; this is best neutralized by the addition of sodium acetate (Pl. XIII. Fig. 2).

Chlorine.—This test is important for the minerals of the sodalite group. The powdered substance is put in a small hemispherical platinum dish and covered with somewhat concentrated sulphuric acid. The dish is covered with a small glass cover, on the under surface of which is a drop of water, while on the upper surface another drop of water serves to keep it cool. The dish is warmed moderately, and the escaping chlorine is caught in the drop on the glass cover. After the upper drop has been removed the glass cover is taken off, and the drop beneath it containing the distillation is placed on an object-glass. If a grain of thallium sulphate is put in it there form octahedrons, and the combinations of $O . \propto O$ of thallium chloride. These are strongly refracting, and with low magnifying-powers are almost opaque in consequence of the total reflection. Or by adding silver nitrate flocculent silver chloride may be obtained, to which if strong ammonia be added and the whole dried, isometric crystals (O and $\propto O \propto$, rarely with $\propto O$) of silver chloride are formed, which are strongly refracting.

8

Sulphur is tested as sulphuric acid; upon adding to the solution containing it a calcium salt the characteristic gypsum crystals are produced.

Phosphorus only occurs in the phosphates among the rock-making minerals. The soluble phosphates may be treated with the nitric-acid solution of ammonium molybdate; the solution when dried gives rhombic dodecahedral crystals of the well-known precipitate (Pl. XIII. Fig. 5), which are yellow by incident light and green in transmitted light. If the mineral substance tested is not pure, and there be any soluble silica present, it may be rendered insoluble by being dried on an object-glass; the residue is again dissolved in nitric acid and the reagent added. Insoluble phosphates are first decomposed with soda. The reaction with ammonium chloride and magnesium sulphate is just as sharp; by this method the phosphoric acid is obtained as crystals of ammonium magnesium phosphate (Pl. XIII. Figs. 3 and 4).

Titanium.—If a titaniferous mineral be carefully melted on platinum wire with a grain of potassium bisulphate, and then placed in a porcelain dish and moistened with a drop of hydrogen superoxide in water, the bead and the solution will be colored yellow or orange-yellow, according to the amount of titanic acid present.* The reaction is extremely sharp, even for the smallest amounts.

---

* Schöner, Zeitschr. f. analyt. Chemie, 1870, IX. 41.

SPECIAL PART.

SYSTEM OF CLASSIFICATION.

THE determination of minerals under the microscope is based primarily on the investigation of their optical properties in connection with their crystal forms, as these are indicated by outline and cleavage (Blätterdurchgänge). Hence the minerals treated in the descriptive part of this book are arranged according to their system of crystallization. From the introductory or general part it is evident that mineral bodies may be classified according to their optical and crystallographic characters in the following groups:

1. Isotropic minerals.................. .... { Amorphous substances.
   { Isometric.

2. Anisotropic minerals. .... { a. With one optic axis { Tetragonal.
   (optically uniaxial). { Hexagonal.

   b. With two optic axes { Orthorhombic.
   (optically biaxial). { Monoclinic.
   { Triclinic.

Besides these there is a small number of cryptocrystalline substances, which are definitely characterized as such by their double refraction, but which always occur in such imperfect forms and in such microscopical aggregation that their crystal system cannot be determined with certainty. These will be placed under the head of *aggregates*.

The method of procedure which should be adopted in a microscopical determination will be briefly given. The question which first arises is whether the substance is optically a unit (single individual), or an aggregate, the crystal system of whose component individuals cannot be determined by their form nor by their behavior toward polarized light. A substance is recognized as a unit or individual by the fact that it shows the same optical behavior throughout its whole extent, so far as this is not modified by twinning. The substance is therefore first studied in parallel light between crossed nicols to see whether the extinction of the light takes place at one and the same time throughout the whole extent of the substance, or of the parts of the twins, when it is rotated with the stage of the microscope. If this is not the case, but if the substance which appeared as a unit in ordinary light is separated into a number of individuals, which are irregularly bounded, or are combined in a fibrous or laminated spherical form and are not separately determinable, and for which, collectively, the extinction is never synchronous during a rotation between crossed nicols, then the substance is an aggregate (cf. p. 88).

If the substance is found to be optically a unit, it is then necessary to observe whether it belongs to the isotropic or to the anisotropic division. An isotropic substance, in which the elasticity of the luminiferous ether is the same in all directions is characterized (p. 52) by the fact that it remains dark in parallel light between crossed nicols during a complete rotation, while it shows the same color and intensity of light in all positions between parallel nicols. An isotropic substance can never show interference colors whatever may be its position or the direction of the nicols to one another. Since, now, doubly refracting, uniaxial bodies may behave in the same manner when they are intersected at right angles to an optic axis, it is necessary, after establishing the isotropic behavior of a substance in parallel light, to test it in convergent light. If the interference figure of a uniaxial body is not obtained by this latter test, the substance is isotropic, and is either amorphous or of the isometric system. In order not to overlook a very slight double refraction which may lead to confusion when the investigation is only made between crossed nicols, and the field of view is not completely dark, but is somewhat gray, one of the stauroscopic methods of testing in parallel light described on page 63 should be used.

Whether an isotropic substance is amorphous or is crystallized in the isometric system is, in general, easily determined by the form of the body, and by the presence or absence of cleavage. Amorphous substances exhibit no independent boundary, their form always depending on those of the surrounding substances; they do not possess cleavage; and the lines of internal parting when present are not straight.

If the substance under investigation is anisotropic, it is next necessary to find whether it belongs to the optically uniaxial division or to the optically biaxial. In order to do this, those crossed sections are sought which show the least difference of color during rotation in parallel light between crossed nicols, or, when possible, such as remain dark or uniformly clear for all positions during rotation. The first will show the axial cross of uniaxial substances (page 68) in convergent light; the second, the axial bar of biaxial substances (page 72) with or without isochromatic curves. All other sections will exhibit a maximum extinction four times during a complete rotation between crossed nicols, which maxima are 90° apart; and in the intermediate positions they will show an interference color whose maximum lies exactly in the middle between the directions of maximum extinction. If the substance is found to be optically uniaxial, the distinction between the tetragonal and hexagonal systems is determined by the cleavage and the crystal form. Isotropic sections of tetragonal substances which show axial figures in convergent light are bounded by quadratic or octagonal outlines, or show two cleavages at right angles to one another. For hexagonal substances these outlines are hexagonal, trigonal, or nine-sided, and the cleavage forms equilateral hexagons or triangles. These sections of tetragonal or hexagonal minerals which are not isotropic show very different outlines according to the crystal form of the substance. However, all sections in the prism-zone generally exhibit parallel edges and parallel cleavage, that is, bisymmetric outlines and cleavage lines respectively, and the extinction will always take place when their directions of symmetry lie parallel to the principal sections of the nicols In sections in other zones, also, the directions of extinction lie parallel or at right angles to the directions of symmetry, whenever the sections possess any.

If the substance is anisotropic and optically biaxial, it remains to be found whether the crystal system is orthorhombic, monoclinic, or triclinic. This may be done in many cases by investigating the axial dispersion in sections which lie at right angles to the acute bisectrix. From the nature of things these sections are rare, consequently the position of the axes of elasticity must be investigated.

In the orthorhombic system these axes lie parallel to the crystal axes. Hence all sections in a zone $oP : \infty P \breve{\infty}, oP : \infty P \bar{\infty}$, $\infty P \breve{\infty} : \infty P \bar{\infty}$ will extinguish parallel to the axis of the zone. This zonal axis is easily recognized by the form of the outlines which are bisymmetric or monosymmetric, or by the cleavage. The cleavage in the orthorhombic rock-making minerals is pinacoidal or prismatic, and the extinction lies parallel and at right angles to the cleavage cracks when these are parallel, and bisects the angle between them when they intersect one another. In sections which do not belong to one of the principal zones just named, the extinction lies parallel and at right angles to the directions of symmetry of the outline, or of the cleavage cracks, whenever they are present. Stauroscopic methods must often be employed for the exact determination of the directions of extinction. If the sections in the three principal zones are tested in convergent light, some will be found exhibiting the point of emergence of a bisectrix; and when these are observed on cleavage plates the bisectrices will emerge perpendicularly (for pinacoidal cleavage), or will be inclined to one side (for prismatic cleavage).

In the monoclinic system only one axis of elasticity coincides with a crystal axis, which is always the axis b. The outline and cleavage can only form symmetrical figures in the zone $oP : \infty P \bar{\infty}$ (001 : 100), and in these the extinction lies parallel to the directions of symmetry, that is, parallel and perpendicular to the axis b. In the vertical zones as well as in the zone $oP \,; \infty P \dot{\infty}$ (001 : 010), the extinction no longer lies parallel to the outlines or to the cleavages, which in this system run parallel $(oP, \infty P \dot{\infty})$ to the axis b or in the projection, at least, perpendicular $(\infty P \dot{\infty}, \infty P)$ to this axis. This inclined extinction is an important characteristic. Cleavage plates of monoclinic minerals parallel to the pinacoids can only show a bisectrix perpendicular to their planes when the pinacoid is the plane of symmetry; the dispersion is then crossed: on the other pinacoids a bisectrix either does not appear at all, or emerges at one side of the field with considerable inclination.

In the triclinic system all sections are asymmetric. In general the direction of extinction lies oblique to the outline or to the cleavage lines.

These statements may be tabulated as follows:

1. The substance has the same optical orientation throughout its whole extent, or the parts having different optical orientation are bounded rectilinearly (twinned).   HOMOGENEOUS and UNIFORM.

  (1) All sections of the same substance remain dark during a complete rotation in parallel light between crossed nicols, and give no interference figure in convergent light.   ISOTROPIC.

    (1a) All independent form and cleavage is wanting.   *Amorphous.*

    (1b) Independent form or cleavage is present.   *Isometric.*

  (2) The sections generally show an interference figure between crossed nicols and become dark four times during a rotation between crossed nicols. The sections, which remain dark, or nearly so, during a complete rotation between crossed nicols, show a dark cross in convergent light, with or without isochromatic circles, the arms of the cross remaining un-

changed or moving parallel to themselves during the rotation.

ANISOTROPIC, OPTICALLY UNIAXIAL.

(2a) The sections showing uniaxial interference figures are quadratic (octagonal) or show rectangular cleavage.

Tetragonal.

(2b) The sections showing uniaxial interference figures are hexagonal, trigonal or nine-sided, or show systems of cleavage cracks intersecting at 60°.

Hexagonal.

(3) The sections in general show an interference color between crossed nicols and become dark four times during a complete rotation. No section remains dark during a whole rotation. Those sections showing no interference color, but having a uniform illumination for all positions, show the locus of an optic axis.

ANISOTROPIC, OPTICALLY BIAXIAL.

(3a) Sections in the three principal zones are symmetrical with respect to outline and cleavage, and the extinction lies parallel and perpendicular to the directions of symmetry.

Orthorhombic.

(3b) The outline and cleavage is only symmetrical in the zone $oP : \infty P \infty$, and this is the only zone in which the extinction lies parallel and perpendicular to the directions of symmetry.

Monoclinic.

(3c) The outline, cleavage, and extinction are unsymmetrical in all zones.

Triclinic.

II. Different parts of the same substance show different optical behavior, and the individual parts are not regularly bounded nor regularly optically oriented.

AGGREGATE.

The crystal system of the few minerals which remain opaque in thin section must necessarily be determined by their outline and cleavage alone. There remains, however, a considerable number of completely transparent minerals whose optical behavior in part is in evident contradiction to their crystallization (optically anomalous minerals), or whose optical behavior in certain instances strikingly approaches that of a crystal system other than that to which they belong. Optically anomalous bodies are frequent among the isometric minerals as well as among the tetragonal and hexagonal; many garnets, perofskite, leucite, apophyllite, and tridymite are familiar examples. Minerals of the mica family are among the substances which in many instances show an optical behavior approaching that of another crystal system; from their action in polarized light these have in some cases nearly a uniaxial character, in others nearly an orthorhombic character. Of triclinic minerals oligoclase and andesine show an approximately monoclinic orientation of their direction of extinction in the zone of their principal cleavage.

AMORPHOUS MINERALS.

AMORPHOUS minerals are produced by the chilling of melted bodies or by the solidification of gelatinous ones. The first may be called *glasses*, the second *opals* (hyaline and porodine substances).

The glasses occur either as independent geological bodies, in which case they are considered with the rocks, as obsidian, pearlite, pumice, pitchstone, tachylite, hyalomelan, palagonite, sideromelane, sordawalite, and wichtisite; or they appear as the residuum of crystallization in certain porphyritic rocks. In the latter case, since their composition varies greatly according to the manner and extent to which crystallization has advanced previous to their consolidation, they cannot be treated as minerals, but will be described under those rock groups in which they are found.

Of the little-known porodine amorphous minerals only the different varieties of amorphous silica have a general distribution among rocks—in most cases as products of the leaching out and of the alteration of silicates; in others it occurs in the form of concretionary masses, where silica has accumulated through the action of organisms on hydrous solutions; and to a very small extent, if at all, as the last residuum of crystallization of acid eruptive masses.

OPAL.

Literature.

H. BEHRENS, Mikroskopische Untersuchungen über die Opale. Sitzber. d. Wien. Akad. LXIV. 1871. 1. Abthl.

E. REUSCH, Ueber einen Hydrophan von Czerwenitza. Pogg. Ann. CXXIV. 1865. 431.

MAX SCHULTZE, Verhandlungen des naturhist. Ver. d. preuss. Rheinlande und Westphalens. XVIII. 1861. 69.

Sir DAVID BREWSTER, On the cause of the colors in precious opal.—Edinb. New Phil. Journ. by Jameson. XXXVIII. 1845. 385.

Opal is wholly amorphous and singly refracting; its varieties are known as precious opal, fire-opal, and common opal, according to the color and to the presence or absence of iridescence. Opal forms irregularly bounded patches, strings, and veins, as well as pseudomorphs after feldspar, augite, and other minerals in decomposed eruptive rocks of the trachyte and andesite series, and in related massive rocks of coarsely crystalline texture. The substance of the opal is often rendered impure by remnants of the constituents of the rocks mentioned,

as well as by flakes of hematite or by very small and indeterminable
dust-like particles. Spots which are dull by incident light and also
appear cloudy in transmitted light belong to hydrophane according to
Behrens. They absorb water eagerly, and become completely transparent, air-bubbles escaping during the process.

When the opal in rocks or along crevices in them assumes the
spherical form it often exhibits in parallel polarized light the interference cross of amorphous spherical substances with centripetal condensation (p. 89).

The energetic double refraction which is occasionally shown, especially by the more precious varieties of opal, is occasioned by strains
which are supposed to arise from the unequal drying of the gelatinous silica. The beautiful play of colors of the precious opal was explained by Brewster by the presence of a succession of cavities whose
varying dimensions gave rise to different colors. Reusch observed
that the color phenomena of precious opal and hydrophane are complementary by incident and transmitted light, and explained this by
the assumption of cracks which are parallel to the plates of the mineral,
or are slightly inclined to their surface, and act like thin plates.
Behrens could find no cavities in the opals examined by him, and
explained the color phenomena as the result of thin lamellæ of an opal
of differing index of refraction, which may be inclosed in the normal
opal. Since the power of refraction of opal changes with the amount
of water present, one may assume that the variable amount of water
contained in the mass of the opal gives rise to the play of colors. For
opal $n_\rho = 1.442$ to 1.450.

From the small value of the index of refraction of opal compared
with that of Canada balsam it may be easily overlooked, and the opal
portion of a thin section mistaken for a cavity filled with balsam.

Colorless hyalite usually forms botryoidal and reniform aggregates
along crevices and in cavities in phonolitic, tephritic, and basaltic
eruptive rocks. $n_\rho = 1.437$ and 1.455. Hyalite is doubly refracting,
and under certain conditions exhibits the interference cross of a uniaxial substance with isochromatic curves. This appearance is referred
to conditions of strain occasioned by the concentric shell-like structure
of the mineral. The character of the double refraction is negative.
The hyalite cross often separates into hyperbolas during a rotation of
the section between crossed nicols, which would necessarily be the
case if the layers were not regular spherical shells.

Common opal and half opal are distinguished from precious opal
and hyalite by a greater admixture of foreign inclusions; among these

tridymite is of special interest. G. Rose * first discovered this mineral in the opal of Kosemütz, Silesia, Iceland, Kaschau, Persia, and Zimapan, Mexico. It occurs in the form of round or hexagonal plates, and in concretions of these forms.

All silica hydrates are soluble in caustic potash; the presence of finely divided opal substance in clay slates and sandstone can be proven by this reaction. The specific gravity of opal varies with its impurities between 1.9 and 2.3. For the pure varieties sp. gr. = 2.2.

Carbonaceous Matter

occurs in sedimentary rocks of widely different formations in finely disseminated particles, occasionally in somewhat larger accumulations, or else in such a finely divided state that it is only recognized as the coloring of the other minerals. The carbonaceous flakes are without regular boundary, and are opaque, lustreless, gray to grayish black by incident light. They are unaffected by acids; when heated to redness on platinum foil they burn up. They are often intimately mixed with the iron ores (pyrite, limonite, magnetite), and then they are consumed with difficulty; if the heated section be treated with acids and again heated to redness, the dark-colored spots will become light. It is not definitely known whether these dark combustible particles are always carbon or sometimes a carbon compound. They color mineral substances black to gray, and when very finely divided, bluish.

---

* Monatsber. d. Berl. Akad. 1869. Sitzung vom 3. Juni. p. 449.

MINERALS OF THE ISOMETRIC SYSTEM.

MINERALS of the isometric system differ from all other crystalline substances by the absence of double refraction, and from amorphous bodies by the presence of a regular boundary or by the occurrence of rectilinear cleavage cracks intersecting one another at uniform angles. They remain dark during a complete rotation in parallel light between crossed nicols, give no interference figure in convergent light, and exhibit no interference colors in polarized light. Optical anomalies are frequent.

Pyrite.

Literature.

Fr. Becke, Aetzversuche am Pyrit. T. M. P. M. VIII. 239–337, 1887.

Pyrite is occasionally present in all kinds of rocks as an accidental accessory constituent, and is widely distributed in small quantities. It forms cubes, pentagonal dodecahedrons, or combinations of these forms striated by the oscillatory combination of $\infty O \infty$ (100) with $\frac{\infty O2}{2} \pi$ (210); less frequently, irregular grains and aggregates of grains.

It is opaque and yellow by incident light, with strong metallic lustre. This color distinguishes it from other opaque iron ores.

Sp. gr. = 4.9–5.2. Chemical composition = FeS_2. Soluble in nitric acid with the separation of sulphur; not noticeably acted on by hydrochloric acid.

Pyrite is often intimately associated with magnetite, hematite, and ilmenite. It is often peripherally or completely pseudomorphosed into limonite, more rarely into red transparent hematite. Its presence may be detected during the grinding of the section by its dark grayish-black streak.

Magnetite.

Magnetite forms crystals whose predominant form is O (111), and whose dimensions vary greatly in one and the same rock; twins, ac-

cording to the spinel law, often very much shortened in the direction of the twinning axis; and simple crystals grown together in parallel posi-
tions. Cross-sections of such forms
are shown in Fig. 51. Skeleton crystals (Pl. III. Fig. 2) are frequent in
highly ferruginous eruptive rocks.
It is sometimes quite uniformly disseminated through the rock in the
form of grains and dust-like particles,

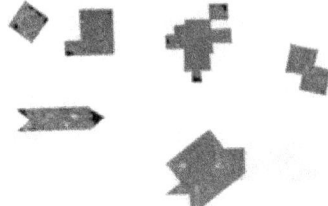

Fig. 51

or is clustered in small aggregates. The latter occur in certain eruptive rocks (phonolites, trachytes, andesites, tephrites), where they often exhibit a more or less close approximation to the crystal forms of hornblende, biotite, hypersthene, more rarely of augite; or they form in combination with other substances complete pseudomorphs after the minerals just named. When the original mineral was hornblende, biotite, or hypersthene, the aggregate is composed of magnetite with augite in very small grains and crystals (Pl. XIV. Fig. 1). Such pseudomorphs appear to have been caused by the resorption of the older secretions which crystallized at a particular period in the development of the magma, but which at a later period could no longer exist. The cleavage parallel to O (111) is not generally perceptible under the microscope.

Magnetite is opaque in rock sections, and has a bluish-black color by incident light with strong metallic lustre, easily recognized except with very small dimensions.

Sp. gr. $= 4.9–5.2$. Chemical composition $= FeO, Fe_2O_3$, often containing a variable amount of titanium, rarely with a small amount of chromium. Dissolves without difficulty in hydrochloric acid. It remains unattacked in a rock powder treated with hydrofluoric acid. It is easily isolated from the powder by means of a weak magnet. This property may be used to distinguish it from hematite, ilmenite, chromite, and graphite.

Magnetite, which contains no titanium, when weathered becomes coated with a yellowish-brown, non-metallic, earthy limonite, which then impregnates the surrounding rock, forming a halo about the magnetite. When there is considerable TiO_2 present, a fibrous or granular, whitish or yellowish substance forms around the magnetite, which is strongly doubly refracting, and is the same mineral which often accompanies ilmenite, titanite, or leucoxene. Magnetite is frequently crystallized with pyrite and ilmenite, more rarely with chromite and rutile.

No other mineral is so generally distributed in eruptive rocks, crystalline schists, and phyllites as magnetite. In the eruptive rocks it belongs to the oldest crystalline secretions from the magma, and therefore frequently appears as inclusions in all other constituents, especially in those which immediately followed its formation, as olivine, biotite, hornblende, and pyroxene. Less frequently magnetite is of younger origin than the ferruginous bisilicates, when it very probably arises from the re-solution of older basic constituents.

Chromite.

Literature.

E. DATHE, Olivinfels, Serpentine und Eklogite des sächsischen Granulitgebietes. N. J. B. 1876. 247–249.

J. THOULET, Note sur le fer chromé. Bull. Soc. min. Fr. 1879. II. 34–37.

M. E. WADSWORTH, Lithological Studies. Cambridge, Mass., 1884, pp. 176–186.

Chromite occurs in small crystals in the form of octahedrons, seldom of cubes or irregular grains and aggregates; its cross-sections are quadratic, rhombic, triangular, or irregular. Cleavage not noticeable; irregular cracks frequently occur, along which alteration products are often located.

In sufficiently thin sections it is transparent with brown to reddish-brown color. Its highly roughened surface arises from its strong index of refraction, $n = 2.0965$. Chromite has a weak metallic lustre in reflected light when its color is grayish black to black, and it is not completely transparent; but there is no metallic lustre on transparent portions which are of a gray or lilac-gray color.

Sp. gr. $= 4.8$ and over; hardness, 5.5. These are important in distinguishing it from picotite. Chemical composition $= FeO, Cr_2O_3$ when pure, not attracted by an ordinary magnet, and not noticeably attacked by acids. The grains of chromite in the rocks are often surrounded by a green halo of chrome ochre.

Chromite is common in crystalline rocks rich in magnesia in these it belongs to the oldest secretions, like magnetite, and is therefore usually enclosed in the next oldest constituents, especially in olivine. Chromite is also widely disseminated in the magnesian rocks of the Archæan formation, particularly in serpentine. Chromite can only be distinguished from picotite (chrome spinel) by its hardness, specific gravity, or quantitative analysis. It is readily distinguished from all other substances by its chemical behavior, especially its reaction in blow-pipe beads.

Spinel Group.

Literature.

H. Fischer, Kritische, mikroskopisch-mineralogische Studien. Freiburg i. B. 1869. 18; 1. Fortsetzung 46, 60; 11. Fortsetzung 66, 88.

E. Kalkowsky, Ueber Hercynit im sächsischen Granulit. Z. D. G. G. 1881. XXXIII. 533–539.

W Prinz, Les enclaves du saphir, du rubis et du spinelle. Ann. Soc. Belge de Microscopie. — Bruxelles 1882.

H. Schulze und Alfr. Stelzner, Ueber die Umwandlung der Destillationsgefässe der Zinköfen in Zinkspinell und Tridymit. N. J. B. 1881. I. 120–161.

A. Stelzner, Zinkspinell-haltige Fayalitschlacken der Freiberger Hüttenwerke. N. J. B. 1882. I. 170–176.

J. Thoulet, Étude microscopique de quelques spinelles naturels et artificiels. Bull. Soc. Min. Fr. 1879. II. 211–213.

M. E. Wadsworth, Lithological Studies. Cambridge, Mass., 1884.

F. Zirkel, Basaltgesteine. — Bonn 1870. 97.

All the *spinels* which occur as rock constituents crystallize in simple forms, O (111), and twins of such, in which the twinning axis is normal to the octahedral plane; more rarely in combinations of O with ∞ O (110), and 3O3 (311). Their cross-sections are mostly quadratic. Certain spinels, as hercynite, only occur in irregular grains; all of them occur in this form very frequently. Cleavage not noticeable microscopically.

The spinels have high indices of refraction (in precious spinels from Ceylon $n_{na} = 1.7150$), and exhibit rough surfaces in Canada balsam. Their color in transmitted light varies with their chemical composition.

H. = 7.5–8; sp. gr. = 3.6–4.5. No spinels are attacked by hydrochloric and hydrofluoric acids; they are therefore, in general, easily isolated because of their density and ability to withstand chemical action.

Spinel proper, $MgO. Al_2O_3$, is colorless to light red by transmitted light; generally in sharply defined crystals. It occurs sparingly in the Archaean, especially in gneiss, and comes into river sands through the mechanical and chemical disintegration of these rocks.

Pleonaste $(MgO, FeO) (Al_2O_3, Fe_2O_3)$ is green in transmitted light, but by incident light or when opaque it is black without metallic lustre. It occurs sometimes with magnetite among the oldest secretions of many eruptive rocks. It is extremely common in gneisses, especially in those intercalated beds bearing cordierite or garnet, in which minerals pleonaste is often included. In these occurrences its forms are usually irrregular, while in the eruptive rocks it is mostly well crystallized.   Its occurrence in regions of contact metamorphism is specially interesting.

Hercynite, FeO, Al$_2$O$_3$, is dark green by transmitted light, and can only be distinguished from pleonaste through isolation and chemical analysis.   It occurs in the granulites of Saxony, and in norite at Cruger's Station, on the Hudson River.[*]

Gahnite or *automolith*, a zinc spinel, is also green when transparent ; it occurs sparingly in octahedrons or grains in crystalline slates under circumstances analogous to those accompanying pleonaste, from which it can only be distinguished chemically.

Picotite, a chrome spinel, is yellow or brown by transmitted light, and can only be distinguished from chromite chemically, or by means of its density and hardness.   It occurs as inclusions in the olivine of basaltic rocks in minute individuals with sharp crystal form ; it occurs very rarely as an independent constituent in such rocks, as was observed in the basalt of Mount Shasta, California, by Wadsworth.[†]   In the lherzolites and olivine rocks it is more often found in irregular grains than in crystals, but attains greater dimensions ; it is the same in the serpentines.

The spinel minerals remain perfectly fresh in rocks, all the rest of whose constituents are altered and decomposed.

Fluorite.

Fluorite as a rock constituent does not appear in crystals, but in irregular grains, often of considerable size.   The cleavage parallel O (111) forms quite sharp systems of lines, which intersect at angles changing with the position of the section.

Transparent, clear, bluish or violet colored, often with quite unequal distribution of the pigment.   Its very low index of refraction is highly characteristic; $n_{na} = 1.4332$–1.4340.   The color appears to come from intermolecular organic matter, hydrocarbons; it is lost when the fluorite is heated to redness.   Traces of double refraction are rare in rock-making fluorite.

Sp. gr. $= 3.18$–3.20.   Its specific gravity in connection with its resistance to all acids but sulphuric facilitate its separation from a

* G. H. WILLIAMS: The Norites of the "Cortlandt Series," etc.   Am. Journ. Sci., Vol. XXXIII., March, 1887.
† Harvard University Bulletin, 1882, No. 22, p. 259.

rock powder. Decomposition products are wholly wanting. Fluid inclusions are frequent.

Fluorite occurs only as an accessory constituent in rocks; in gneis, granite, quartz porphyry, syenite, elæolite syenite and in the crystalline schists.

Garnet Group.

Literature.

C. KLEIN, Optische Studien am Granat. Nachr. d. kön. Ges. d. Wiss. zu Göttingen. 1882. No. 16. 457–564 and N. J. B. 1883. I. 87–163.

A. VON LASAULX, Ueber die Umrindungen von Granat. Sitzungsberichte d. niederrhein. Ges. zu Bonn. 1882. 3. Juli.

E. MALLARD, Explication des phénomènes optiques anomaux que présent un grand nombre de substances cristallisées. Paris. 1877. (Ann. des Min. ?, X 60-203. 1876.)

A. RENARD, Les roches grenatifères et amphiboliques de la région de ... gris. Bulletin du Musée Roy. d'hist. nat. de Belgique. Bruxelles. T. I. 1882.

A. SCHRAUF, Beiträge zur Kenntniss des Associationskreises der Magnesia i. s. Z. X. 1882. VI. 321–388.

A. WICHMANN, Ueber doppelbrechende Granate. Pogg. Ann. CLVII. 282–299. 1876. — Z. D. G. G. 1875. XXVII. 749.

All members of the garnet family exhibit very simple crystal forms; those found in rocks have mostly the forms ∞O (110) and 2O2 (211), alone or in combination. Their cross-sections are then erratic, hexagonal, or eight-sided. Irregular grains and aggregates occur exclusively in many rocks; in others they are accompanied by well defined crystals. The outlines of the garnet grains are exceedingly irregular in some Archæan rocks. Cleavage is not noticeable in thin sections; the great brittleness of the mineral gives rise to irregular fracturing.

The high index of refraction of all garnets is a distinguishing characteristic. $n_{na} = 1.7468{-}1.8141$. The dispersion is strong, especially in pyrope $n_\rho = 1.7776, n_\nu = 1.8288$. The rock-making garnets are in general completely isotropic, or exhibit very faint traces of double refraction. Nevertheless there are certain occurrences, especially the lime-silicate hornstones and the garnet rocks, which show very double refraction, usually connected with a zonal structure in garnets. These optical anomalies have been thoroughly studied, particularly by E. Mallard and C. Klein.

H. = 7–7.5, sp. gr. = 3.4–4.3, varying with the chemical composition. Garnet which has not been heated to redness is almost wholly unacted on by acids, including hydrofluoric, and it is decomposed by

9

alkali carbonates only after long fusion with very fine powder. Fused garnet is decomposed by hydrochloric acid with the separation of gelatinous silica. Its chemical behavior and high specific gravity greatly facilitate its separation from a rock powder.

From the great tendency of all garnets to form isomorphic laminæ or shells, which is shown by a difference in color between the centre and the margin, or between alternating shells (Pl. V. Fig. 4), the chemical composition seldom corresponds to one of the simple combinations which are treated in mineralogy as the varieties of garnet. Moreover, analytical investigations of rock-making garnets are too few to permit of giving the distribution of the different varieties in the rocks with certainty. The following statements therefore may undergo more or less modification:

Grossular, $3CaO$, Al_2O_3, $3SiO_2$, transparent and colorless or nearly so in sufficiently thin sections. Sp. gr. = 3.4–3.6. Easily fusible. Zonal structure and optical anomalies frequent; sometimes well crystallized, ∞ O (110), sometimes in grains and aggregates. Occurs especially in lime-silicate hornstones, and as inclusions in the granular limestone belts of the Archæan, also in garnet rock often combined with common garnet and allochroite. If there is a zonal difference in its color, the centre is darker than the margin. It frequently encloses fluid inclusions, calcite, quartz, wollastonite, epidote, vesuvianite, and graphite. Decomposition products unknown.

Almandine, $3FeO$, Al_2O_3, $3SiO_2$, is red by transmitted light. Sp. gr. = 4.1–4.3. Easily fused to a dark magnetic bead. It occurs as grains in many granitic rocks, seldom in crystals, 2O2 (211); it also occurs in the Hungarian andesites in grains and crystals. It is most abundant in the gneisses, granulites, and in those Archæan rocks free from feldspar; generally as grains, more rarely in crystals, which have 2O2 (211) predominant in rocks rich in feldspar, and ∞ O (110) predominant in those poor in feldspar. It usually encloses the minerals associated with it. Its substance is generally extremely fresh. It is found altered to chlorite at Spurr Mountain Iron Mine, from which locality it has been described by Pumpelly,[*] and more recently it has been thoroughly investigated by Penfield [†] and Sperry, who also described a similar alteration of iron-alumina garnet from Salida, Chaffee County, Colorado.

Common garnet, an isomorphic mixture of the grossular, alma-

* Amer. Journ. Sci. (3) X. 17. July, 1875.

† Amer. Journ. Sci. (3) XXXII. Oct. 1886.

dine, and melanite molecule, occurs in certain garnet rocks, in a metamorphosed eruptive rock of the diabase and gabbro series, in Archæan rock, especially in kinzigite, eulysite, amphobolite eclogite. pyroxene rocks, and their derivatives, as well as in the phyllite formations. It is reddish brown to yellowish red by transmitted light, often nearly colorless. Zonal structure is very common, often accompanied by optical anomalies. The crystal form is generally wanting. Inclusions of the associated minerals and fluid inclusions centrally accumulated are frequent; so is also a micropegmatitic intergrowth with the associated minerals, and their radial arrangement about the garnet as a centre. Decomposition products are not uncommon. One form is the pseudomorph of chlorite after garnet, described by Hawes * from the phyllites of the Connecticut Valley in New Hampshire. Its alteration into hornblende has been quite often observed, and in one instance into scapolite.

Lime-iron garnet. 3CaO, Fe_2O_3, $3SiO_2$, in velvet-black crystals with the form ∞O (110) $2O2$ (211), called melanite, frequently occurs as an accessory constituent in those basic eruptive rocks rich in alkali (phonolite, leucitophyre, nephelinite, tephrite). It is brown in transmitted light with various depths of color, and forms one of the oldest secretions. Optical anomalies are rare. Decomposition phenomena are wanting. Melanite accompanied by wollastonite and fassaite has been described by Fouqué † as a volcanic contact phenomenon.

Green lime-iron garnet occurs in many serpentines; it sometimes shows a zonal structure of green and red layers. It is brown in many iron-ore beds in the crystalline schists. The lime-iron garnets have sp. gr. = 3.4–4.1, and fuse to a strongly magnetic bead.

Spessartine, essentially 3MnO, Al_2O_3, $3SiO_2$, occurs occasionally in granitic rocks in the form of grains of considerable size. Its color is sometimes blood-red, sometimes yellowish red to colorless. Its decomposition processes are not known. It occurs with topaz in the lithophysæ of rhyolite from Nathrop, Colo., according to Cross.‡

Pyrope, principally 3MgO, Al_2O_3, $3SiO_2$, with some chromium and a variable admixture of the almadine molecule, never forms crystals, but angular to rounded grains of red or blood-red color by transmitted light, and mostly of very pure substance. It fuses with difficulty to a non-magnetic bead. Sp. gr. = 3.7–3.8. Pyrope appears

* Mineralogy and Lithology of New Hampshire. Concord, 1878. 75.
‡ Compt. rend. 1875. 15 Mars.
‡ Amer. Jour. Sci., Vol. XXXI., June, 1886, p. 432.

to be confined to rocks rich in magnesia: the peridotites and their derivatives, the serpentines.   In these rocks the pyrope is often surrounded by a radial grouping of the other constituents, especially the pyroxenes and their alteration products.   Very frequently the pyrope in serpentine is surrounded by a radially fibrous, light grayish-brown shell, which is an alteration product of the pyrope, probably with the co-operation of the olivine substance.   This shell has been called *kelyphite* (Pl. XIV. Fig. 4); its composition is not constant, and its mode of formation is still doubtful.   Diller * has described kelyphite shells of biotite and magnetite around pyrope in the peridotite from Elliott Co., Ky.

Leucite.

Literature.

H. BAUMHAUER, Studien über den Leucit. Z. X. 1877. I. 257–273.

A. DES CLOIZEAUX, Nouvelles recherches sur les propriétés optiques des cristaux naturels ou artificiels et sur les variations que ces propriétés éprouvent sous l'influence de la chaleur. Paris. 1887. 513–515.

J. HIRSCHWALD, Zur Kritik des Leucitsystems. T. M. M. 1875. IV. 227 ff.

— Ueber unsere derzeitige Kenntniss des Leucitsystems. T. M. P. M. 1878. II. 85–100.

C. KLEIN, Optische Studien am Leucit. Göttinger gelehrte Nachrichten 1884. No. 11. 421–472. — N. J. B. 1885. Beil.-Bd. III. 522–584.

E. MALLARD, Explication des phénomènes optiques anomaux que présentent un grand nombre de substances cristallisées. Paris. 1877. 24–39. (Ann. des Mines (7). X. 1876.)

G. VOM RATH, Ueber das Krystallsystem des Leucits. M. B. A. 1872. 1 Aug. — Pogg. Ann. Ergänzungsband VI. 1872. 198 ff. and Sitzungsber. der niederrhein. Ges. Bonn. 4. Juni 1883.

A. WEISBACH, Zur Kenntniss des Leucits. N. J. B. 1880. I. 143–150.

F. ZIRKEL, Ueber die mikroskopische Structur der Leucite. Z. D. G. G. 1868. XX. 97 sqq. — Basaltgesteine 1870. pg. 44 ff.

The crystal system of leucite has been the subject of much discussion and great uncertainty for many years.   Its habit, without exception, is that of an isometric crystal, exhibiting the icositetrahedron, 2O2 (211), alone, or with ∞ O (110) and ∞ O ∞ (100) less strongly developed.   And notwithstanding its anomalous optical behavior, it was considered an isometric body by Brewster,† Biot,‡ and Des Cloizeaux (l. c.) until the year 1872.

The polarization phenomena were explained as lamellar polariza-

---

* Bull. 38, U. S. Geol. Surv. 1887, p. 15.
† Edinburgh Phil. Journ. 1821. V. 218.
‡ Memoire sur la polarisation lamellaire. 1841. 669.

tion, or as the effect of intercalated lamellæ. G. vom Rath, in 1872, placed it among the tetragonal minerals as the result of his study of the twinning striæ on the crystal faces and of the values of the angle of the apparent 2O2 (211) edges. Hirschwald, on the ground of its crystal habit and of the twinning, which is parallel to the 6 faces of ∞ O, restored it to the isometric system; while Weisbach considered it orthorhombic. Investigating the mineral by physical methods, Baumhauer concluded that the result of etching its crystal faces indicated its tetragonal nature, or at least did not militate against such an assumption. Mallard's investigation of the optical behavior of sections parallel to the cubical faces (supposing leucite isometric) led him to refer it to the monoclinic system. C. Klein was convinced by his study of leucite that under the physical conditions acompanying its formation it had crystallized isometrically, but that its molecular structure at ordinary temperatures may be considered orthorhombic. Summing up the results of all leucite studies down to the present time, it may be confidently stated that all leucite crystallized in the isometric system, but that the isometric molecular arrangement, at least of the larger crystals, cannot obtain for the temperatures and pressures at the earth's surface; that it therefore experiences a molecular displacement, in consequence of which there arises a more or less complicated, apparent twinning. This molecular displacement has not only an optical effect, but leads to a more or less profound deformation of the crystal form. It is not yet possible to decide from the goniometric and optical behavior of leucite to which crystal system this molecular displacement tends to change it.

Leucite furnishes an excellent example of the group of minerals in which optical anomalies are occasioned by dimorphism.

The cross-sections of leucite crystals are six-sided, eight-sided, or rounded, according to the greater or less development of 2O2; there also occur quadratic, triangular, or rhombic sections from the surface of the crystals parallel to (100), (111), (110). The larger individuals frequently exhibit irregularities of outline due to the corrosion of the crystals, and sometimes appear to be composed of a number of smaller crystals. Leucite crystals vary greatly in size; massive leucite appears to be extremely rare.

Cleavage is not noticeable in thin section; but a cracking of the crystals along irregular faces is very often present, as the result of molecular shifting.

The very small crystals of leucite appear wholly isotropic when investigated optically; in the larger individuals a very complicated twin lamination is observed between crossed nicols, especially upon the insertion of a gypsum plate; these laminæ are doubly refracting, and intersect at angles which change with the position of the section. The index of refraction of leucite is low, and differs very little from that of a rock glass. The double refraction is weak, and positive according to Des Cloizeaux ($\omega = 1.508$, $\epsilon = 1.509$) and Klein. Tschermak[*] found its character negative in one instance.

In very thin sections it is necessary to use a sensitive tint in order to perceive the anisotropism and to study the twinning. The interference colors in thin sections do not exceed grayish blue of the 1st order. This double refraction, and with it the twin lamination, disappear when the crystals are exposed to a temperature of about 500° C., when they appear isotropic as first observed by Klein,[†] and subsequently by Penfield.[‡] Rosenbusch [§] showed that at this temperature the twinning striæ on the crystal faces disappear, so that they reflect light with perfect uniformity, from which it is very probable that they also return to isometric symmetry in a goniometrical sense.

The twinned-like structure of leucite has been thoroughly studied by Klein, whose results are given with illustrations by Rosenbusch, but are here omitted.

The larger crystals of leucite usually enclose the older secretions associated with them; these minerals are magnetite, picotite, apatite, olivine, augite, haüyne, nepheline, and melanite. More frequently the inclusions are prismatic microlites, which are in part green and most likely augite, and in part colorless and indeterminable. These are either crowded at the centre of the leucite, or are arranged in concentric zones, Fig. 52; they then lie with their longer axis parallel to the boundary of the enclosing mineral. These are often accompanied by glass inclusions and gaseous interpositions, and more rarely by those of a fluid (Capo di Bove, Monte Vulture, Ölbrück). The glass inclusions take the form of the enclosing mineral. Sometimes all the inclusions in one crystal are of the same kind; sometimes the different kinds

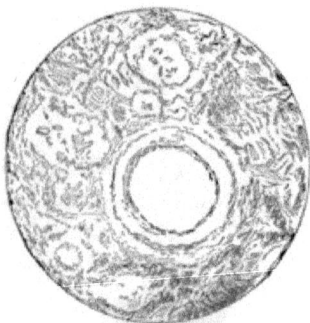

Fig. 52

lie together indiscriminately; at others they are so arranged that zones

---

[*] T. M. M. 1876. 66.   [‡] N. J. B. 1884. II. 224.

[†] N. J. B. 1884. II. 50.   [§] N. J. B. 1885. II. 59.

of different kinds alternate with one another. Very rarely there is a
radial arrangement of the interpositions, or a combination of radial and
zonal arrangement in the same crystal (Pl. XIV. Figs. 5 and 6).
Very small leucite crystals are usually free from interpositions. Leucite
crystals are often encircled by a veil or shell of augite microlites (Pl.
XV. Fig. 1), which is at times a means of recognizing the leucite when
the characteristic twinning is not noticeable.

Sp. gr. = 2.45–2.5. Leucite, K_2O, Al_2O_3. $4SiO_2$. mostly with no very
considerable percentage of Na_2O, is very slightly attacked even by hot
hydrochloric acid when in thin section, but as powder it is strongly
attacked with the separation of pulverulent silica. The mineral is
therefore better isolated from the rock by specific-gravity methods
than by chemical ones.

Leucite alters quite frequently into analcite without the form of
the crystal being changed in any way. Nevertheless there is formed
as a side-product radially and confusedly fibrous, double refracting
aggregates of an indeterminable nature, which are often in considerable
quantity. The analcite, in turn, is altered to a mixture which is principally feldspar and light-colored mica.

Leucite is a mineral wholly confined to Tertiary and recent eruptive rocks and their tuffs; it accompanies sanidine and nepheline in
rocks of the phonolite series, plagioclase and nepheline in the tephritic
rocks, and nepheline alone in the leucite basalts and leucitites.

Zirkel * has described its occurrence at Leucite Hills, Wyoming
Ter.; and von Chrustschoff † has found it in leucite porphyry from
Cerro de las Virgines, in Lower California.

Sodalite Group.

Literature.

L. L. HUBBARD, Beiträge zur Kenntnis der Nosean-führenden Auswürflinge des
    Laacher Sees. T. M. P. M. VIII. 1887. 356–399.
B. MIERISCH, Die Auswurfsblöcke des Monte Somma. T. M. P. M. VIII. 1887.
    113–189.
G. VOM RATH, Mineralogisch-geognostische Fragmente aus Italien. Z. D. G. G.
    1866. XVIII. 620–624.
— Skizzen aus dem vulkanischen Gebiet des Niederrheins. Z. D. G. G. 1860. XII.
    29; 1862. XIV. 663; 1864. XVI. 73.
A. SAUER, Untersuchungen über phonolithische Gesteine der canarischen Inseln.
    Halle. 1876. Zeitschr. f. d. ges. Naturw. XLVII.

* Microscopic Petrography Washington, 1876.
† T. M. P. M. Vol. VI. 1885, pp. 160–171.

H. Vogelsang, Ueber die natürlichen Ultramarinverbindungen. Kon. Akad. van
   Wentesch. Amsterdam (2) VII. 1873.
F. Zirkel, Untersuchungen über die mikroskopische Zusammensetzung und Struk-
   tur der Basaltgesteine. Bonn. 1870. 79 ff.

The sodalite group includes a number of minerals crystallizing in
the isometric system, sodalite, haüyne, with the varieties nosean, ittnerite and skolopsite, and lapis-lazuli.   They are characterized chemically
by the fact that they present isomorphous combinations of a silicate
molecule with the salt of another acid, or with a haloid compound.
The first-named substances are widely spread rock-making minerals of
extremely characteristic geological position.

Sodalite.

Sodalite forms simple crystals, more rarely twins according to the
spinel law.   When they occur in porphyritic rocks they are rhombic
dodecahedrons and octahedrons, either alone or in combination with
one another.   In granular rocks their forms become more indistinct
or are lost altogether ; between the sodalite and the younger rock constituents the former shows its crystal outline ; between it and the older
constituents the outline is that of the older secretions.   Cross-sections
of the crystals are mostly quadratic or hexagonal, but are often distorted by the crystals being strongly developed in the direction of a
trigonal secondary axis.

In the freely crystallized sodalite crystals of Monte Somma, the
cleavage parallel to ∞O is very clearly perceptible, even in thin section ; the sodalites crystallized within rocks exhibit much less typical
cleavage.

In transmitted light sodalite is colorless ; also bluish, greenish, light
pink, red, and yellowish.   Its index of refraction is low ; $n_{na} = 1.4827$ -
1.4858.   Optical anomalies have been observed occasionally in the
vicinity of inclusions.

Sp. gr. $= 2.28$-2.34. Chemical composition $= 2(Na_2O, Al_2O_3, 2SiO_2)$
$+ NaCl$. Easily and completely soluble in hydrochloric and nitric
acids ; upon standing gelatinous silica separates out ; it is even acted
on in thin section by acetic acid.   In order to observe the gelatinization well in thin section it should be moistened with only a very thin
coat of acid.   If the mineral has been treated with hydrochloric acid
there will be an abundance of common salt crystals when the jelly dries.

Sodalite is quite a constant constituent of elæolite-syenites ; in these
rocks it occurs in crystals, grains, and massive forms, or in veins and

streaks within other minerals, especially feldspar. The formation of sodalite in these rocks followed the secretion of the iron-bearing constituents and preceded that of the feldspar. Its age relative to that of elæolite appears to vary. The primary nature of sodalite in these rocks is in general beyond question. Only the vein-like massive occurrences are possibly of secondary origin. In the rocks mentioned sodalite often encloses the ores, and needles of pyroxene and hornblende associated with it, besides fluid-inclusions ; these when abundant are usually accumulated centrally.

Upon the alteration of sodalite, which is sometimes very far advanced in these rocks, there arise tufted aggregates of zeolites ; spreustein, according to Brögger, is a pseudomorph of natrolite after sodalite. In other cases there arise aggregates of muscovite and kaolin. Carbonates which are not infrequently found along cracks in sodalite are evidently infiltration products.

Sodalite occurs only in sharp and distinct crystals in the younger rocks of the trachyte and phonolite families. It is wide-spread in the trachytes of the island of Ischia . it is found sparingly in the place of haüyne in many phonolites of Northern Africa and of the Cantal (Pas de Compains), as also in the granular tephrite of the Crazy Mountains, Montana Territory.*

Its formation follows that of augite without exception, and precedes that of nepheline or feldspar. Inclusions are rare ; they consist of the older minerals which accompany it, especially augite, together with glass and rarely fluid inclusions.

Its isotropism in connection with the low index of refraction and its ready solubility in acids, as well as its low specific gravity, distinguish sodalite readily from all other minerals, with the exception of haüyne and nosean. These are characterized by their chemical behavior, giving reaction for sulphuric acid.

Haüyne and *Nosean.* — All those members of the sodalite group are classed under the name *haüyne*, which are considered isomorphous mixtures of $2(Na_2O, Al_2O_3, 2SiO_2) + Na_2O, SO_3$ and $2(CaO, Al_2O_3, 2SiO_2) + CaO, SO_3$. The first of these compounds is found in a nearly pure condition in the volcano Siderno, in the Cape Verde Islands.† The second compound is not known to occur alone. The analyses as well as the micro-chemical reactions very frequently give a slight percentage

---

* J. E. Wolff. Notes on the Geology of the Crazy Mountains. Northern Transcontinental Survey. 1885. c.f. N. J. B. 1885. I. 69.

† C. Doelter, Die Haüyne der Capverden. T. M. P. M. 1882. IV. 461.

of chlorine, which indicates the presence of an isomorphous admixture of the sodalite compound.

The members of this group rich in soda are called *nosean*, in distinction to those rich in lime, called *haüyne*. All the members of this series gelatinize easily with acids: if the gelatinous silica be allowed to dry under the microscope, or better, if the solution obtained through the action of acids be removed and allowed to evaporate, there arise in the case of haüyne abundant characteristic gypsum crystals; if the mineral was nosean, these would appear but sparingly, or not at all. The hydrochloric acid should not be too concentrated, nor the temperature too high, as in this case orthorhombic cube-like crystals of anhydrite will be produced in place of gypsum. The color is changed in many occurrences by heating to redness, or the crystals may become colored if they were colorless before. They are colored blue when heated to redness in vapor of sulphur.

The sp. gr. varies from 2.27–2.50, according to the chemical composition and to the greater or less abundance of interpositions. Nosean is lighter than haüyne; but this may be obscured by the presence of included hematite or ilmenite plates. The cleavage parallel ∞O (110) is rarely observed microscopically.

The haüynes crystallized in rocks, when fresh, are generally perfectly isotropic; but there are occurrences which exhibit optical anomalies (Vesuvius, Lake Laach). These are of two kinds: in the first case there is a local double refraction, which only occurs around inclusions, especially about gas inclusions; it is characterized by a dark cross between crossed nicols, the arms lying parallel to the principal sections of the nicols, and remaining so during a rotation of the section. The gas inclusion is at the centre of the cross, and gives rise to the phenomenon by the pressure which it exerts on the surrounding mineral. The elasticity is greater in radial directions about the inclusions than in tangential directions. Similar anomalies are often observed along cracks in the haüyne.

The second kind of optical anomaly is a double refraction throughout the whole extent of the mineral (Vesuvius; Lake Laach, Niedermendig, Rhine Province): this is always weak, often only noticeable by the use of a gypsum or quartz plate.

The index of refraction of haüyne is low, but somewhat higher than that of sodalite. $n_{na} = 1.4961$. The color is extremely manifold: haüyne is colorless, blue, gray, brownish, red, yellow, and green; and the color is often irregularly distributed in spots, streaks, and stripes, or in concentric zones in thin section. Occasionally the color is most intense along the cracks, and is most easily developed here in colorless individuals. This led Vogelsang to conclude that the color is of secondary nature. Very strong heating destroys the original or artificial color of haüyne.

The haüynes, with the exception of ittnerite and skolopsite, never occur in irregular masses, but always in crystals or crystal fragments or in grains (rounded crystals). The forms and cross-sections are the same as those of sodalite. The microstructure of haüyne is very variable. Most all occurrences abound in inclusions. These are usually the iron-ores, oxides, gas and glass inclusions; fluid inclusions in great numbers and of very different forms are confined to particular localities (Nieder-mendig). All the foreign bodies, for the most part, are present at the same time, either scattered irregularly through the mineral substance, or regularly arranged, especially when their quantity is considerable. In the latter case they are sometimes aggregated in the centre or periph-erally; sometimes they are aggre-gated in concentric shells. More-over, the interpositions are often arranged in lines parallel to the octahedral axes. Cross-sections par-allel to $\infty O \infty$ (100) and ∞O (110) exhibit two systems of lines inter-secting at right angles, while in sec-

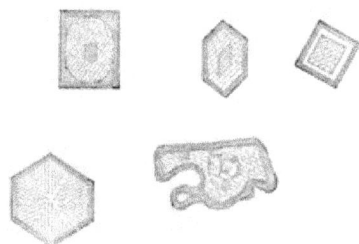

Fig. 53

tions parallel O (111) there are three systems intersecting at 60° (Fig. 53). The glass and gas inclusions, when large, frequently have the crystal form of the enclosing mineral.

The noseans in the leucite porphyries of the volcanic territory of the lower Rhine often have a broad opaque border, having a bluish-black or brownish-black color. This probably arises from the conver-sion of the iron-bearing compounds in a zone rich in interpositions into limonite, and the possibly contemporaneous kaolinization of the haüyne substance.

The haüyne minerals are very often found in nature in a more or less advanced stage of alteration. There are probably two processes which can be distinguished—the zeolitization and the weathering proper. The zeolitization, which takes place through the addition of water and the loss of the sulphates from the molecule, shows itself very quickly by the formation of a fibrous structure. These grayish to yellowish fibres penetrate the clear mineral substance in the form of bundles, starting from cracks, from the surface, and from the larger interpositions. In consequence of the anisotropic nature of the zeolite fibres they are very noticeable between crossed nicols.' When the alteration is complete the whole section consists of bundles of fibres radiating from different points. The zeolite resulting from haüyne is usually natrolite. The cloudy coloring of this natrolite aggregate is not due to a pigment, but is mostly the result of its extraordinarily fine fibrous aggregation. That other zeolites than natrolite, especially stilbite and chabazite, may result from the alteration of the haüynes rich in lime is very probable, both from the microscopical habit and also from the microscopical occurrence of these minerals in rocks rich in haüyne; but the direct proof of it has not yet been produced. Very often the lime component of the original mineral is secreted in the form of calcite when it is altered to natrolite. Zeolitized haüyne becomes clouded when heated to redness through the loss of water.

The distribution of haüyne in rocks is very great. However, it is confined to rocks of the youngest geological periods which are poor in silica and rich in alkali: in this it is distinguished from sodalite, whose formation goes back to the Palæozoic period. The true phonolytes and leucite porphyries contain it almost without exception; it is found with less regularity in the tephrytes, the nepheline and leucite rocks, and in their varieties free from feldspar. Nosean occurs in the phonolite at Black Hills, Dakota. In many nepheline rocks the amount of haüyne can become so excessive that it forms the next principal constituent to pyroxene, and supplants the nepheline: such rocks have been called haüynophyres. In all these varieties of rocks the formation of haüyne in the molten magma followed that of the older pyroxenes, and preceded that of nepheline; it is therefore the oldest of the feldspathic components. In all the rocks above-named haüyne is associated with nepheline, or with nepheline and leucite; the only rocks in which it occurs without these minerals are certain andesites of the Canary Islands.

Ittnerite and *skolopsite*, which correspond microscopically and chemically to haüyne, have only been found in one locality in the Kaiserstuhl.

Analcite.

Literature.

A. BEN SAUDE, Ueber den Analcim. N. J. B. 1882. I. 41–79.
C. KLEIN, Analcim vom Table Mountain bei Golden, Col. N. J. B. 1884. I. 250.
E. MALLARD, Explication des phénomènes optiques anomaux que présentent un grand nombre de substances cristallisées. Paris. 1877. 57–61.

Analcite is never an original rock constituent, but is always a product of secondary secretion or alteration; in the first case it occurs in cavities as freely developed or attached crystals with the forms $\infty O \infty$. 2 O 2 (100) (211), or 2 O 2 (211), or else it completely fills the cavities without a regular crystal form; in the second case it occurs in pseudomorphs, and therefore has the form of the original mineral, consequently its cross-sections are not characteristic. The most frequent pseudomorphs are after nepheline. They also occur after leucite. The cleavage parallel to $\infty O \infty$ (100) is usually quite noticeable, or may be developed by a rapid heating of the section.

The index of refraction is low, as for all zeolites. According to Des Cloizeaux, $n_\rho = 1.4874$. Colorless in transmitted light. The optical anomalies, which are very common in the freely crystallized analcites, are comparatively rare in those crystallized in mass, if the investigation is not carried on under a sensitive tone of color. It is highly probable that the double refraction is a result of internal strain. A gentle warming in a water or paraffine bath, according to A. Meriam, diminishes the strength of the double refraction very considerably; and according to C. Klein it disappears altogether upon stronger heating in an atmosphere of steam. On the other hand, heating to redness, by which the analcite begins to lose water, increases the double refraction, or gives rise to it if not previously present. This latter characteristic may be used in the diagnosis of analcite.

Sp. gr. = 2.15–2.28. Chemical compositions = Na_2O, Al_2O_3, $4SiO_2 + 2aq$. Soluble in all mineral acids with the separation of gelatinous silica; covered with a thin coat of hydrochloric acid, the surface gelatinizes, and may be readily colored. The clouding due to the loss of water when heated to redness is very marked, often complete.

Analcite with strong double refraction may be distinguished from leucite, which it closely resembles, by its gelatinization with hydrochloric acid and the treatment with hydrofluosilicic acid. Analcite furnishes almost exclusively the characteristic hexagonal crystals of sodium fluosilicate; leucite, a preponderance of the isometric crystals of potassium fluosilicate. They may also be distinguished by their specific gravities.

Perofskite.

Literature.

A. BEN SAUDE, Ueber den Perowskit. Göttingen. 1882.

EM. BOŘICKY, Ueber Perowskit als mikroskopischen Gemengtheil eines für Böhmen neuen Olivingesteins, des Nephelinpikrites. Sitzungsber. der k. böhm. Ges. d. Wissensch. 13. Okt. 1876.

C. KLEIN, Perowskit von Pfitsch in Tyrol. N. J. B. 1884. I. 245–250.

A. SAUER, Erläuterungen zu Section Wiesenthal der geol. Specialkarte des Königr. Sachsen. Leipzig. 1884. 54.
A. STELZNER, Ueber Melilith und Melilithbasalte. N. J. B. B.-B. II. 1882. 390 ff.

Perofskite appears in the eruptive rocks in octahedrons (Pl. XV. Fig. 2), and forms microscopic crystals, 0.02–0.03 mm. in diameter, which are usually quite sharp, though sometimes rounded. They are occasionally gathered together in groups. Incipient forms of growth also occur, which appear like intersection twins and irregularly ramifying skeleton crystals; to these may be added the jagged plates mentioned by several authors. On the other hand, the perofskite crystals, which occur sparingly in the Archæan rocks or their intercalations, almost always have the cubical form.

By incident light perofskite is grayish yellow to gray-brown, the minute crystals appearing like fine powder; in transmitted light it is grayish white, violet-gray, gray-brown, brownish yellow to red-brown, seldom with greenish tones, more rarely with a slight zonal change of color. The index of refraction has not yet been measured; it is, however, very strong, and over 1.7. Consequently, the total reflection is considerable, and the surface of intersected crystals is strongly wrinkled. The dark borders due to the total reflection have given rise to numerous illusions as to the crystal forms and the presence of a shelly structure.

Perofskite does not appear isotropic between crossed nicols, but doubly refracting, so that in the larger crystals the parts with different orientation of the axes of elasticity penetrate one another in the form of a complicated twinning, with optically biaxial striæ arranged in intersecting systems. These are not noticeable in very small crystals. The above-cited studies of C. Klein and A. Ben Saude render it highly probable that the double refraction of perofskite is an anomaly, and not the result of mimetic structure.

H. = 5.5. Sp. gr. = 4.1. Chemical compositions, CaO, TiO_2. Part of the lime is not infrequently replaced by FeO in considerable quantity. Perofskite is not attacked by hydrochloric acid nor by hydrofluoric acid in water. It is dissolved by concentrated sulphuric acid upon being heated. Perofskite is easily isolated from rocks by the combined use of its specific gravity, its resistance to chemical action, and its very indifferent behavior towards a strong electro-magnet.

Perofskite is generally quite free from inclusions, and is undecomposed in rocks. But Sauer (l. c.) observed in the nepheline basalt of Oberwiesenthal its alteration into a substance (leucoxen) quite analogous to the alteration product of ilmenite and rutile.

Perofskite, when opaque, is easily mistaken for the iron ores and spinels; from the first of these it is distinguished by the lack of metallic lustre and its insolubility in hydrochloric and nitric acids.

The transparent crystals and grains of perofskite are easily confounded with spinel, chromite, garnet, and titanite. A definite determination can only be made on isolated material by proving the presence of titanic acid and the absence of silica and chromium.

Perofskite is an almost constant ingredient of the younger basic eruptive rocks, especially melilite basalt. It occurs in leucite and nepheline rocks; more rarely in elæolite syenite (Ditró). It occurs in serpentinized peridotite in the Onondaga salt group at Syracuse, N. Y.* It belongs to the oldest secretions in the eruptive magmas of these rocks, and therefore occurs as inclusions in most of the other constituents. It is accompanied by magnetite and chromite, with which also it grows together and often surrounds later secretions with a kind of wreath.

---

* Geo. II. Williams, Amer. Journ. Sci. Vol. XXXIV. Aug. 1887.

MINERALS OF THE TETRAGONAL SYSTEM.

TETRAGONAL minerals are doubly refracting, with one optic axis. The latter coincides with the principal crystallographic axis, and is the axis of greatest or least elasticity. In the first case, the character of the double refraction is said to be negative, and the ordinary ray is more strongly refracted ($\omega > \epsilon$) ; in the second case the substance is optically positive, and the extraordinary ray is more strongly refracted ($\omega < \epsilon$). Each of the two rays is differently absorbed : thus tetragonal minerals, when colored, exhibit a more or less noticeable pleochroism in all sections which do not lie parallel to oP (001). Sections at right angles to c have quadratic or octagonal outlines or cleavage lamellæ, or the regular outline is wanting and the cleavage is parallel to oP (001); such sections behave like isotropic substances in parallel polarized light—they remain dark during a complete rotation between crossed nicols ; in convergent polarized light they show an interference cross with or without colored rings, whose arms lie parallel to the principal sections of the nicols, and which do not change their position during a rotation of the section. Sections parallel or inclined to c exhibit outlines varying with the position of the section and the form of the crystal ; the cleavage lamellæ are recognized by parallel or intersecting systems of cracks. In parallel polarized light the sections are doubly refracting ; for a complete rotation between crossed nicols they appear dark and light four times, and the position of darkness is always reached when a principal section of the nicols is either parallel to the cleavage cracks or bisects the angle between them. In convergent polarized light the interference figure appears at one side of the field of view and moves around the margin during a rotation in such a manner that the arms of the cross move parallel to themselves, so long as the section is not too greatly inclined to c ; in sections parallel to c the interference figure separates into hyperbolas which lie symmetrically with respect to the principal axis.

Optical anomalies show themselves in basal sections when examined in convergent polarized light by the interference cross opening into hyperbolas during a rotation of the section, and presenting the appearance of a biaxial body with small axial angle, cut at right angles to the acute bisectrix. Different portions of such an abnormal plate usually show different sizes of the apparent axial angle, and a varying position of the apparent axial plane. In parallel polarized light the basal section sometimes exhibits a division of the field into irregularly lighted parts.

Rutile.

Literature.

A. CATHREIN, Ein Beitrag zur Kenntniss der Wildschönauer Schiefer und der Thonschiefernädelchen. N. J. B. 1881. I. 169–183.
A. von LASAULX, Ueber Mikrostruktur, optisches Verhalten und Umwandlung des Rutil in Titaneisen. Z. X. 1883, VIII. 59–75.
A. SAUER, Rutil als mikroskopischer Gesteinsgemengtheil. N. J. B. 1879. 569–576.
— Rutil als mikroskopischer Gemengtheil in der Gneiss- und Glimmerschiefer-formation, sowie als Thonschiefernädelchen in der Phyllitformation. N. J. B. 1881. I. 227–238.
L. van WERVEKE, Rutil im Ottrelithschiefer von Ottrez und im Wetzschiefer der Ardennen. N. J. B. 1880. II. 281–283.

Rutile appears in rocks under a great variety of forms. Where its dimensions are large it usually takes the form of grains, or has its edges and corners rounded; on the other hand, the extremely minute micro-scopic individuals in certain schists possess crystal forms of almost ideal sharpness. $111 : 11\overline{1} = 84° 40'$. Their habit is always prismatic, and the forms recognizable are the same as on the macroscopic crystals; they also possess the same striation parallel c. Twinning is extremely common, and follows either the law, the twinning plane is $P\infty$ (101); or, as is less frequent among the macroscopic individuals, the twinning

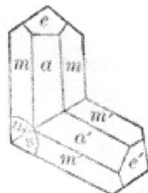

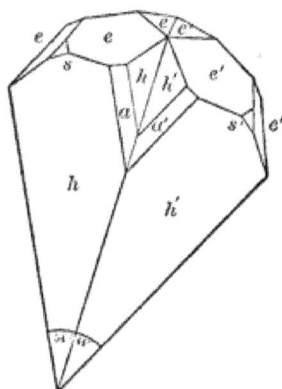

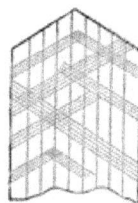

Fig. 54   Fig. 55   Fig. 56

plane is $3P\infty$ (301). By the first method the principal axes of the twinned individuals form an angle of 65° 35', by the second an angle

10

of 54° 44′. Twins of the first kind are usually genicular (Fig. 54,
$\infty P . \infty P \infty . P \infty$ (110) (100) 101); those of the second kind are heart-
shaped (Fig. 55, $\infty P3 . \infty P \infty . P \infty . P$. (310) (100) (101) (111)).
Not infrequently there is no indication of twinning in the outline of
the larger grains and crystals, yet in polarized light they prove to be
twinned through and through, so that in one individual a greater or
less number of lamellæ are intercalated in twinned position after the
first-named method.   These lamellæ cut the principal axis either at an
angle of 65° 35′ or 57° 12.5′, and stand in twinned position to this
or to themselves, and are generally arranged parallel to all faces of
$P \infty$ (101).   In a section parallel to $\infty P \infty$ (100) this lamination
would appear as in Fig. 56; in a basal section it would be as in Fig. 57.
The lamellæ may traverse the crystal completely or only in part.

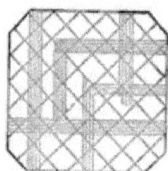

Fig. 57

Fig. 58

O. Mügge has shown that this twinned lamination is probably a result
of pressure, the face $P \infty$ (101) serving as a gliding-plane. (cf. N. J. B.
1884, I. 216.)   The very small individuals of rutile frequently occur
as intergrown twinned needles, and form net-shaped groups, called
sagenite by De Saussure, represented in Fig. 58. The meshes of such
a sagenite web are apparently 60° and 120°; in actual fact the angles
correspond to the two laws just given. Such structures are usually
very small, the length of the single individuals being $\frac{1}{100}$ to $\frac{1}{1000}$ mm.

The cleavage of rutile parallel ∞P (110) is very perfect, and
shows itself as fine and very straight cracks (Figs. 56 and 57); that
parallel $\infty P \infty$ (100) is less perfect: the cracks are rough, irregular, and
frequently interrupted; they are only distinct in very thin sections.
In basal sections, therefore, there are two systems of cracks intersecting
at right angles: the first of these cuts the other at 45°; in sections
inclined to the principal axis the cleavage cracks intersect in rhombic

figures, which are more acute as the inclination is less ; in sections parallel to *c* all the cleavage cracks are parallel.

Rutile is strongly refracting, and is optically positive; among rock-making minerals there are none with higher index of refraction nor more strongly doubly refracting. Bärwald determined on rutile from the auriferous sands of Syssert, in the Urals,

$$\omega_{li} = 2.5671 \qquad \epsilon_{li} = 2.8415$$
$$\omega_{na} = 2.6158 \qquad \epsilon_{na} = 2.9029$$
$$\omega_{le} = 2.6725 \qquad \epsilon_{le} = 2.9817$$

From this it arises that the surface of its sections is very distinctly wrinkled, the total reflection at the margin is strong, and the interference colors even of the minutest needles are very brilliant. For a thickness of hardly $\frac{1}{1000}$ mm. it shows red of the first order; as soon as the dimensions are somewhat increased, the colors are the indistinct ones of a higher order, which are no longer recognizable when the rutile is strongly colored.

The pleochroism is changeable in those rutiles which in transmitted light appear yellow, reddish brown, or fox-red, according to their thickness. The thicker microscopic crystals and the extremely minute needles do not appear at all pleochroic; in those of medium size *O* is sometimes yellow to brownish yellow, *E* brownish yellow to yellowish green, the absorption is $E > O$. In consequence of this pleochroism twinned crystals often appear to be colored green and yellowish in stripes.

In consequence of the twinning structure, also, basal sections of rutile frequently do not exhibit the simple interference figure of a uniaxial mineral, but phenomena which suggest more or less strongly that of a biaxial crystal cut perpendicular to a bisectrix. Between crossed nicols the section is dark throughout its whole extent only when the diagonals of the cleavage parallel ∞P (110) coincide with the principal sections of the nicols. In all other positions the portions of the crystals free from lamellæ remain dark; but those containing twinned lamellæ are variously illumined according to the number and position of those lamellæ, as may be seen from Fig. 59, which represents the cross-section of a basal section of rutile with twinned lamellæ.

Fig. 59

The high sp. gr. = 4.20–4.25 of rutile, whose chemical composition is TiO_x, and its resistance to hydrochloric and hydrofluoric acids facilitate its isolation from rocks even when of the smallest dimensions. It is strongly attacked by treatment with hydrofluoric and sulphuric acids, or with sulphuric acid alone. To distinguish the isolated powder from zircon and cassiterite, with which it may be confounded, a small quantity is melted to a bead on platinum wire with dehydrated bisulphate of potassium, and this is tipped with a drop of hydrogen superoxide. When titanic acid is present the bead and the liquid are colored yellow or orange, according to the amount of titanium present.

Rutile is frequently found altered into a fibrous or granular substance, strongly refracting, of white, yellowish or greenish color (Pl. XV. Fig. 3), which is identical with leucoxene, an alteration product of ilmenite. It has been called titanomorphite by von Lasaulx, and wrongly supposed to be calcium bititanate. Sauer has taken the same substance in similar rocks for titanic acid. Cathrein proved it to be titanite. Rutile is also altered into ilmenite.

Rutile may also occur as a secondary product; it has been found as the alteration of titanite in an elæolite syenite of the Serra de Monchique, and as an alteration product from ilmenite in altered diabase.* It is probable that the sagenite webs found in the decomposed micas of the kersantites and minettes, and those in altered phlogophites, are products of the leeching out of the minerals containing them. The rutile needles often lie in three directions, intersecting at angles of 60°, and parallel to the rays of the pressure figure. Rutile is also a secondary product in the hornblendes of many diorites.

Rutile occurs as a primary constituent both in eruptive rocks and in the schists, but more frequently in the latter. G. W. Hawes† considered the long hair-like interpositions widely disseminated in the quartzes of granite as rutile, although he was not able to definitely determine them, their breadth being so minute that they appear opaque.

M. Maclay-Miklucko‡ isolated and measured rutile crystals from the mica of the topaz-bearing granite of Greifenstein, which were accompanied by microscopic cassiterite.

G. H. Williams§ determined rutile as interpositions in the mica of a porphyritic diorite out of the gneiss from the region of Triberg, in the Black Forest. K. A. Lossen (l. c.) discovered it as a primary constituent in peculiar concretionary mineral masses, which occur as inclusions or like older secretions in the kersantite of Michaelstein in the Hartz.

---

* G. H. Williams, N. J. B. 1887. II. 263.

† Mineralogy and Lithology of New Hampshire. Concord. 1878. 45.

‡ N. J. B. 1885. II.

§ N. J. B. B.-B. II. 1883. 617.

Rutile is very common in grains and crystals in the gneisses and mica schists and the masses intercalated in them, especially in the rocks rich in hornblende and augite. It is very widely disseminated in the phyllite formation. And it is in the phyllitic slates that the sagenite webs are most perfectly developed:

The extremely minute microlitic needles which F. Zirkel first called attention to in the clay slates and roofing slates, and which have been known as clay-slate needles (thonschiefernadeln) (Pl. XV. Fig. 4), were shown by A. Cathrein to be rutile. For further notes on the distribution of rutile the student should consult the work of H. Thürach.*

In all rocks rutile is characterized by the more or less complete absence of interpositions.

Anatase.

Literature.

J. S. DILLER, Anatas als Umwandlungsprodukt von Titanit im Biotitamphibol-granit der Troas. N. J. B. 1883. I. 187–193.

A. STELZNER, Studien über Freiberger Gneisse und ihre Verwitterungsprodukte. N. J. B. 1884. 1. 271–274.

H. THÜRACH, Ueber das Vorkommen mikroskopischer Zirkone und Titanmineralien in den Gesteinen. Würzburg. 1884.

Anatase always occurs in crystals, and is never massive; its habit is predominantly pyramidal, more rarely tabular, and never prismatic. Of the pyramidal forms, which are known macroscopically, a considerable number also occur in the microscopic crystals. P (111), however, predominates, and determines the habits of the crystals, $111 : 111 = 136° 36'$. Less frequently a pyramid $\frac{1}{m} P\infty$, not yet completely identified, or the base determines the type of the crystals.—The fundamental pyramid is always striated parallel to the edge $P : oP$ (111) : (001).—On account of the minute dimensions of the crystals they are generally seen microscopically as complete bodies; their outlines in cross-sections are quadratic parallel to oP, acutely rhombic parallel to c.

The cleavage parallel P (111) and oP (001) is sharp and clear in cross sections, but is not noticeable when the crystals are not cut by the surfaces of the thin section.

---

* Ueber das Vorkommen mikroskopischer Zirkone und Titan mineralien in den Gesteinen. Würzburg. 1884.

Schrauf * determined

$$\omega_R = 2.511 \text{ and } 2.515 \qquad \epsilon_R = 2.476 \text{ and } 2.477$$
$$\omega_D = 2.534 \qquad\qquad \epsilon_D = 2.496 \text{ and } 2.497$$

Miller found

$$\omega = 2.554 \qquad\qquad \epsilon = 2.493$$

Among rock-making minerals rutile alone has a higher index of refraction; the double refraction in anatase is lower than in dolomite and calcite, but somewhat higher or at least as high as in zircon. The crystal form is seldom easily recognized because of the strong total reflection along the margin of the crystals; the relief is strong, and the interference colors, even for very small thicknesses, are of the second and third order. With somewhat thicker crystals the colors are those of the fourth and fifth order, which approach white. In basal sections the interference cross is accompanied by several rings. The character of the double refraction is negative. The pleochroism generally is not strong, and varies considerably with the color: in blue crystals the color for E is deep blue, for O light blue; in yellow crystals E is light yellow, O orange, according to von Lasaulx.

Optical anomalies, which show themselves by causing the interference cross to separate into hyperbolas during a rotation between crossed nicols, have led Mallard to consider anatase mimetic. The interference cross of the blue crystals is not black, but apparently blue.

Anatase exhibits an adamantine lustre by incident light; by transmitted light it is at times colorless, or yellow of different intensity, or brown; at times, blue in different shades; seldom green. The color often varies in one crystal, either parallel to the pyramidal cleavage cracks in concentric bands, or parallel to the diagonals of this cleavage. The colorless or yellow portions usually behave optically normal, while the blue portions appear more often to exhibit anomalies.

Sp. gr. = 3.9. Chemical composition = TiO_2. Its chemical behavior is the same as that of rutile, from which it is easily distinguished by the form, cleavage, and optical character. Anatase is usually free from interpositions.

Anatase has not yet been observed as a primary constituent of rocks. In all its occurrences it must be considered an alteration product of titaniferous minerals. Thus Diller found it as a probable alteration product of titanite in the hornblende biotite granite of the

---

* S. W. A. XLII. 1860.

Troad, and of ilmenite in the Schalstein of Redwitz, near Hof, Bavaria. It has been found in gneiss, diabase, quartz porphyry. Thürach observed it in various granites, diorites, and crystalline schists; also in numerous grauwackes, sandstones, shales (Schiefer thonen) and limestones of all formations, from the Silurian to the Tertiary.

Anatase in some cases is probably derived from rutile, as the latter is sometimes an alteration product of anatase. Titanic iron is also found among the cleavage cracks of the blue Brazilian anatase, in the same manner as was described for rutile.

Cassiterite.

Cassiterite forms short prismatic or pyramidal crystals and twins; or its crystallographic boundary may be entirely wanting. $111 : 111 = 87°$ $7'$. Twinning is not so general as for rutile, with which otherwise the cross-sections and their angles correspond closely. It sometimes occurs in radially columnar aggregates; the single individuals are then long, slender prisms.

The cleavage parallel $\infty P \infty$ (100) is not noticeable, or at most is only in traces, on the microscopic individuals and on the cross sections; this is one of the most important means of distinction from rutile.

Optically positive. Index of refraction high. Double refraction strong; consequently the interference colors are only recognizable on very thin lamellæ. Grubemann has determined on the cassiterite of Schlaggenwalde:

For the red part of the spectrum $\omega = 1.9793$ $\epsilon = 2.0799$
For the yellow part of the spectrum $\omega = 1.9966$ $\epsilon = 2.0934$
For the green part of the spectrum $\omega = 2.0115$ $\epsilon = 2.1083$

In transmitted light yellowish to brown or red in different shades, seldom almost colorless, often variously colored in bands and stripes; by incident light almost metallic adamantine lustre. The interference cross not infrequently separates into hyperbolas upon rotation.

Sp. gr. $= 6.87$. Chemical composition $= SnO_2$. Unattacked by acids. Distinguished from rutile and zircon with certainty only by means of its specific gravity, measurements of angles on isolated crystals, or by its chemical reaction. The coloring ruby-red of a borax bead, previously colored blue by copper vitriol, may be accomplished after sufficient practice even with extremely minute particles.

Up to the present time cassiterite has only once been unquestionably shown to exist as a microscopic rock constituent, and then it oc-

curred with rutile as inclusions in the lithia mica of the granite of Greifenstein.* It appears to be locally abundant in gneisses and granites
specially those of Cornwall, and in the contact zones of the schists in
the immediate vicinity of the granites. It can only be definitely determined by isolating the crystals from these rocks.

Zircon.
Literature.

K. VON CHRUSTSCHOFF, Beitrag zur Kenntniss der Zirkone in Gesteinen.   T. M.
   P. M. 1886. VII. 423.
H. ROSENBUSCH, Sulla presenza dello zircone nelle roccie.   Atti della R. Accad.
   delle Sc. Torino. XVI. 1881.
A. E. TÖRNEBOHM, Om Zirkonens utbredning in bergarterne.   Geol. För. i Stock-
   holm Förhandl. 1876. III. No. 34.
TH. VON UNGERN-STERNBERG, Untersuchungen über den Finnländischen Rapa-
   kiwi-Granit. Inaug.-Diss. Leipzig.1882.

Zircon occurs as a primary constituent of rocks only in the form
of crystals, never massive; $111 : 11\overline{1} = 84°\ 20'$. The habit of the microscopic crystals is almost always short prismatic, seldom long, and very
rarely pyramidal. The forms are the same as those found on the macroscopic crystals. It is to be remarked that the pyramid $3P3$ (311)

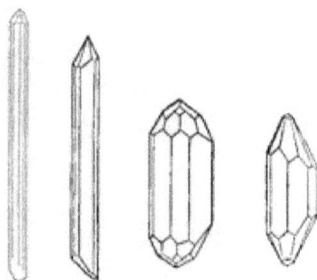

is very often the predominant form
on the poles of the principal axis of
the microscopic crystals, and that not
infrequently other biquadratic pyramids
occur. Fig. 60 shows some of the
rarer forms; Pl. XV. Fig. 5 shows
some of the commonest. The form of
the cross-sections is readily derived from
a consideration of the crystal forms.
The crystals are seldom shorter than 0.01
mm. Twinning has not yet been noted.

Fig. 60

The imperfect cleavage parallel to the prism and pyramid is not
noticeable microscopically on the crystals, and but rarely on cross-sections; on the other hand, in basal sections of large individuals the
cleavage parallel to ∞P (110) is very distinct, and that parallel to
$\infty P\infty$ (100) is observed in traces. One must be careful not to mis-
take the shell-like (zonal) structure, so common in this mineral, for
cleavage.

The index of refraction of zircon and its double refraction are both
very strong; its optical character is positive. On the hyacinth of
Ceylon the following constants have been determined : $\omega = 1.960, \epsilon =$

----
* M. Maclay-Miklucko.   N. J. B. 1885. II. 88.

2.015 (Brewster), $\omega_\rho = 1.92$, $\epsilon_\rho = 1.97$ (Senarmont), $\omega_{na} = 1.9239$, $\epsilon_{na} = 1.9682$ (Sanger), and on the zircon from Miask, Urals, $\omega_{na} = 1.9313$, $\epsilon_{na} = 1.9931$ (Sanger). These numbers explain the high relief, the broad total reflection borders, the wrinkled surface of cross-sections and the brilliant red and green interference colors with crossed nicols, which are exhibited by the smallest individuals. Sections parallel to the base $_oP$ (001) yield several rings about the dark cross in convergent light. Here and there in zircon, also, the interference cross separates into hyperbolas, especially when the shelly structure is quite strongly developed.

A pleochroism which is often very distinct in macroscopic crystals is very faint in the microscopical occurrences, and is more generally not noticeable at all. Haidinger observed in the brownish pearl-gray crystals of Ceylon: O, clove-brown; E, asparagus-green; in light clove brown crystals, O, grayish violet-blue; E, grayish olive-green; in yellowish white crystals from the same locality, O, light blue; E, light yellow.

The microscopic zircons are mostly colorless or very light yellow and pink or violet; very seldom reddish to brownish from a coating of limonite, which may be removed with hydrochloric acid. When color is present it is not usually uniformly distributed, but is arranged in zones, or at the centre or along the axes of the crystals. Shelly or zonal structure is very common; it generally repeats the outward form of the crystal, when this is made up of $\infty P . P . (110) (111)$. It rarely indicates different forms from those developed on the crystal. However, it frequently happens that the lines of the zones are straight so long as they follow the prisms, but appear rounded at the poles of the crystals.

Inclusions in zircon are not infrequent, but are difficult to determine on account of the high index of refraction of their matrix. Of the unindividualized interpositions, fluid ones may be recognized with certainty by the movement of their bubbles; inclusions without bubbles, which always have very dark borders and occur in various forms, may be either gas or glass inclusions. Indeterminable acicular microlites also occur.

II. $= 7.5$. Sp. gr.$=4.4-4.7$. Chemical composition $= ZrO_2, SiO_2$. Not appreciably attacked by acids; it is easily isolated from rocks on account of its specific gravity, its resistance to acids, and its indifference to magnetic attraction. The crystals, then, though of very minute dimensions, may be measured by a goniometer. As a more certain test, a small portion may be fused with bicarbonate of soda in a platinum dish in the proportion of 1 : 4 to 1 : 10, in order to obtain hexagonal plates of ZrO_2.

Zircon occurs very constantly and often abundantly in the acid eruptive rocks of the granite series, syenite, diorite, and gabbro, as well as in their porphyritic equivalents; it occurs less abundantly, but quite constantly, in the more basic eruptive rocks of the diabase family, and in all the younger eruptive rocks.   It is also a constant and often an abundant constituent of Archæan rocks, especially of the gneisses. Formerly it was very generally mistaken for rutile. Zircon was recognized by G. W. Hawes[*] in granites.

A. E. Törnebohm was the first to comprehend the wide distribution and petrographical importance of zircon. It has been isolated from the rapakiwi of Finland, and determined qualitatively and goniometrically by H. Rosenbusch and Th. von Ungern-Sternberg. Its distribution in different eruptive and schistose rocks has been noted by H. Thürach.[†] It is found in the sand derived from granites and Archæan rocks; and also as a foreign constituent in fossiliferous rocks, limestones, shales, marls, sandstones, etc.

In all cases zircon belongs to the oldest constituents of the rocks it occurs in ; consequently its crystal form is always perfect, and it may be enclosed in any of the other minerals associated with it. It is undoubtedly older than all the silicates, but its relative age as compared with apatite and the ores is not so certain. When enclosed in mica, pyroxene, hornblende, cordierite, etc., it is often surrounded by a pleochroic halo.[‡]

Scapolite Group.

Literature.

Fr. Becke, Die Gneissformation des niederösterreichischen Waldviertels. T. M. P. M. 1882. IV. 369 et passim.

V. Goldschmidt, Ueber Verwendbarkeit einer Kaliumquecksilberjodid-Lösung bei mineralogischen und petrographischen Untersuchungen. N. J. B. B.-B. 1. p. 225 ff. 1880.

A. Michel-Lévy, Sur une roche à sphène, amphibole et wernérite granulitique des mines d'apatite de Bamle, près Brevig (Norvège). Bull. Soc. min. Fr. 1878. 1. 43-46.

---

* Mineralogy and Lithology of New Hampshire. Concord. 1878. 75.

† Ueber das Vorkommen mikroskopischer zirkone und Titan mineralien in den Gesteinen. Wurzburg. 1884.

‡ A Michel Lévy, "Sur les noyaux à polychroïsme intense du mica noir." C. R. 1882. XCIV. IIj. Gylling, Nagra ord om Rutile och Zirkon, etc. Geol. Fören. i Stockholm Förhdl. 1882. VI. No. 74. 162. sqq. G. H. Williams, N. J. B. B.-B. II. 1883.

Hⱼ. Sjögren, Om de norska apatitförekomsterna och sannolikheten att anträffa apatit i Sverige. Geol. Fören. i Stockholm Förhandl. 1882. VI. No. 81. 469 ff.

A. E. Törnebohm, Ett par skapolitförande bergarter Geol. Fören. i Stockholm Förhdl. 1882. VI. No. 75. 193 ff.

G. Tschermak, Die Skapolithreihe. S. W. A. LXXXVIII. Nov. 1883.

The rock-making *scapolite* minerals usually show no outward crystal shape, but form irregularly defined grains, or fibrous aggregates, mostly with a confused arrangement. Except those which occur in limestone, whose outline in the prism zone is generally formed by ∞P (110), $\infty P \infty$ (100) rarely $\infty P2$ (210) or $\infty P3$ (310), and much more rarely by the pyramidal termination P (111). $111 : 11\bar{1} = 63° 42'$. Consequently for aggregates the sections are irregular, but for imbedded crystals they are quadratic or octagonal in cross-sections, and rectangular or long lath-shaped in longitudinal sections.

The cleavage parallel to $\infty P \infty$ (100) is recognized by distinct parallel cracks in longitudinal sections, and by rectangularly intersecting ones in cross-sections (Pl. IX. Fig. 6); it is generally more noticeable in those occurrences which are no longer entirely fresh. A transverse parting by which the prisms separate into several members is very common to the lath-shaped individuals, and plays an important part in the process of alteration of the crystals. $H. = 5.5$.

In a fresh condition the scapolite minerals are colorless and transparent, more rarely gray to brown because of needle-shaped interpositions which are as yet indeterminable. All scapolites are optically negative, with an index of refraction which is not high, but with strong double refraction. There has been determined, on Vesuvian meionite,

$$\omega_{na} = 1.594\text{–}1.597 \qquad \epsilon_{na} = 1.558\text{–}1.561 \text{ (Des Cloizeaux)};$$

on dipyre from Pouzac,

$$\omega_{na} = \quad 1.5673 \qquad \epsilon_{na} = \quad 1.5416 \text{ (Lattermann)};$$
$$\omega_{\rho} = \quad 1.558 \qquad \epsilon_{\rho} = \quad 1.543 \text{ (Des Cloizeaux)};$$

on scapolite from Arendal, Norway,

$$\omega_{\rho} = \quad 1.566 \qquad \epsilon_{\rho} = \quad 1.545 \text{ (Des Cloizeaux)}.$$

The index of refraction apparently sinks with a decrease of the Ca percentage, and an increase of the alkali percentage. The interference colors in longitudinal sections in consequence of the great difference, $\omega - \epsilon$, are more brilliant than for most of the colorless minerals, especially for the feldspars and feldspar-like substances, as well as for quartz; even in very thin sections they seldom fall below orange and yellow of the 1st order. Cross-sections in convergent light yield a distinct cross, and with sufficient convergence the first ring may be plainly seen.   On the other hand, the relief is slight, and sections in Canada balsam show no roughened surface.

Scapolite is distinguished from feldspar and cordierite by its uniaxial character and the cleavage; from quartz by the cleavage and the character of the double refraction; from apatite by the index of refraction, the cleavage, and the chemical reaction with phosphoric acid.

The specific gravity rises with the percentage of lime from 2.569 in marialite to 2.735 in meionite.   According to Tschermak's investigations, the scapolite group presents a continuous series of isomorphous mixtures of two silicates, which are not known to occur by themselves. One of them, $8CaO, 6Al_2O_3, 12SiO_2 = Si_{12}Al_{12}Ca_8O_{46}$, predominates in meionite at 88 per cent, and is called the meionite molecule $= Me$; the other, $3Na_2O, 3Al_2O_3, 18SiO_2 + 2NaCl = Si_{18}Al_6Na_8O_{46}Cl_2$, occurs in marialite at 84 per cent, and is called the marialite molecule $= Ma$. The less siliceous mixtures of Me to Me_2Ma_1 are completely decomposed by acids, or nearly so, without gelatinization; the mixtures of Me_2Ma_1 to Me_1Ma_2 are only slightly attacked by acids, and the more acid mixtures from Me_1Ma_2 to Me_2Ma_1 completely resist the attack of acids.

The rock-making scapolite minerals have not as yet been sufficiently investigated chemically to refer them with certainty to one of these three groups, but it is the mixtures containing a medium and higher percentage of silica which appear especially widespread.

There are no inclusions which particularly characterize the minerals of this family; besides the minerals associated with them in the rocks, especially epidote, calcite, diopside, actinolite, magnetite, pyrite, and feldspar, there are fluid inclusions of irregular shapes or in negative crystal forms.   If the scapolites have been formed epigenetically from other minerals (feldspar) they sometimes contain the interpositions of the parent mineral.

The scapolites withstand but slightly the action of the atmosphere and of surface waters; altering easily from the cross fractures and cleavage cracks into a fibrous substance, not yet definitely determined, but not unlike zoisite from its low double refraction, or into a lamellar aggregation of kaolin or muscovite.   It also weathers into carbonates.

Meionite forms attached crystals, and does not appear to occur as an actual rock-making mineral.

With the exception of dipyre, all the rock-making minerals of the scapolite group are here designated as *scapolites*.   They never occur as primary constituents in eruptive rocks, but are sometimes developed in them at the expense of the feldspars.   As primary minerals they abound in the Archæan rocks, where they occur not only in the limestone layers, but also as constituents of the gneisses, especially those rich in lime, and in the epidote and pyroxene-bearing varieties. Such occurrences are treated of in the works of Becke and Törnebohm, cited above.

Dipyre and *couzeranite* belong to the scapolites in which the marialite molecule predominates. Both minerals are to be considered as identical, the difference arising mainly from the want of pure material for the investigation of the second variety. Dipyre accompanies the contact metamorphism of limestone and schists in the Pyrenees. In granular limestone it is usually well crystallized in the vertical zone, while in the schists it furnishes irregularly rounded and elliptical sections. In limestone it is quite free from inclusions, with the exception of calcite. In the schists it is often completely filled with carbonaceous particles, muscovite plates, rutile needles, and quartz grains; and by incident light and even in transmitted light with a low magnifying power it appears yellowish, reddish, bluish, or almost opaque. These inclusions are still more abundant in couzeranite, which also forms irregular grains, or quadratic, sometimes octagonal (∞P . $\infty P \infty$) prisms without terminal faces. It is easily confused with andalusite, which may be avoided by observing the cross-section in convergent polarized light. It is also a contact mineral in limestones and schists. The microscopical characteristics of the dipyres have been given by Zirkel,[*] Fischer,[†] and v. Lasaulx.[‡]

Vesuvianite.

The rock-making *vesuvianite* occurs more frequently in irregular pieces or prismatic aggregates than in crystals; well-crystallized individuals only occur in granular limestone, and then have the faces ∞P . $\infty P \infty$ (110) (100) in the prism zone; as terminal faces oP (001) appears to predominate, P (111) to be subordinate. $111 : 11\bar{1} = 74° 27'$.

The imperfect cleavage parallel to the prism faces is indicated microscopically by a few cracks, usually short; they only follow one prism, probably $\infty P \infty$ (100). H. = 6.5.

By transmitted light vesuvianite is nearly colorless, yellowish to greenish yellow, rose-red, very seldom dark reddish brown or blue. The colors often vary in concentric shells or more rarely in

[*] Z. D. G. G. 1867. XIX. 202.
[†] Kritische, mikroskop-mineralog. Studien. I. Fortsetzung, Freiburg i. B. 1871. 52.
[‡] N. J. B. 1872. 848.

irregular patches, especially toward the centre of the crystal. The index of refraction is high (on a crystal from Ala it was $n_{na} = 1.7258$), the surface of the section is therefore rough. The double refraction is very small, with negative character, the difference $\omega - \epsilon$ scarcely exceeding 0.0015.

On idocrase from Ala, Tyrol,

$$\omega_{na} = 1.719\text{--}1.722 \qquad \epsilon_{na} = 1.718\text{--}1.720 \text{ (Des Cloizeaux)};$$

consequently the interference colors are very low. The double refraction varies in intensity in one and the same crystal, sometimes in concentric zones; therefore there may be stripes of different interference colors in the crystal between crossed nicols, though it is of absolutely uniform color in ordinary light. Moreover, the character of the double refraction may change in the different-colored stripes, so that with predominating optically negative stripes there may occur optically positive ones (Hammrefjeld in Norway). The extinction in longitudinal sections remains parallel to the prism edges for all the stripes. In the cross-sections in thin section the interference cross is very faintly seen without the slightest trace of rings. The pleochroism is generally weak to very weak; the ordinary ray colorless or yellowish; the extraordinary, reddish, yellowish, or greenish. Vesuvianite may be easily confounded with pistacite on account of its pleochroism and index of refraction; but the small double refraction is a sure means of distinction without considering the cleavage, position of the axes of elasticity and the phenomena in convergent light.

Sp. gr. $= 3.40\text{--}3.47$. Vesuvianite is a lime-alumina silicate, whose formula is not exactly known: it contains some water among its bases, and in certain occurrences fluorine, according to Jannasch. Besides CaO it contains MgO and MnO; besides Al_2O_3, Fe_2O_3. and also small quantities of alkalies. It is not unattacked by acids; it fuses easily with intumescence to green or brown glass, which is then soluble without difficulty in HCl with the separation of SiO_2.

Vesuvianite occurs principally, if not exclusively, in metamorphic rocks; it is widely disseminated in the limestones and lime-silicate hornstones of many granite contact zones.

It frequently forms porphyritic crystals in the granular limestones of the Archæan, as well as a constituent of the closely related inclusions of lime silicates. It also occurs to a limited extent in the gneisses. It is most frequently accompanied by wollastonite, diopside and other pyroxenes. garnet, epidote and titanite.

Alteration products are not known in rock-making vesuvianite. It occasionally encloses the substances accompanying it, especially calcite and pyroxene; it also contains fluid inclusions; but characteristic interpositions of any kind are wanting, and the mineral is most frequently completely homogeneous.

Melilite.

Literature.

A. STELZNER, Ueber Melilith und Melilithbasalte. N. J. B. B.-B. 1882. II. 369–387. cf. N. J. B. 1882. I. 229.

F. ZIRKEL, Untersuchungen über die mikroskopische Zusammensetzung und Structur der Basaltgesteine. Bonn. 1870. 77 sqq.

Rock-making *melilite* often occurs in perfectly crystallized individuals, and then has the form of quadratic, octagonal, or rounded plates, according to whether the prism ∞P (110) alone, or with $\infty P \infty$ (100) or $\infty P3$ (310), is combined with the basal plane. More rarely it is in the form of short prisms. When, as is most frequently the case, the individuals are not well crystallized, it is the prism zone which is least developed, producing thin plates with irregular boundaries, the basal plane also being uneven. Hence sections of melilite are lath-shaped parallel to *c*, and quadratic, octagonal, or rounded at right angles to *c*. In some rocks it occurs in quite irregular grains, and receives its outline from that of the other constituents, which existed in the rock previous to its crystallization.

The cleavage parallel oP (001) is poorly developed; and sections inclined to the basal plane exhibit but few cleavage lines, often only one, to which the extinction is parallel. Irregular cracks sometimes traverse the longitudinal sections.

In transmitted light melilite is either colorless or yellowish to brownish; and even the apparently colorless sections, when compared with actually colorless substances (apatite, nepheline, leucite) in the same thin section are found to be dull yellowish, with a tinge of green or gray. The index of refraction is higher than for the associated colorless silicates; the double refraction is very weak. In very thin sections its cross-sections appear almost isotropic, and the double refraction can only be noticed by observation in sensitive colors (gypsum plates, etc.). Even in thicker sections the interference colors do not exceed gray-blue of the 1st order. But the degree of its double refraction appears to be somewhat different in different occurrences. The character of the double refraction is negative.

On humboldtilite from Vesuvius—

$$\omega_p = 1.6312 \qquad \epsilon_p = 1.6262 \text{ (L. Henniger)}$$
$$\omega_{na} = 1.6339 \qquad \epsilon_{na} = 1.6291$$

A. Michel Lévy[*] determined on the same occurrence $\omega - \epsilon = 0.0058$. In convergent light basal sections yield a very faint cross, and the optical character has to be determined in parallel light. Pleochroism is entirely absent from the colorless and slightly colored melilite; for the decidedly yellow variety Stelzner found E dark yellow, O light yellow.

Melilite possesses a peculiar and constant micro-structure in rocks, which may be used to advantage with its essential characteristics as a means of identification. Longitudinal sections ($\|c$) exhibit either fine lines quite like cleavage cracks, which traverse the section parallel to the principal axis (Pl. XIV. Fig. 6), or there starts out from the basal terminal plane curious forms shaped like pegs, spears, spatulæ, oars, or pipes (Pl. XV. Fig. 6) which extend to a greater or less depth into the crystal, and sometimes widen out to spherical or funnel-shaped forms, or less frequently send out arms parallel to oP (001). The substance filling these forms appears to be isotropic glass. Melilite also encloses the older minerals associated with it, especially augite and leucite; less frequently the iron ores, oxides, perofskite, and apatite. The arrangement of the inclusions is mostly central or irregular, more rarely zonal.

Its specific gravity, 2.90–2.95, allows it to be easily separated from the rock powder. Melilite is a lime-alumina silicate, $12CaO, 2Al_2O_3, 9SiO_2$, in which besides lime there may be magnesia and alkalies, and besides alumina there may be sesquioxide of iron. It gelatinizes very readily in hydrochloric acid, and in the solution sulphuric acid precipitates a great quantity of gypsum needles. It is frequently found decomposed in nature; almost always altering to a fibrous aggregate, possessing strong double refraction, and probably belonging to a zeolite; the fibres generally stand perpendicular to the basal plane of the crystal, and penetrate it from both these faces. In other cases the fibres are arranged in delicate radial groups. By incident light such more or less altered melilite appears chalk-white or ochre-colored and earthy. Törnebohm[†] observed an alteration of melilite to garnet in the basalt of Alnö, Sweden.

Melilite is confined to the younger volcanic rocks. It is widely disseminated in the leucite and nepheline rocks. It also forms rocks in which it takes the place of the feldspars. Besides nepheline, leucite, and augite, perofskite is a constant accompaniment of melilite.

---

[*] Bull. Soc. Min. Fr. VII. 1884. 46.
[†] Geol. Fören. i. Stockholm Förhandl. 1882. VI. No. 76. 243.

MINERALS OF THE HEXAGONAL SYSTEM.

HEXAGONAL minerals are anisotropic and uniaxial, like tetragonal ones. The optic axis coincides with the principal crystallographic axis, and is at the same time the axis of greatest or least elasticity. In the first case the character of the double refraction is negative, and $\omega > \epsilon$; in the second case it is positive, and $\omega < \epsilon$. The two rays are differently absorbed; and hexagonal minerals, if colored, show a more or less distinct pleochroism in all sections except those parallel to the basal plane. Sections at right angles to c have hexagonal or triangular (in tourmaline nine-sided) outlines, or directions of cleavage. Or else there is no regular outline, and the cleavage lies parallel to the base oP (001). Such sections act like isotropic media in parallel polarized light, that is, they remain dark between crossed nicols during a complete rotation. In convergent light they exhibit a dark interference cross with or without colored rings, which remains unchanged during a rotation of the section, the arms of the cross lying parallel to the principal sections of the nicols. Sections which are inclined or parallel to c show outlines which vary with the position of the section and the form of the crystal; the cleavage appears as systems of cracks running parallel to or intersecting one another. These sections are doubly refracting in parallel polarized light; for a complete rotation of the section between crossed nicols they become dark and light four times, and the position of darkness is always reached when the cleavage cracks are either parallel to the principal sections of the nicols or the latter bisect the angles made by intersecting cleavage lines. In convergent polarized light the basal interference figure sometimes appears at one side of the field of view, and moves in the margin of the field during a rotation of the section in such a manner that the arms of the cross move parallel to themselves. Finally, in sections parallel to the principal axis there appear hyperbolic curves lying symmetrical to the principal axis.

Optical anomalies are recognized in convergent light by the interference cross in basal sections opening into hyperbolas, and thus presenting the interference figure of a biaxial body with small optic angle cut at right angles to the first bisectrix. Different parts of such an abnormal section generally show different sizes of apparent axial angle, and different positions for the apparent axial plane. In parallel polarized light such basal sections usually exhibit a structure resembling twinning.

From the foregoing it is evident that tetragonal and hexagonal minerals can only be distinguished by their form and cleavage, but not by their optical characters.

Graphite.

The *graphite* which occurs in rocks generally has quite irregular forms : it constitutes flakes and grains of very variable shape, as well as disk-like bodies ; occasionally it shows a more or less distinct approximation to crystalline outline, and then possesses hexagonal to rounded sections, and rectangular, staff-shaped longitudinal sections. For the most part it is disseminated in minute particles, only recognized as such with high magnifying powers. Graphite is opaque ; by incident light black to brownish black with metallic lustre. It is not acted on by acids ; is consumed only with difficulty in thin sections on platinum foil, even after the iron oxides accompanying it have been removed by acids.

Graphite $= C$ is widely disseminated as a constituent or a pigment of the oldest formations, more especially of the phyllitic kinds, where it is evidently the residuum of organic carbonaceous substances. It also occurs under similar conditions in rocks of the Archæan, extending down deep into the gneiss. It only occurs in rocks of more recent formations, when these have assumed a more or less crystalline character through metamorphic processes.

Magnetic Pyrites—Pyrrhotite.

Pyrrhotite never forms crystals in rocks, but always in irregular masses. Its outlines are therefore irregular. Cleavage is wanting. It is opaque ; by incident light bronze yellow, with distinct metallic lustre. It is distinguished from pyrite by its color, its attraction by an electro-magnet, as well as by its solubility in hydrochloric acid. Chemical composition $= Fe_n S_{n+1}$. It occurs occasionally in old erruptive rocks, is especially frequent in gabbro, more rarely in schists.

Hematite.

Hematite occurs in three different forms in rocks—as specular iron, micaceous hematite, and red hematite.

As specular iron it forms rhombohedral or tabular crystals, with a parting parallel to the faces $R\pi$ $(10\bar{1}1)$ $(86° 10')$ which is often distinct, and which is probably to be referred to the twinning according to the rhombohedral faces, described by Bauer.[*] This twinning is probably a mechanical phenomenon, as Mügge[†] is inclined to think. The sections

---

[*] Z. D. G. G. 1874. XXVI. 186.   [†] N. J. B. 1884. I. 216.

parallel to the base are then triangular or hexagonal, and perpendicular to the base are mostly lath-shaped.

As micaceous hematite, it always has the form of thin plates with hexagonal outlines; the sides of the hexagons are often of very unequal length. The outlines are also ragged, or quite irregular. The plates are often aggregated to delicate forms of growth of many shapes, especially when they occur in laminated minerals (mica), when their position and arrangement are dependent on the crystallization of the matrix.

As red hematite it is massive, and forms a very finely divided pigment in rocks, recognized with high magnifying-powers as flakes and grains, or as loose aggregates.

Hematite does not exhibit any cleavage microscopically; but the parting parallel to the fundamental rhombohedron gives rise to lines, which can scarcely be distinguished from cleavage cracks microscopically.

As specular iron, hematite is opaque, with metallic lustre by incident light, iron black to grayish black, with a tinge of reddish, which is noticeable in a strong light.

Micaceous hematite is submetallic in lustre, and is transparent, with a color varying with the thickness of the plates from deep red through yellowish red to yellowish gray. Pleochroism is not noticeable. Isolated plates when thin enough yield a uniaxial interference figure in convergent light.

As red hematite it is opaque, reddish by incident light, without metallic lustre.

Specific gravity $= 4.5–5.3$. Chemical composition $= Fe_2O_3$. Soluble in hydrochloric acid, but considerably slower than magnetite. Not attracted to a simple magnet unless attached to grains of magnetite, which character may be used as a means of separation between the two.

Hematite in the form of specular iron is a very widely disseminated constituent in the acid eruptive rocks, such as granite and syenite, trachyte, rhyolite, and andesite; also in many phonolites, as well as in many crystalline schists of the Archæan. In the eruptive rocks it belongs to the oldest individuals. As micaceous hematite it occurs in the same eruptive rocks, but chiefly as inclusions in other minerals which are colored red by it: thus in the quartz, feldspar, and mica of granites; in the haüynes of phonolites, and of nepheline or leucite rocks. In the crystalline schists it occurs both independently and as inclusions in the other constituents. The red color of the phyllitic schists is almost universally due to the presence of plates of micaceous hematite. More-over, this form of hematite is the most widely distributed pigment in the mineral world. It is extremely common in the micas of the peg-matitic forms of granite and gneiss, where it is often combined with tourmaline to form the asterism of the mica. Such occurrences have been described by G. Rose,[*] from New Providence in Pennsylvania, and from Grenville in Canada. As finely divided, loose, red hematite, it is present in the acidic porphyritic rocks, quartz porphyries, rhyo-lites, quartz porphyrites, and dacites. It colors the ground-mass of these rocks, especially when they assume a microfelsitic development. Finally, hematite is very common, partly in the micaceous form, partly as earthy red hematite, in pseudomorphs after pyrite in the phyllitic schists, as well as in pseudomorphs after olivine and bronzite in the basic eruptive rocks (melaphyres, basalts, etc.); lastly, after garnet in eruptive and schistose rocks.

Ilmenite.

The development of *ilmenite* in rocks is exactly similar to that of hematite. It is most frequently found in irregular masses without crystallographic outline, or in rhombohedral crystals, or tabular ones parallel to oR (0001). The sections, therefore, are either irregular or triangular and hexagonal when parallel to the basal plane, or often have very jagged and irregular contours; perpendicular to the base they are lath-shaped. Frequently ilmenite plates produce incipient forms of growth by arranging themselves in three parallel groups, which cut one another at 60° in cross-sections. Another kind of occur-rence strongly resembles the micaceous variety of hematite, being in very thin plates. It may be designated as micaceous titanic iron. Finally, ilmenite is found in an ochre-like form, and appears as minute particles and aggregates, and serves as a pigment to minerals enclosing it in a finely divided state. When perfectly fresh, ilmenite exhibits no microscopic cleavage cracks; as soon, however, as chemical alteration sets in, a system of stripes and lines appears in cross-sections, which may follow the cleavage parallel to $R \pi$ (10$\bar{1}$1). But since there are striations noticeable by incident light on the natural basal plane of even the freshest individuals, which appear parallel to the intersection of $R \pi$ (10$\bar{1}$1), and arise from twinned lamellæ, so the apparent cleavage noticeable in partly decomposed sections is to be explained as a shelly structure parallel to R, which becomes more noticeable through decom-

---

* S. B. A. 1869, 19 Apr., p. 352 sq.

position. Moreover, a shelly structure parallel to $o R$ (0001) is sometimes observed microscopically in cross-sections.

Ilmenite is opaque, with metallic lustre; by incident light iron-
black, with a tinge of brownish. Micaceous titanic iron, as K. Hof-
mann first showed, becomes transparent with a clove-brown color, and
is quite strongly doubly refracting, mostly with a sub-metallic lustre.
The ochre-like, finely divided ilmenite loses the metallic habit, and by
incident light is brownish black and dark brown.

Specific gravity $= 4.3–4.9$. Chemical composition $= FeTiO_3$.
when pure; but there appears to be quite a complete series between
specular iron $FeFeO_3$ and $FeTiO_3$, in which there also occur members carrying $MgTiO_3$. Hot hydrochloric acid attacks ilmenite
somewhat more slowly than it does specular iron. The solution when
heated with tin-foil becomes violet. Hot concentrated sulphuric acid
yields a blue solution. Pure ilmenite, like hematite, is somewhat indifferent toward the magnet; a distinct attraction toward the magnet
indicates an admixture of magnetite. It is with difficulty distinguished
from titaniferous magnetite, when neither the crystal form nor shelly
structure nor cleavage permits the determination of its system of crystallization.

Ilmenite, as it occurs in rocks, is very frequently more or less
completely altered into other substances. In most cases this alteration
commences with the formation of a strongly refracting substance, only
slightly transparent, which when sufficiently thin is strongly doubly
refracting. Its color is white, yellow, or brown by incident light; and
its structure is sometimes granular, sometimes distinctly radiating, the
fibres standing perpendicular to the ilmenite. This alteration product,
which may also arise from titaniferous magnetite and from rutile, was
called *leucoxen* by Gümbel, who, however, considered its substance as
of primary nature. Its chemical composition is not the same in all
cases where it has been investigated, and it has been considered the
equivalent of a variety of minerals (titanite, anatase, and siderite) by
different observers. Since leucoxen grows at the expense of the ilmenite during the process of alteration, its outlines possess similar geomet-
rical forms to those of ilmenite (Pl. XVI. Fig. 1). When there is a
shelly structure parallel to $R \pi$ (1011) or $o R$ (0001), the pseudomorph follows these shells (Pl. XVI. Fig. 2), working from their faces
inward until the whole is transformed. A. Cathrein has shown that
the brown and yellow color of many leucoxens is due to the mechanical mixture of rutile in the form of sagenite, which already existed
intergrown with the ilmenite.

This alteration of ilmenite to leucoxen takes place both in eruptive rocks and in schistose ones. But so far as experience goes, the alteration of ilmenite into carbonates rich in iron, in which either the previously existing rutile remains as such, or the titanic acid contained in the ilmenite is converted into rutile, is confined to schistose rocks of the phyllite series and to phyllitic schists of more recent formations. On the other hand, according to von Lasaulx it is very probable that ilmenite may at times be derived from rutile.

The distribution of titaniferous iron in the form of ilmenite is very great. It accompanies or replaces specular iron in granites, syenites, etc.; it belongs to the essential ingredients in diorites, but especially in diabase, gabbro, and related rocks, as well as in their mesozoic and more recent equivalents, augite porphyrites, melaphyres, basalts, etc. In these rocks ilmenite, together with magnetite, which often accompanies it, belongs to the oldest secretions from the magma; its formation precedes that of olivine and pyroxene, and seldom appears in the later stages of the development of the rocks. Ilmenite frequently forms a constituent of gneiss and mica-schist, of the labradorite rocks of Norway and of Canada, and of amphibolite from many localities. In the form of micaceous titanic iron it occurs in the basalts of southern Bakony, Hungary; in the augite-porphyrites and melaphyres of the Saar Nahe region, Lower Rhine; as well as in the nepheline basalts and pyroxenites of Kaisersthul. Ochreous titanic iron probably forms the dust-like pigments which often give to the plagioclases of gabbros and ophites their peculiar brown color; the globulites of the basic rock glasses (augite porphyrite and basalt) are very likely titanic iron.

Corundum.

The forms of rock-making *corundum* vary greatly. At times it crystallizes in long prismatic shapes, at times in sharp pyramids or in thinly tabular forms, and several such types may occur together (Pl. XVI. Fig. 3). Cross-sections parallel to the base are hexagonal or rounded, those parallel to $\overset{1}{c}$, lath-shaped; but the longitudinal direction of the section corresponds in some cases to that of the principal axis, in others to that at right angles to it. It is often only possible to distinguish between these types by means of an optical determination. Furthermore, corundum forms irregular grains and masses. Cleavage is only observed in the larger individuals of corundum; it is parallel to $R\,\pi\,(10\bar{1}1)$. The concentric structure also, which is determined by the twinning parallel to $R\,\pi\,(10\bar{1}1)$, is seldom observed in microscopic individuals.

Corundum is generally almost colorless or transparent blue, seldom brown or red. The color is not disseminated uniformly, but is usually in quite irregular patches and streaks, or in concentric zones. Corundum has a high index of refraction, but a low double refraction; Osann found for corundum from Ceylon $\omega_{na} = 1.7690$, $\epsilon_{na} = 1.7598$. The character of the double refraction is negative. The relief and the dark border of total reflection as well as the rough surface are strongly marked; the interference colors are low, and in good thin sections do not exceed red of the first order. Basal sections show the cross, but without rings in thin sections of the normal thickness, and the arms of the cross are somewhat indistinct. Optical anomalies common to the larger individuals are mostly wanting in microscopic individuals. Pleochroism is only strong when the coloring is quite deep; for the blue corundums (sapphire and emery), O is blue, E is sea-green to colorless.

H. $= 9$. Sp. gr. $= 3.9$–4.0. Chemical composition $= Al_2O_3$. Corundum is not attacked by acids, and is not dissolved by fused soda; it is therefore easily separated out of the rocks.

Rock-making corundum possesses no constant microstructure; most frequently it encloses gas and fluid inclusions, which latter are often found to be liquid carbon dioxide. It has often grown in contact with ilmenite, and also encloses it. Rutile crystals and sagenite webs also occur frequently in the larger crystals, seldom in the microscopic ones.

Corundum never occurs as an essential constituent of rocks, with the exception of emery, which, together with iron oxides, forms independent bodies in the crystalline schists. It appears only as an accessory constituent in granites, gneisses, granular limestones and dolomites, and is constantly accompanied by spinel, rutile, and sillimanite. Corundum occurs with magnetite and pleonaste as a contact mineral in connection with the norites of the Cortlandt series at Stony Point, on the Hudson, N. Y.,[*] and in the dunite and serpentine of Pennsylvania and North Carolina.

Brucite.

Brucite in rocks forms irregular as well as hexagonal plates, seldom fibrous aggregates. When in the form of plates, the basal sections are six-sided, rounded or irregular, and also ragged; sections parallel to c give narrow lath-shaped figures.

The perfect cleavage parallel to oR (0001) is very clearly expressed microscopically by fine cracks, which run parallel to one another and to the longitudinal direction of the lath-shaped sections. Basal sections exhibit no cleavage, but sometimes show cracks and curves which do not appear to possess any regular course.

---

[*] G. H. Williams, Am. Journ. Sci. XXXIII. Feb. 1887. 198.

Brucite is transparent and colorless, its index of refraction about the same as that of Canada balsam ; the double refraction is strong and positive. Bauer determined $\omega_\rho = 1.560$, $\epsilon_\rho = 1.581$. Basal sections exhibit a very distinct interference cross in convergent light; upon rotation this cross very often opens into hyperbolas, the position of whose poles varies for different parts of the section. The figure corresponds to that of a biaxial medium with small optic angle cut at right angles to the acute bisectrix. The fact that not only the apparent axial angle but often the position of the axial plane varies in the same section, indicates that the source of the phenomena is due to strains. In parallel polarized light such basal sections are not homogeneous and isotropic, but exhibit streaked or striped doubly refracting areas with very dull polarization colors. Longitudinal sections in parallel polarized light, when the cleavage cracks are inclined 45° to the principal planes of the nicols, are brightly colored, and may be easily mistaken for muscovite and talc, which, however, are still more strongly doubly refracting. They may be distinguished by the fact that for brucite the elasticity of the ether parallel to the cleavage is greater than that perpendicular to it, while for muscovite and talc the reverse is true.

Sp. gr. $= 2.3–2.4$. Chemical composition $= MgO$, H_2O. It is soluble in acids ; upon being heated to redness on platinum foil it is sometimes colored pink through the oxidation of a trace of FeO to Fe_2O_3. If a thin section containing brucite be moistened with a solution of silver nitrate after it has been slightly heated to redness on platinum foil, the brucite quickly turns brown through the deposition of oxide of silver.

Brucite is disseminated in small quantities in phyllites containing magnesite or dolomite, as well as in similar crystalline schists, in actinolite schists, and also in serpentines and many decomposed diabases. In these cases it has evidently been derived from the carbonates and silicates of magnesia. It also occurs as a contact mineral in some granular limestones.

Quartz.

When rock-making quartz possesses crystal form it appears in dihexahedrons $\pm R$ (10$\bar{1}$1), on which the prism ∞R (1010) is but rarely developed, and then only to a slight extent. Hence sections parallel to the base give regular hexagons ; those in the prism zone rhombs, with angles of 76° 26′ and 103° 34′ ; inclined sections have triangular or trapezoidal outlines. On account of the very general rounding of

the edges and corners, the sections often appear round; in consequence of mechanical deformation of crystals which were originally
regularly bounded, the outlines of the fragments are irregular and
sharply angular; curved and looped contours are due to the corrosion
of completed crystals (Pl. V. Fig. 1). Under certain conditions the
quartz of later generation in many porphyritic (granophyric) eruptive
rocks appears to crystallize in peculiar forms of growth, which are
analogous to the quartz of graphic granite; the habit then is apparently prismatic or trapezohedral. The individuals are intergrown in
the most intimate manner with feldspar (orthoclase, albite, or oligoclase), and within one and the same feldspar crystal they lie exactly
parallel to one another (Pl. VIII. Fig. 3).

In by far the greatest number of rocks the quartz is massive without crystal form, and its outline is consequently of no determinative
value. This sort of quartz sometimes forms single individuals, sometimes aggregates; the latter are almost always granular. Only in the
spherulites of porphyritic rocks or where the quartz is pseudomor-
phous after a fibrous mineral is it fibrous. In the first case it
seems to be necessary to refer it to chalcedony, which hardly occurs
in any other form than fibrous.

In the rock-making quartzes there is no trace of the interpenetration twinning so characteristic of attached crystals of quartz. This
may be due to the fact that the ordinary twinning in which the systems
of axes are parallel could not be detected optically on account of the
extreme thinness of the rock sections.

The imperfect cleavage parallel to the rhombohedron is very rarely
met with in microscopic quartz, and cleavage cracks are almost entirely
absent. A rhombohedral shelly structure is occasionally recognized
by the mode of arrangement of inclusions, especially fluid inclusions.
The absence of cleavage is one of the most characteristic negative criteria of quartz under the microscope.

In thin sections rock-making quartz is transparent and colorless,
even when it appears colored by incident light. The milky clouding
in incident light is mostly due to inclusions of fluids and gases. The
source of the blue color of many granitic quartzes is not yet definitely
known; the red color of the quartzes of silicious rocks, seen by incident light, arises from minute plates of hematite and ilmenite; the
green color in those of many hornblendic schists is due to needles of
amphibole; the jet-black and blue-black color of the quartz in many
porphyroids and phyllitic rocks is caused by carbonaceous substances
(graphite, coal), rarely by magnetite. The index of refraction of

quartz is almost the same as that of Canada balsam, consequently the surface of the section is perfectly plain and without relief. The double refraction is weak and its character positive. Rudberg found $\omega_{na} = 1.54418$, $\epsilon_{na} = 1.55328$. The interference colors of quartz in thin sections scarcely ever exceed those of the 1st order; in good thin sections it shows the bluish or yellowish tints of the 1st order. Basal sections give the interference cross without rings, and permit the positive character to be easily determined with the quarter undulation mica plate. This is the most important optical means of determination for quartz, taken in connection with its weak double refraction and transparency. Circular polarization does not appear in thin sections because of their thinness.

Sp. gr. = 2.65. Chemical composition = SiO_2. Not attacked by ordinary acids; is dissolved by hydrofluoric acid, which acts slowly on thin sections. The resistance which quartz offers to all the reagents occurring in nature accounts for the fact that it never appears weathered in thin sections, but is always completely fresh.

The following varieties of rock-making quartz may be conveniently distinguished:

Granitic quartz. Massive, and with its outline determined by those of the minerals associated with it. It forms the youngest primary constituent of the acid granular eruptive rocks, either as an essential or an accessory ingredient; thus in granites, syenites, diorites, and certain diabases. Granitic quartz is highly characterized by an abundance of fluid inclusions. These are mostly in irregular swarms and streaks, but are sometimes arranged in planes parallel to the rhombohedral faces. The fluidal cavities are sometimes completely filled with fluid (water, more rarely liquid carbon dioxide, or both), or there may be a gas bubble present. The relative sizes of the bubble and fluid vary within the widest limits. Besides the fluid inclusions occur gas inclusions, whose contents are probably water vapor. In the fluid inclusions there are sometimes crystalline secretions, usually of cubical form, which may in some cases be sodium chloride. Besides these inclusions, granitic quartzes often contain extremely fine opaque microlites, which Hawes referred to rutile.*

Massive granitic quartz often bears the traces of mechanical deformation in the peripheral shattering of the larger grains, as well as in the wavy extinction due to a continuous change in the direction of

---

* Mineralogy and Lithology of New Hampshire. Concord. 1878. 45.

the principal axis in one and the same grain. This deformation is undoubtedly the result of mountain-making forces.

Closely related to granitic quartz is the *quartz of the crystalline and phyllitic schists;* this also is massive, and destitute of outward crystalline boundary. But it does not receive its form from the minerals associated with it to the same extent as the granitic quartz does. They rather mutually penetrate one another, especially in the case of feldspar. The forms are rounded to lenticular, and range from microscopic grains to those of very considerable dimensions. The mutual intergrowth with feldspar (occasionally also with garnet, hornblende and other minerals) is similar to the granophyric structure of certain eruptive rocks. The inclusions correspond in all points to those in granitic quartz, and the phenomena of mechanical deformation are still more widespread.

Porphyritic quartz should properly exhibit a well-developed crystal form, which, however, may be more or less completely lost through chemical corrosion or mechanical deformation. Its forms never show a dependence on those of the associated minerals, and it is evident that this quartz was formed at a time when more or less of the rock existed in the condition of magma. Porphyritic quartz is an essential constituent of quartz porphyry, quartz porphyrite rhyolite (liparite), and dacite (quartz andesite). Gas and fluid inclusions are found here as in granitic quartz, but generally not in such quantities; they are accompanied by the very characteristic glass inclusions of round and dihexahedral form (Pl. VII. Figs. 2 and 3). Phenomena of mechanical deformation are quite rare, except those in the closely related quartzes of the porphyroides which are without glass inclusions; the greater part of them are from strains produced by glass inclusions. Fracturing caused by the fluid movement in the plastic rock mass is common. The chemical corrosion of porphyritic, quartzes (Pl. V. Fig. 1) is highly characteristic and distinguishes them from those of granites and schists. Not infrequently in the porphyritic rocks the quartz assumes spherical forms, which sometimes consist of a single individual, sometimes of two or three, seldom of more, which then appear as spherical sectors. This has been called *quartz globulaire* by Michel-Lévy. The substance of these forms is often mixed with more or less microfelsite.

The *clastic quartz* of sandstones, graywackes, and related rocks is usually without crystal form, being angular or rounded; the shape of the grains is not determined by its aggregation with other minerals, but by the mechanical processes which took part in its deposition.

The microstructure of the separate grains is that of the particular variety of quartz to which it originally belonged. Naturally the microstructure of the granitic and gneissic quartzes predominates. By the secondary deposition of silica in crystallographic orientation around the clastic grains the latter may assume the crystal form* (so-called crystallized sandstones, etc.). Such regenerated quartzes are not uncommon in the clay slates. The so-called quartz of the siliceous slates and of related rocks requires more exact investigation, and probably does not belong to quartz, but to chalcedony.

Gangue quartz forms irregular masses in granular aggregates, whose microstructure closely resembles that of the gneissic quartz on account of the abundance of fluid and gas inclusions. To this variety belong the secondary quartz in rocks of all classes, which arises from the decomposition and weathering of the silicates. It often occurs in pseudomorphs after these and other minerals (feldspar, mica, hornblende, pyroxene, etc.). Occasionally the quartz in these pseudomorphs is fibrous, but only when the original mineral was fibrous, as fibrous calc spar, asbestus, chrysotile, crocidolite, etc.

Chalcedony.

Chalcedony forms concretionary crystalline masses, mostly with a radially fibrous structure and shelly parting, rarely with a parallel fibrous structure. The fibres always appear to stand perpendicular to the surface of the shells or layers; they have very variable dimensions transversely, but are always extremely thin. Within the solid rock chalcedony is generally in the form of spherulites (Pl. IX. Fig. 1), while in cracks and cavities it occurs as a coating or in stalactites. The crystal system of chalcedony has not yet been definitely determined; however, it appears to be optically uniaxial. The index of refraction is smaller than for quartz, $n_\rho = 1.537$; the double refraction is somewhat stronger. The character of the double refraction is negative, which may be detected by examining the spherulites with a quartz wedge or a gypsum plate. This characteristic permits of its being readily distinguished from quartz. Tangential sections exhibit a fine-grained aggregate polarization, and the small dimensions of the fibres prevent their possible uniaxial nature from being determined. Central sections through the concretions give the spherulitic interference cross.

Sp. gr. = 2.59–2.64, somewhat lower than for quartz. Chemical composition = SiO_2, with the same chemical behavior as for quartz.

---

* R. D. Irving, Am. Journ. Sci. June, 1883, and Irving and Van Hise, Bulletin, No. 8., U. S. Geol. Survey. 1884.

Chalcedony appears as an original constituent of the ground mass of very silicious porphyritic eruptive rocks which have a microfelsitic development; thus in many quartz porphyries, rhyolites (liparites), quartz porphyrites, and dacites. Moreover, the spherulites of silica in quartz slates and related rocks appear to belong to chalcedony. As a decomposition product it occurs in all kinds of silicate rocks.

Tridymite.

Literature.

A. VON LASAULX, Ueber das optische Verhalten und die Krystallform des Tridymits. Z. X. 1878. II. 253-274.

A. MERIAN, Beobachtungen am Tridymit. N. J. B., 1884. I. 193-195.

M. SCHUSTER, Optisches Verhalten des Tridymit aus den Euganäen. T. M. P. M. 1878. I. 71-77.

F. ZIRKEL, Ueber den mikroskopischen Tridymit. Pogg. Ann. 1870 CXL. 492.

Tridymite forms tabular crystals, sometimes with rounded outlines which are bounded by the planes oP (0001) and $\bar{\infty}P$ (10$\bar{1}$0). In the attached crystals in cavities there occur in addition to the derived pyramids the prism of the 2d order, and a dihexagonal prism often developed hemihedrally. Moreover, the attached crystals are almost always juxtaposition or penetration trillings along the faces $\frac{1}{6}P$ (10$\bar{1}$6) and $\frac{3}{4}P$ (30$\bar{3}$4), while in the rock-making crystals this twinning does not seem to occur. The dimensions of the rock-making tridymites are always microscopic; consequently they almost never appear as sections, but as complete bodies. The microscopic individuals occur almost without exception in tile-like aggregates, in which the faces oP overlap one another (Pl. XVI. Fig. 4). Through the suppression of one pair of prism faces the outline is often rhombic.

Cleavage is not known in rock-making tridymite, though there is sometimes a parting parallel to oP (0001) due to the growing together of several plates along this face.

Tridymite as a rock constituent is free from inclusions, with the exception of gas interpositions; it is transparent and pellucid, with a weaker refraction than Canada balsam; and with moderate double refraction whose character is positive. The plates which have grown in the rock behave isotropic when they lie on their basal plane, and weakly doubly refracting in other positions. The larger attached crystals often exhibit in basal planes a very complicated division into areas which are doubly refracting, and which show the locus of optic axes or a bisectrix in convergent light, which would indicate that these areas belong to the triclinic system. Such tridymite plates become isotropic upon being heated, but resume their doubly refracting character when allowed to cool. From this it is inferred that the tridymite plates crystallized originally in the hexagonal system, but that under the physical conditions existing at the surface of the earth they become subjected to strains in an attempt to assume another molecular arrangement.

Sp. gr. $= 2.28$–2.33. Chemical composition $= SiO_2$; chemical behavior the same as for quartz, except that tridymite is soluble in boiling caustic soda.

Tridymite is chiefly a volcanic mineral; it is a frequent constituent of rhyolite (liparite), trachyte, and andesite. It is particularly abundant in the lithophysæ of obsidian and rhyolite in the Yellowstone National Park. It has also been found by G. Rose in the opals of Kosemütz, Silesia; Kashan, Persia; and Zimapan, Mexico; and in the cacholong of Iceland. In an augite andesite from Grad-Jakán in Java it occurs probably through the decomposition of feldspar. Tridymite has also been found in meteorites.

Calcite.

As a rock-making constituent *calcite* never occurs in crystals: it is either in irregularly bounded grains and plates or in aggregates, or it occurs in parallel or radiating fibrous aggregations, or else it presents that peculiar concretionary, crystalline form called oölitic. Hence its crystal form is of no importance for its microscopical determination. Still the grains and plates of calcite found in rocks are mostly characterized by the twinning parallel to $- \frac{1}{2} R \pi$ (01$\bar{1}$2), by which each grain is converted into a polysynthetic individual (Pl. XVI. Fig. 5). This polysynthetic twinning is of very common occurrence in crystalline limestone, and may very likely have been produced by pressure; it can also be produced during the process of grinding when the section is sufficiently thin.

The calcite cleavage parallel to $R \pi$ (10$\bar{1}$1) shows itself in thin section by numerous sharp cracks (Pl. X. Fig. 1), whose inclination to one another changes with the position of the section. This cleavage is one of the most important means of distinguishing calcite from other minerals, with the exception of dolomite, magnesite, etc. Along the cleavage cracks, which do not cut the calcite sections perpendicularly, there often appear Newton colors produced by the interference of the light reflected back and forth from the sides of the cracks.

Calcite when pure is colorless, but appears dark gray, bluish, almost opaque, brownish or yellowish in transmitted light in consequence of organic pigments. The mean index of refraction is not high, consequently the surface of the sections is plain, and the relief small. On the other hand, the double refraction is very strong, with negative character; $\omega_{na} = 1.6585$, $\epsilon_{na} = 1.4864$. Hence the interference colors between crossed nicols are high, even for very thin sections; during a rotation between crossed nicols darkness alternates with clear white or pale green and bright pale green, and the more precise colors of the lower orders are wanting. Basal sections in convergent polarized light give an interference cross with several colored rings, even for very thin sections. The same thing is obtained from radial aggregates which sometimes occur as secondary minerals in eruptive rocks, when the section is tangential and the objective of the microscope is not focused on the surface of the section, but on a point in which the rays of like phasal difference intersect. Oölitic aggregates often give in parallel polarized light the interference figure of spherical uniaxial aggregations (Pl. IX. Fig. 2). Calcite does not show pleochroism, but the strong absorption of the ordinary ray is distinctly noticed when the sections are not too thin.

Sp. gr. $= 2.72$, which serves to distinguish it from dolomite and aragonite (2.95). Chemical composition $= CaO, CO_2$; it is easily attacked by acids. It is distinguished from other isomorphous carbonates by the readiness with which it is attacked by weak acids even at ordinary temperatures. A part of the Ca in the formula may be replaced by Mg, Fe, or Mn, without affecting the solubility. To distinguish magnesia-bearing calcite from normal calcite it is advisable to employ the method of G. Linck given on page 112.

Calcite frequently contains fluid inclusions and rhombohedrons of dolomite or magnesite. Mechanical deformation is recognized by the curving of the cleavage cracks and the undulating extinction, and is specially common in granular limestone.

The distribution of calcite is very great, even when we leave out of consideration its prevalence in the sedimentary formations, where it occurs as marble, limestone, oölite, chalk, calcareous tufa, and in marl, calcareous sandstones, calcareous clay slate, and calcareous mica-schist, etc. In all kinds of eruptive rocks, more especially in those poor in silica, it appears as the filling of cavities and cracks, and partly within the compact rock mass. It is very often a product of atmospheric decomposition, and then at times forms complete alteration pseudomorphs after lime silicates (plagioclase, augite, etc.), or replacement pseudomorphs after silicates poor in lime or free from it (olivine, biotite, etc.). Moreover, it occurs in many eruptive rocks (minettes, ker-

santites) in apparently primary grains, which are nevertheless secondary, occupying the spaces once filled by silicates. In other cases the
presence of calcite in eruptive rocks is due to infiltration from neigh-
boring calcareous rocks.

Dolomite.

In contrast to calcite, *dolomite* occurs in rocks chiefly as crystals,
and even when in dense homogeneous aggregates there is an evident
tendency toward outward crystalline boundaries, so that it assumes
a saccharoidal structure.   The form of the crystals occurring in
rocks appears to be almost universally the fundamental rhombohedron
R π (1011), seldom more acute rhombohedrons.   Hence the cross-sections are triangular, six-sided, and rhombic.   From the tendency to
curved surfaces which characterizes this mineral, the outlines are often
crooked, bent, and distorted.   Twinning has not been observed on
rock-making dolomite; the lamination parallel to $-\frac{1}{2}R$ π (01$\overline{1}$2),
so common in calcite, is wanting.   Oölitic structure occurs with
dolomite as with calcite.   The cleavage parallel to R π (1011) is just
as distinct in dolomite as in calcite, and the difference in the rhombohedral angles of both substances cannot be used as a means of distinguishing them from one another

The optical behavior of dolomite is the same as that of calcite;
the character of the double refraction is negative; its amount is considerable: $\omega_{na} = 1.6817$, $\epsilon_{na} = 1.5026$ (Fizeau).   Interference colors and
axial figure the same as for calcite.   In transmitted light dolomite is
colorless or yellowish to brownish in consequence of the alteration of
the ferrous carbonate to limonite; it is gray, brownish, or blackish
through organic pigments.   Pleochroism not noticeable; the absorption $O > E$.

Sp. gr. varies with the iron percentage from 2.85 to 2.95.   Chemical composition, $CaO. CO_2$, MgO, CO_2, in which varying amounts of Mg
may be replaced by Fe and Mn.   Acetic acid and cold dilute hydrochloric acid attack dolomite but slightly; in heated hydrochloric acid
it is dissolved rapidly with strong effervescence.

Dolomite occurs as an independent rock among the crystalline
schists, and in close geological connection with limestone in the palæozoic and mesozoic sedimentary formations.   As occasional scattered
crystals it is found in limestones, silicious slates, clay slates, and phyllites, especially where the latter show regional metamorphism.   It
may amount to an essential component of these rocks.

Magnesite and Breunnerite.

Magnesite forms suspended crystals in the form of the fundamental rhombohedron $R\,\pi$ ($10\overline{1}1$), when it occurs as an accessory constituent. When it is an essential component of the rock, it appears as isolated grains or in granular aggregations without crystalline boundaries to the separate grains. Twinning is absent, as in dolomite.

The cleavage parallel to the fundamental rhombohedron is manifested, as in calcite and dolomite, by numerous cracks which are mostly straight, less frequently slightly curved.

Its behavior toward light is the same as that of calcite and dolomite; the double refraction is strong and negative. In transmitted light magnesite is colorless to grayish; also yellowish for a high iron percentage.

The sp. gr. is about 3.0–. Chemical composition, $MgO\,CO_2$; with the introduction of the isomorphous iron carbonate in variable proportions it passes into breunnerite. The corresponding Ca and Mn combinations are only present to a slight extent. Cold hydrochloric acid does not attack magnesite in thin sections and in fragments.

Breunnerite often becomes yellow to reddish brown by the separating out of limonite. Magnesite and breunnerite occasionally occur as accessory minerals in chloritic and talcose schists, as well as in Swedish olivine schists. With talc they form certain crystalline schists, and with bronzite, sagvandite.

Apatite.

Literature.

F. Zirkel, Untersuchungen über die mikroskopische Structur und Zusammensetzung der Basaltgesteine. Bonn. 1870. 72–74.

In eruptive rocks *apatite* is mostly in the form of long, slender hexagonal columns which are terminated by the base or by the fundamental pyramid ($10\overline{1}1:10\overline{1}\overline{1} = 80°\,12'$ to $80°\,36'$), sometimes by both forms; less frequently the crystals appears as short, thick columns bounded by the same faces. The latter is specially noticeable in rocks of the gabbro family. On the other hand, in the crystalline schists apatite occurs quite often in rounded or long oval grains, with slight indications of crystalline boundaries, if any. Hence sections parallel to the base are regularly hexagonal, parallel to the principal axis more or less elongated rectangles, which are sometimes pointed at the ends or have the corners truncated, or the cross-sections may be round or oval. Crystals occur grown together in parallel position, but twinning is absent. The cleavage parallel to the base and prism is seldom observed microscopically, but the long columnar crystals almost always exhibit a transverse jointing so that they fall into distinct pieces which not infrequently have been more or less dislocated.

Apatite of itself is clear and transparent, but sometimes in rocks possesses a gray, violet-blue, yellowish or brownish color of different intensity. Its index of refraction is higher than that of the other colorless minerals generally associated with it: hence its bright white color, and not inconsiderable relief. The double refraction is weak and negative. In apatite from Jumilla, Spain, $\omega_{na} = 1.6388$, $\epsilon_{na} = 1.6346$ (Lattermann). Therefore the interference colors in thin section scarcely exceed white of the 1st order, being mostly in the grayish-blue tones. In convergent light basal sections give only a cross, without rings. Colorless apatite has no pleochroism; the colored apatite is always distinctly and often strongly pleochroic, the absorption being $E > O$, which is a convenient microscopical means of its determination from tourmaline. This strong absorption of the extraordinary ray is noticeable by careful observation even in colorless apatite. The optical anomalies frequently noticed in large attached crystals are scarcely ever observed in the microscopical individuals found within rocks.

On account of its high specific gravity, 3.16–3.22, apatite, when separated mechanically from the rocks, falls with the minerals having heavy metallic bases, and may be generally separated from these with the electro-magnet without trouble. From non-magnetic minerals (zircon, titanite, rutile, etc.) it may be separated by means of Klein's solution.

Chemical composition $= 3Ca_3P_2O_8 + Ca(Cl, Fl)_2$. It is readily soluble in acids; from the solution upon the addition of ammonium molybdate there is precipitated yellow octahedral or rhombic dodecahedral crystals or groups, which are formed even in the cold (Pl. XIII. Fig. 5). In another part of the solution dilute sulphuric acid precipitates crystals of gypsum. Many crystals contain a noticeable percentage of manganese; the solution of these upon treatment with hydrofluosilicic acid yields rhombohedral crystals with prismatic habit of fluosilicate of manganese. Although apatite is so easily attacked by acids, it is remarkable that it is found perfectly fresh in rocks which are completely decomposed. In some instances it is of ideally pure substance, in others it is more or less filled with interpositions, of which gas and fluid inclusions predominate, while glass inclusions are rarer. These are often arranged in a very orderly manner, and mostly show a central accumulation parallel to the principal axis ; in other cases they are in concentric shells parallel to the outward form of the crystal, or may be scattered generally through the whole mineral. Very rarely the interpositions are massed peripherally or are arranged parallel to the principal vertical section, so that they form six-rayed stars in cross-sections. The surface of the apatite is often rough through corrosion, and covered with irregular depressions.

Apatite is present in all rocks, and in the eruptive rocks appears as one of the oldest, if not the oldest secretion of the magma. The needles of this mineral pass uniformly through all the other constituents. Though mostly disseminated quite uniformly throughout the whole rock mass, it is sometimes crowded together with the older secretions (iron ores, zircon, mica). It appears to be more abundant in the older granular eruptive rocks and in the feldspathic crystalline schists, than in the younger eruptive rocks and in the feldsparless schists ; the basic eruptive rocks also appear to contain more apatite than the acid ones. It is found particularly associated with biotite and nepheline.

Nepheline and Elæolite.

Literature.

H. ROSENBUSCH, Der Nephelinit vom Katzenbuckel. Freiburg i. B. 1869. 46–59.
F. ZIRKEL, Mikroskopische Untersuchungen über die Zusammensetzung und Structur der Basaltgesteine. Bonn. 1870. 38.
— Ueber die mikroskopische Zusammensetzung der Phonolithe. Pogg. Ann. 1867. CXXXI. 303.

Nepheline and elæolite bear the same relation to one another as sanidine and orthoclase do. The first includes the glassy colorless occurrences in the younger volcanic rocks ; the second the massive occurrences, often somewhat colored, in the older plutonic rock and their pegmatitic secretions. Identical in substance and in all essential physical characters, they are nevertheless rightly separated on account of their diverse habit and different geological position.

Nepheline and elæolite show themselves in the rocks partly as completely developed, short prismatic crystals of the form ∞P (1010) . oP (0001), whose basal edges are sometimes truncated by small faces of P (1011). The angle over the edge of the prism is 88° 10′. The nepheline individuals in general are considerably smaller than those of elæolite when compared with the grains of the containing rock. But the elæolite indivduals also sink to microscopic dimensions. The outline of the sections are naturally hexagonal parallel to the base, and short rectangular to quadratic in longitudinal sections, occasionally with the corners truncated. An outward crystalline boundary is often completely wanting in elæolite, as would naturally be the case in granular rocks; but nepheline seldom occurs massive. Their boundary in this case, then, is in no way characteristic of the minerals.

The cleavage parallel to ∞P $(10\bar{1}0)$ and oP (0001) is seldom observed under the microscope in glassy nepheline; it is more common in elæolite. The cleavage becomes more evident after decomposition has attacked these minerals, as the alteration products are first deposited along the cleavage cracks.

Nepheline and elæolite become transparent and colorless; their index of refraction corresponds very closely to that of Canada balsam, and the double refraction is weak. Hence the absence of relief in thin section and their low interference colors (grayish blue or at most white of the 1st order). The low index of refraction may be used as a means of distinguishing these minerals from apatite. The character of the double refraction is negative. In nepheline from Vesuvius, J. E. Wolff determined $\epsilon_{na} = 1.5376$, $\omega_{na} = 1.5416$; M. E. Wadsworth, $\epsilon_{na} = 1.5378$, $\omega_{na} = 1.5427$; in elæolite from Hot Springs, Arkansas, S. L. Penfield found $\epsilon_{na} = 1.5422$, $\omega_{na} = 1.5469$. In convergent light thin sections give a broad interference cross without rings. In parallel light the double refraction of very microscopic individuals is not at all noticeable, except by using a gypsum plate or quartz wedge.

Sp. gr. $= 2.55-2.61$; it is generally somewhat higher for elæolite than for nepheline, probably on account of the difference in their interpositions. It lies between that of the triclinic and orthorhombic feldspars, and permits their mechanical separation. The chemical composition is $4Na_2O$, $4Al_2O_3$, $9SiO_2$, in which one quarter of the Na is generally replaced by K, while only a very small amount of Ca occurs in these minerals. Nepheline and elæolite gelatinize quite easily and quickly with hydrochloric acid, but more difficultly than the minerals of the sodalite group. This gelatinization and the method of staining it already described (p. 65) are the best means of recognizing and determining nepheline and elæolite under the microscope. The absence of calcium in the solution prevents a confusion with melilite.

Elæolite very commonly carries microscopic interpositions of augite and hornblende needles, fluid and gas inclusions. In the fluid inclusions cubes of NaCl are sometimes secreted. Nepheline is also rich in inclusions of the minutest dimensions, which are often scarcely determinable; they are augite microlites, fluid, gas, and glass inclusions. In both minerals the arrangement of the interpositions is mostly in zones (Pl. XVI. Fig. 6, and Pl. XVII. Fig. 1); they are seldom crowded together at the centre.

Elæolite and nepheline are easily altered to zeolites, of which natrolite appears most frequently to form pseudomorphs after them. The process commences from the cracks and margin, and leads to the formation of parallelly fibrous, confusedly fibrous or radiating aggregates, with brilliant double refraction. Nepheline and elæolite are also known to alter into analcite and thomsonite.

While the zeolitization of these minerals has taken place through the action of hot waters soon after the solidification of the rock, there is produced from them through the ordinary atmospheric influences, muscovite and kaolin (liebenerite and gieseckite).

Nepheline is only known in volcanic rocks; it occurs with sanidine in phonolites[*] and leucite porphyries, with triclinic feldspars in teschenites and tephrites, without feldspars in nepheline basalts and nephelinites, and with leucite in leucite basalts and leucitites.

Elæolite occurs with orthoclase as an essential ingredient of elæolite syenite,[†] and occurs in a rock without feldspars at Mt. Jivaara in Finland. It is an accessory mineral in the augite syenites of Southern Norway. The frequent occurrence of nepheline and elæolite with minerals of the sodalite group is to be noted.

In all the eruptive rocks the formation of nepheline and elæolite follows the secretion of the bisilicates and the micas, and in the first generation at least precedes that of the feldspars. When minerals of the sodalite group occur with them as primary crystals they are the older secretions.

Eucryptite is a lithia nepheline described by Brush and E. S. Dana[‡] as an alteration product of spodumene from Branchville, Conn.

Cancrinite.

Literature.

A. KOCH, Petrographische und tektonische Verhältnisse des Syenitstockes von Ditró in Ostsiebenbürgen. N. J. B. B.-B. I. 1881. 144.

H. RAUFF, Ueber die chemische Zusammensetzung des Nephelins, Cancrinits und Mikrosommits. Z. X. 1878. II. 456–468.

A. E. TÖRNEBOHM, Om den s. k. Fonoliten från Elfdalen, dess klyftort och förekomstädt. Geol. Fören. i. Stockholm Förhdl. 1883. VI. No. 80. 383.

---

[*] Am. Journ. 1880, XX. 259.

[†] J. H. Caswell has described phonolite from the Black Hills, Dakota. Microscopic Petrography of the Black Hills of Dakota. Washington, 1876. 492.

[‡] Elæolite syenite occurs near Deckertown, N. J. (B. K. Emerson, Am. Journ. Sci., April 1882. 302). also at Litchfield, Me., and Magnet Cove, Ark.

Cancrinite as a rock-making mineral sometimes occurs in long columnar crystals with the faces ∞P (10$\overline{1}$0), P (1011); more frequently in staff-like individuals developed only in the prism zone; occasionally it is in irregular grains whose outlines are dependent on the other rock constituents.

The cleavage parallel to ∞P (10$\overline{1}$0) appears microscopically in distinct and sharp cracks, both in transverse and longitudinal sections; longitudinal sections also exhibit a distinct cleavage parallel to oP (0001). An imperfect cleavage appears to run parallel to $\infty P2$ (11$\overline{2}$0).

Cancrinite when fresh is transparent and colorless. The index of refraction is lower than that of Canada balsam, the double refraction is negative, and considerably stronger than that of nepheline. The interference colors in thin section generally range from orange of the 1st order upwards, and are similar to those of scapolite. In cancrinite from Miask, $\epsilon_\rho =$ 1.4955, $\omega_\rho = 1.5244$ (Osann). Cross-sections in convergent light give a sharp cross with rings.

The sp. gr. is about 2.45, which greatly facilitates its mechanical separation from the minerals associated with it. The chemical composition $= 4Na_2O, 4Al_2O_3, 9SiO_2 + 2CaO, CO_2 + 3H_2O$. It is decomposed by cold hydrochloric acid with the separation of gelatinous silica and the liberation of bubbles of carbonic acid. When heated to redness in thin section it becomes clouded, and may thus be distinguished from nepheline. Cancrinite has no constant microstructure: it is sometimes quite free from interpositions; at others it has the same inclusions as elæolite,—especially the pyroxene needles, which are arranged with their longitudinal axes parallel to the principal axis of the cancrinite. The red and yellow color of many occurrences arises from the interposition of plates of hematite. The process of alteration is similar to that in elæolite.

Cancrinite is only known as yet in elæolite syenites (Miask, Brevig, Lichfield) in company with elæolite and sodalite.

Tourmaline.

In many rocks *tourmaline* forms perfectly developed columnar crystals, often with very distinct hemimorphism: on one pole there is usually R, π (10$\overline{1}$1) only, with the terminal angle 133° 10′–133° 20′, rarely with derived rhombohedrons; on the other pole is the base; in the vertical zone there is sometimes only $\infty P2$ (11$\overline{2}$0), sometimes the three-sided ∞R (10$\overline{1}$0) also. Thus the cross-sections are regular hexagons, or hexagons with alternately truncated corners, the longitudinal sections being lath-shaped. More frequently, however, tourmaline appears in staff-like individuals without sharp crystalline development, in bunched or finely radiating aggregates; the cross-sections of separate individuals then are irregularly rounded. More rarely it assumes the form of irregular grains. A shelly structure is quite common, the difference in color of the kernel and shells clearly indicating an isomorphous lamination. Less frequently the layers vary horizontally; and still more rarely they assume the form of an axial cross, differing in color from the main mass of the crystal.

Cleavage is not recognizable microscopically, but irregular transverse and longitudinal cracks are very common, especially in the larger individuals. The tourmaline occurring in rocks is never perfectly colorless, but is always transparent and colored. Moreover, the colors vary extraordinarily both in kind and intensity. Yellow, brown, green, red, and especially violet blue, are those which most frequently appear in transmitted light. The index of refraction is moderate, the double refraction quite strong and negative. Des Cloizeaux found in a crystal consisting of blue and green shells, for both colors, $\epsilon_\rho = 1.6240$, $\omega_\rho = 1.6444$; Miklucho-Maclay found in a colorless tourmaline from Elba, $\epsilon_{na} = 1.6208$, $\omega_{na} = 1.6397$. The relief of the section against the colorless rock constituents is distinctly perceptible; the surface is noticeably rough. Cross-sections give a sharp interference cross with clearly determinable negative character. The pleochroism is stronger as the color is deeper, but is distinctly noticeable even in quite light-colored individuals. The absorption is always strong for the ordinary ray and weak for the extraordinary ray. The colors change with the body color. Similarly strong differences of absorption are only exhibited in rocks by dark mica, hornblende, and allanite; the first two are easily distinguished by their cleavage, while the third is only distinguished by its crystallization.

Optical anomalies are rarely perceptible. They are only shown in the cross-sections, and especially in convergent light through the separation of the interference cross into hyperbolas with very varying intervals between their poles. Apparently it is those individuals composed of isomorphous layers which most frequently possess these anomalies.

The sp. gr. varies with the chemical composition from 3.0–3.24, rising with the increase of the bivalent metals. The chemical composition is a very variable one, and according to Rammelsberg's investigations may be explained as an isomorphous mixture of the molecules:

$$NaHO, \; B_2O_3, \; 3Al_2O_3, \; 4SiO_2,$$
$$5MgO, \; B_2O_3, \; Al_2O_3, \; 5SiO_2,$$
$$5FeO, \; B_2O_3, \; Al_2O_3, \; 5SiO_2 \; ;$$

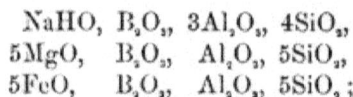

in which in the first molecule, in place of Na, K and Li may also occur, and in the second and third, in the place of Mg and Fe, Mn may occur. Tourmaline is not acted on by acids, including hydrofluoric, and may therefore be easily separated from the rocks by chemical means, even in small quantities. It is then obtained together with rutile, spinel, garnet, zircon, andalusite, sillimanite, etc., and may be separated from these substances by means of Klein's solution, supplemented by magnetic methods.

Tourmaline possesses no distinctive microstructure which can be used in its diagnosis. It occasionally encloses part of the minerals associated with it, or gas and fluid inclusions besides liquid carbon dioxide, but not with any degree of constancy.

As an accessory constituent, tourmaline occurs in the older granular and acid eruptive rocks (granite, syenite, diorite). In massive occurrences of these rocks it is situated most usually on the periphery and in the vicinity of fissures and veins. When these rocks form dikes it is more often scattered through the whole mass. It is particularly frequent in the "greisen"-like modifications of granite, and may in these cases be easily confused with cassiterite, when it forms granular or radial aggregates. It may be distinguished from it in cross-sections by testing the optical character of the interference cross with the quarter undulation mica plate, and in longitudinal sections by means of the quartz wedge. It is also very common in the contact zones of schists near granites, and may become so abundant locally as to form tourmaline hornstone, as in many places in Cornwall, and especially in the White Mountains at Mt. Willard, N. H.[*] It very rarely occurs in the mesozoic pyrophritic acid eruptive rocks (quartz porphyries and quartz porphyrites); it is almost completely absent from the equivalent tertiary lavas. Its whole mode of occurrence in eruptive rocks and their contact zones indicates that it was not directly secreted out of the eruptive magma, but resulted from the action of fumaroles carrying fluorine and boron on the eruptive rock, especially on its feldspar and mica. Tourmaline is very common as isolated crystals, mostly very sharply defined, in the quartz and feldspar-bearing members of the crystalline schists, gneisses, granulites, etc., as well as in

---

[*] G. W. Hawes, The Albany Granite and its Contact Phenomena. Amer. Jour. 1881. XXI. 21-32.

phyllites and clay slates (Pl. XV. Fig. 4). Owing to its resistance to decomposition, tourmaline remains unaltered in the detritus of these rocks and passes into the composition of clastic rocks.

Eudialyte.

Eudialyte forms either crystals of very variable dimensions, which, however, seldom become wholly microscopic and exhibit predominantly the forms oR, (0001), R, π ($10\bar{1}1$), $-\frac{1}{2}R$, π ($10\bar{1}2$), while the forms $\frac{1}{4}R$, π ($10\bar{1}4$), ∞R, (1010) and $\infty P2$ ($11\bar{2}0$) are subordinate; or it forms grains with incomplete crystallographic boundaries. The terminal angle of R is 73° 30′.

The cleavage parallel to oR (0001) is distinctly perceptible microscopically, while the incomplete cleavage parallel to $\frac{1}{4}R$, π ($10\bar{1}4$) and $\infty P2$ ($11\bar{2}0$) is not recognizable.

Eudialyte becomes transparent with light yellowish-red and purplish blood-red color; possesses quite strong positive double refraction, and a high index of refraction, as the decided relief and rough surface in thin section indicates. The pleochroism is weak. The whole appearance suggests garnet.

The sp. gr. varies from 2.84–3.05. The chemical composition approaches Na_2O, $2(Ca, Fe)O$, $6(Si, Zr)O_2$. The chlorine percentage of the analyses is referred to inclusions of sodalite, which are sometimes recognized microscopically. With hydrochloric acid eudialyte gelatinizes quite easily.

The microscopic individuals are usually free from inclusions; the larger massive grains occasionally enclose elæolite, sodalite, arfvedsonite, and numerous fluid inclusions of rhombohedral or rounded form, which are quite large and are arranged in straight lines.

Eudialyte is often a prominent constituent of Greenland elæolite syenite, from which Vrba described it; he also found it accessory to similar rocks from the islands of Langesundfjord in Southern Norway.

Eucolite is chemically and crystallographically identical with eudialyte; it is distinguished, however, from the latter by the distinct cleavage parallel to $\infty P2$ ($11\bar{2}0$), and the negative character of the double refraction. This mineral also is found in the elæolite syenites of Southern Norway.

Chlorite Group.

In the *chlorite* group have been placed a number of minerals which show a relationship because of their chemical composition, their crystallographic development, physical properties, and of their whole habit and geological value, but which, however, partly on account of their optical behavior, partly because of their chemical composition, have been divided into the three varieties: *pennine, clinochlor,* and *ripidolite.* The determination of these varieties, even in comparatively well-developed crystals, is not without difficulties, and in rock-making occurrences often becomes impossible.

Since the subdivisions of this group in present use are very probably provisional, the name *chlorite* will be used to embrace all the minerals of the chlorite group. It has only been placed in the hexagonal crystal system because the observations applicable to most cases practically lead to this system. Nevertheless, it is highly probable that all that is here described as chlorite is to be referred in fact to the monoclinic system. Rock-making chlorite appears mostly in the form of flat leaves of variable size, or of somewhat dense scales of irregular outline. The scales generally lie with their faces upon one another in parallelly laminated aggregations; less frequently they arrange themselves spirally in rosettes. When a crystallographic outline is observed on the scales or plates it is most frequently hexagonal, rarely triangular, or irregularly polygonal, as though the hexagonal prisms of the first and second order had been developed with an incomplete number of faces. The cross-sections of these laminated aggregates are more or less lath-shaped, often with curved and imperfectly parallel edges. Another development of chlorite is the fibrous, in which the fibres are sometimes parallel, sometimes divergent, and at times grouped in uniformly radiating spherulites. Lastly, chlorite frequently occurs in the minutest particles, which exhibit neither laminated nor fibrous structure. This is usually the case when chlorite occurs as finely divided pigments in other minerals.

Chlorite cleaves very perfectly parallel to the flat face, which is considered as the basal plane; the cleavage plates are generally flexible. In cross-sections the cleavage manifests itself by numerous lines running parallel to the edge of the section, which are often twisted in the same manner as this is. The perfection of the cleavage microscopically is scarcely second to that of mica. Plates parallel to the cleavage face never show cleavage lines—a fact which may be used to distinguish it microscopically from chloritoid.

Chlorite is generally green by incident and by transmitted light, although the depth of color varies from greenish white to dark green. The index of refraction is low; the double refraction is very weak. Haidinger determined on pennine $\epsilon = 1.575$, $\omega = 1.576$; Des Cloizeaux, $\epsilon_p = 1.576$, $\omega_p = 1.577$. The character of the double refraction varies, and has been found even for the same occurrence sometimes positive, sometimes negative. Plates parallel to the cleavage face usually show themselves quite isotropic in parallel polarized light, or when rotated between crossed nicols exhibit only a very slight illumination, which often appears to be due to the fact that the plates are not lying perfectly flat. In convergent light they generally give an indistinctly defined interference cross, between whose arms the light quadrants are scarcely visible. The centre of the cross often lies excentrically; moreover, it not infrequently opens into two hyperbolas with very variable polar interval. These phenomena indicate monoclinic chlorite (clinochlore or ripidolite). Cross-sections of plates or fibres of chlorite show themselves as doubly refracting, but mostly with very low interference colors; often the double refraction is only noticeable by careful observation, even for light-colored varieties. The interference colors in thin section do not exceed white of the first order, but these are difficultly distinguishable because of the proper color of the mineral, and often combine with the latter to form a peculiar blue. The extinction lies apparently parallel to the cleavage, or deviates but very slightly from it; consequently in the actually monoclinic chlorites the bisectrix must stand very nearly perpendicular to the base. This is also a good criterion in distinguishing it from most chloritoids. Twinning has not been observed in rock-making chlorite. Radially fibrous aggregates give the interference cross of spherulites; the arms lie apparently parallel to the principal sections of the nicols. Chlorite is distinctly pleochroic, and in the same manner, both in the apparently hexagonal and in the monoclinic varieties, O is green, E pale yellow to red or brown. The difference between the two rays is more perceptible as the chlorite is deeper colored. The pleochroism is generally noticeable, even when the double refraction can only be observed by the insertion of sensitive plates.

. The specific gravity varies between 2.65 and 2.97 with increasing iron percentage. The chlorite minerals are considered as isomorphous mixtures of

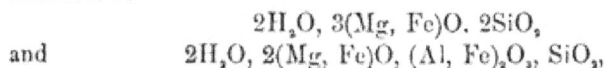

$$2H_2O, 3(Mg, Fe)O. 2SiO_2$$

and $\quad 2H_2O, 2(Mg, Fe)O, (Al, Fe)_2O_3, SiO_2,$

in the proportion of $3:2$ to $1:2$. All chlorites gelatinize in thin section with hot hydrochloric acid, often even with cold; still more easily with sulphuric acid, and may then be colored. Besides magnesia and iron, alumina may also be abundantly detected in the solution by the methods already given. Upon being heated to redness in thin section on platinum foil chlorite loses its water and becomes opaque.   Ferruginous varieties are colored reddish brown to black in this way.

The chlorites are among the most widely disseminated substances in rocks, but in eruptive rocks and their tufas they are always secondary formations and products of alteration.   They are derived from the aluminous members of the biotite and phlogopite micas, of the pyroxene and amphibole families, and of garnets.   To them many eruptive rocks owe their green color.   This secondary origin of the chlorites is undoubtedly proved by their occurrence as pseudomorphs after the above-mentioned minerals, often with the complete preservation of the forms of the original substance.   In other cases chlorite is formed by magnesia- and iron-bearing solutions from non-aluminous substances acting upon solutions from aluminous minerals which are free from magnesia and iron. In this way "replacing" pseudomorphs are formed, for example, after feldspar.   Chlorite has a further distribution in the schistose rocks: it occurs in chlorite schists as the only essential constituent, and is here accompanied by amphibole, magnetite, dolomite, etc., as accessory minerals.   Together with amphibole, it appears in many amphibolites, especially those of phyllite formations; with mica in mica schists and phyllite; with feldspar and mica in phyllite gneisses; with epidote, augite, and actinolite, besides feldspar, and quartz in the so-called green schists.   It also appears in the phyllites proper and in clay slates as quite a regular and often a very abundant constituent.

MINERALS OF THE ORTHORHOMBIC SYSTEM.

SECTIONS of the regularly bounded minerals belonging to the orthorhombic system or the figures made by their cleavage cracks are bisymmetric when they lie parallel to two axes, monosymmetric when they
are parallel to only one axis, and asymmetric when they intersect all
three axes. If the pyramidal faces or cleavage planes are wanting
the sections may apparently possess a higher symmetry than they actually do.

Pinacoidal cleavage furnishes parallel systems of lines in all sections
which are not parallel to the cleavage itself; prismatic cleavage or that
parallel to two pinacoids gives parallel cracks in all sections lying in the
zones of these cleavages, and intersecting cracks in all other sections;
cleavage parallel to three pinacoids or to a pyramid gives in all sections
figures which are enclosed by intersecting lines.

The ellipsoid of elasticity of orthorhombic crystals is triaxial, and
its three axes coincide with the axes of the crystal. Therefore the
plane of the optic axes always lies in a principal crystallographic section, and the optic axes for light of different wave-lengths are disposed
symmetrically with respect to two of the crystallographic axes. If the
bisectrix of the acute angle of the optic axes is the direction of least
elasticity (c), the crystal is called positive; it is said to be negative when
the axis of greatest elasticity (α) bisects the acute angle. Sections perpendicular to an optic axis are uniformly light in all positions in
parallel light between crossed nicols, and yield in convergent light an
axial figure in the centre of the field of view. This figure consists of
approximately circular, concentric curves cut by a dark bar, which is
always straight when the axial plane coincides with a principal plane
of the nicols. Upon rotating the section in its plane the bar turns and
curves slightly to an hyperbola. All other sections become light-
colored and dark four times during a rotation in parallel white light
between crossed nicols. The maximum of darkness occurs when the
line of intersection on the mineral plate of one of the principal sections
bisecting the angle between the optic axes becomes parallel to one of
the principal sections of the nicols. Hence the sections in the three
principal zones extinguish parallel to those boundary lines or cleavage
cracks which run parallel to a crystallographic axis. The maximum

brightness lies at 45° to this position. In convergent light sections perpendicular to a bisectrix exhibit a dark cross whose arms divide the field of view symmetrically when the plane of the optic axes coincides with one of the principal sections of the nicols. Upon a rotation of the section in its own plane the cross opens to hyperbolas whose poles reach their maximum interval after a rotation of 45°. This interval is a measure for the angle between the optic axes. The interference figure is bisymmetrical, and the distribution of the colors shows the kind of dispersion, $\rho < v$ or $\rho > v$. How much of the interference figure is visible depends on the angle, $2E$ or $2H$, according to whether the observation is made in air or in oil. In sections which are inclined to an optic axis, the axial bar when it reaches the straight position passes at length through the locus (point of egress) of the second axis. If the locus of the bisectrix of an interference figure is not in the centre of the field of view, the section is still parallel to a crystallographic axis when one of the bars bisects the field in the crossed position. If this is not the case the section intersects all three of the crystallographic axes.

In general, if orthorhombic minerals exhibit pleochroism, all sections are dichroic which are not at right angles to an axis. The greatest differences of color lie at 90° to one another, and coincide with the positions of darkness of the sections between crossed nicols.

Brookite.

Literature.

H. Thürach, Über das Vorkommen mikroskopischer Zirkone und Titanmineralien in den Gesteinen. Würzburg. 1884. 36-41.

Brookite appears in very small tabular crystals flattened parallel to

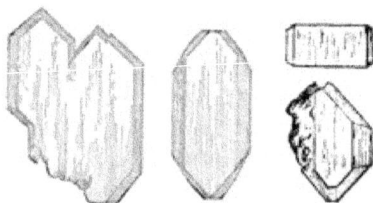

Fig. 61

$\infty P \breve{\infty}$ (100), which are bounded most frequently by ∞P (110) and $P\breve{2}$ (122) besides oP (001) and $2P\breve{\infty}$ (021), more rarely also by $mP\breve{\infty}$ and $\infty P\breve{\infty}$ (010). The tabular face is striated parallel to the vertical axis; the crystals are often greatly distorted, and also combined to form twin-like or irregular groups. Fig. 61 presents some small brookite crystals, according to Thürach.

The cleavage parallel to $\infty P\breve{\infty}$ (010) is not noticeable on the microscopic crystals. Brookite is transparent, with yellow to brownish-red color, according to the thickness of the plates; by incident light it exhibits a strong adamantine lustre, somewhat metallic, and often an ashen-gray color. It is seldom blue or greenish blue. The high index of refraction causes strong total reflections from the faces inclined to the axis of the microscope, and strong relief as well as a rough surface in Canada balsam. The double refraction is strong, the interference colors therefore are high, even in very thin plates. The bisectrix for all colors stands perpendicular to the tabular face, but the plane of the optic axes for red and yellow light is parallel to oP (001), while for more strongly refrangible rays it is parallel to $\infty P\bar{\infty}$ (010). This fact, which may be easily observed in convergent light on every plate by employing colored glasses, is one of the surest means of recognizing this mineral. The character of the double refraction is positive, consequently $a = c$. The pleochroism is weak for rays vibrating parallel to a and c; the absorption is stronger for the first than for the second.

The sp. gr. $= 3.8$–4.15. Chemical composition, TiO_2; reactions the same as those of rutile and anatase.

Brookite possesses no characteristic interpositions; in general it appears to be free from inclusions.

Brookite has not yet been found in fresh eruptive rocks and crystalline schists. Thürach observed it especially in decomposed granites, gneisses, and quartz porphyries; it frequently occurs together with anatase. He also discovered it in many sedimentary rocks, partly alone, partly in company with anatase.

Pseudobrookite.

Literature.

A. Koch, Pseudobrookit, ein neues Mineral. T. M. P. M. 1878. I. 344–350.

Pseudobrookite forms tabular crystals, mostly with rectangular outlines, whose longest dimension is less than 2 mm., and is usually below 1 mm. The boundary according to Koch (Fig. 62) is generally given by $a = \infty P\bar{\infty}$ (100), $m = \infty P$ (110), $b = \infty P\bar{\infty}$ (010), $d = P\bar{\infty}$ (101), with which are often associated a brachyprism $\infty P\breve{2}$ (120) and a derived macrodome $e = \frac{1}{3}P\bar{\infty}$ (103), as well as very small pyramids. The most important angles are $a:m = 154°\ 9'$, and $a:d = 138°\ 41'$. The plates are often only partly bounded by crystal faces, and are always vertically striated in the prism zone. Cleavage parallel to $\infty P\bar{\infty}$ (010) distinct; Törnebohm ob-

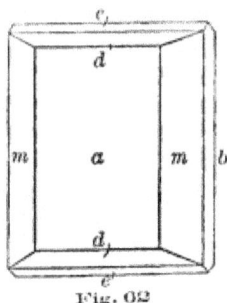

Fig. 62

served another parallel to a brachydome, whose traces on *a* intersect
one another at 60°

Crystals of some thickness are opaque, dark brown to black, and
possess a metallic adamantine lustre; very thin plates are transparent,
brownish, or ruby red, strongly refracting with moderate double refraction. Nothing is known concerning the position and angle of the
optic axes. The pleochroism is weak; the ray vibrating parallel to *c*
is most strongly absorbed.

Sp. gr. = 4.98. The chemical composition is not known exactly;
an incomplete analysis gave Koch 52.74 TiO_2, 42.29 Fe_2O_3 with some
Al_2O_3; the residue was lime and magnesia. The powder is soluble in
concentrated hydrochloric or sulphuric acid after long heating. The
mineral is difficultly fusible, yields an iron bead with borax, and a
titanium bead with salts of phosphorus.

Pseudobrookite has been found in cracks and hollows in the
andesite of Aranyer Berg in Siebenbürgen, Transylvania, in the apatite
of Jumilla, Spain, in the domite-like trachytes of Fayal and San
Miguel, in an augite andesite of Behring's Island, in the basalt débris
of Kreuzberg, in a basalt tuff and in the phonolite débris of Käuling.
It is very abundant in a Central American amphibole andesite of Mira-
valles, Costa Rica.

Aragonite.

Aragonite never forms crystals in rocks, but masses or prismatic
aggregates with parallel, divergent, or radial arrangement of the individuals.

The cleavage parallel to the brachipinacoid is not at all or but
slightly noticeable.

Aragonite is transparent and colorless; the double refraction is
very strong, $\alpha_{na} = 1.5301$, $\beta_{na} = 1.6816$, $\gamma_{na} = 1.6859$. The optic axes
lie in the macropinacoid, and *c* is the negative bisectrix, $2V = 18°$.
Prisms cut at right angles to their axis give the interference figure of
an orthorhombic crystal with $\rho < v$. Pleochroism not noticeable.

Sp. gr. = 2.94. Chemical composition = CaO, CO_2. Reactions the
same as for calcite, from which it is easily distinguished by the absence
of cleavage and the specific gravity. It occurs as a decomposition
product in the basic eruptive rocks.

Anhydrite.

Anhydrite forms granular or columnar fibrous aggregates. The
grains appear to be quite regularly developed in three directions. In

them twin lamellæ often lie diagonal to two cleavages, and are pressure phenomena.

Cleavage parallel to three pinacoids is distinctly shown microscopically by cracks of different degrees of sharpness and of varying frequency. Most perfect cleavage parallel to oP (001), nearly as perfect parallel to $\infty \breve{P}\infty$ (010), less perfect parallel to $\infty P\infty$ (100).

In thin section transparent and colorless. Miller determined $\alpha = 1.571$, $\beta = 1.576$, $\gamma = 1.614$. The axial plane lies in $\infty P\breve{\infty}$; the acute bisectrix coincides with $\breve{a}$. The character of the double refraction is positive. Cleavage plates parallel to the most imperfect cleavage exhibit the axial figure very finely. $2E = 71°\ 10'–71°\ 20'$.

Sp. gr. $= 2.8$–3. Chemical composition $= CaO, SO_3$. But slightly attacked by acids; together with fluorite it melts to a clear bead. Anhydrite changes into gypsum by taking up water under the action of the atmosphere. It forms anhydrite rock, which has been investigated microscopically by Fr. Hammerschmidt (T. M. P. M. 1882, V. 245). It also occurs in gypsum.

Andalusite.

Literature.

H. ROSENBUSCH, Die Steiger Schiefer und ihre Contactzone an den Granititen von Barr-Andlau und Hohwald. Strassburg i. E. 1877. (cf. also N. J. B. 1875. 849.)

Andalusite never occurs massive, but always in more or less well-defined crystals; seldom in rounded grains. The dimensions vary from a length of several centimetres to hundredths of millimetres, but the relation of length to breadth is quite constant—about $1:3$ to $1:4$. The forms of imbedded crystals are very simple, ∞P (110) with an angle of $90°\ 50'$, and oP (001), to which a dome is occasionally added. Hence cross-sections are very nearly quadratic, longitudinal sections elongated rectangles (Pl. XVII. Fig. 2). Twinning does not seem to occur.

The quite perfect cleavage parallel to ∞P (110) shows itself in the cross-sections of the larger individuals, generally in very distinct cracks, which cut each other apparently at right angles and lie parallel to the crystallographic boundaries; in longitudinal sections they run parallel to each other and to the boundaries of the crystal. The cleavages parallel to the dome and to the vertical pinacoid are seldom noticeable. In the very small microscopic individuals the cleavages are frequently not noticeable at all. In the larger individuals of andalusite and chiastolite one may often observe that an apparent twinning line passes diagonally through the cross-section. This is due

13

to mechanical deformation, the face $\infty P \bar\infty$ (100), and never the other pinacoid, having acted as a gliding plane, and the halves of the crystal have been pushed out of an exact parallel position by an extremely thin wedge of the rock mass.

Andalusite is usually colorless, seldom transparent reddish. The index of refraction is quite high, hence the distinct relief; the double refraction is weak, and the interference colors consequently low. The thin section must be at least 0.05 mm. thick to give red of the 1st order under the most favorable circumstances. This is an important means of distinction from sillimanite. The character of the double refraction is negative. The plane of the optic axes lies in $\infty P \bar\infty$ (010); the vertical axis is the bisectrix; thus, $a=c$, $b=b$, $c=a$. Fig. 63 gives the optical scheme in crystallographic orientation. $2 V = 83°-85°$; $\alpha = 1.632, \beta = 1.638, \gamma = 1.643$. Cross-sections show the point of egress of a bisectrix, sections parallel to $\infty P \bar\infty$ (010) give the highest interference colors. The extinction lies diagonally in cross-sections, and parallel to the cleavage and crystal boundaries in longitudinal sections. In unsymmetrical sections a considerable extinction angle may show itself because of the large optic angle. Pleochroism is sometimes completely wanting, while in other cases it is very strong; in rather thick sections $\mathfrak{a}$ = olive-green, $\mathfrak{b}$ = oil-green, $\mathfrak{c}$ = dark blood-red; in thin sections $\mathfrak{a} = \mathfrak{b}$ = quite colorless to very pale greenish, $\mathfrak{c}$ = rose-red. The absorption is $c > b > a$. Sections parallel to a dome and perpendicular to an optic axis exhibit distinctly the polarization brush.

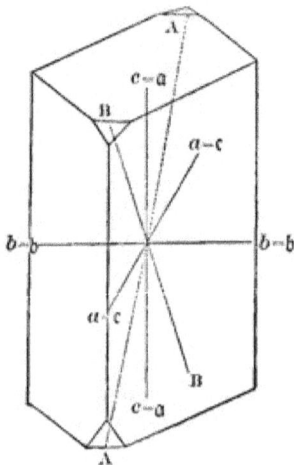

Sp. gr. = 3.16–3.20. Chemical composition = Al_2O_3, SiO_2. Andalusite is not attacked by acids, even by hydrofluoric, and is therefore easily isolated chemically; its isolation by means of separating solutions also is not difficult. The isolated powder when heated to redness with cobalt solution is colored a fine blue. Andalusite is readily altered through the action of the atmosphere into laminated and fibrous aggregates, which may belong in part to muscovite (sericite), in part to kaolin.

Andalusite possesses no constant microstructure: sometimes it encloses the minerals associated with it, such as quartz, biotite, and the ores, as well as fluid inclusions; very frequently carbonaceous particles, about which pleochroic halos often appear. These are yellow when the vertical axis stands parallel to the principal section of the nicol, and disappear upon a rotation of 90°. Glowing destroys them; they may therefore be due to finely divided organic pigments, which also appear to produce the pleochroism of the andalusite.

Andalusite is highly characteristic of metamorphic schists ; its true home is the contact zones of clay slates near granites, syenites, elæolite syenites, and diorites. It occurs far more rarely in the mica schists and gneisses of the Archæan, where also it is probably of metamorphic origin. According to E. Cohen (N. J. B. 1887. B. II.) andalusite is not an uncommon accessory constituent of normal granites, where it occurs in the form of columns, either acicular or with a more compact habit, which are always isolated and not grouped like those in contact zones and in the crystalline schists. The terminations are often incomplete, rounded or jagged ; sometimes they are terminated by two inclined lines, probably derived from a dome or pyramid.

Chiastolite (Pl. XVII. Fig. 3) is chemically, crystallographically, and physically identical with andalusite, and is only distinguished from it by the constancy with which carbonaceous substances are enclosed in it, arranged in the well-known manner.

Andalusite may be confused with diopside upon hasty observation on account of the similar cleavage, with sillimanite (in granulite) because of the same chemical composition, with zoisite and feldspar from a certain similarity in habit and in the decomposition products. Diopside is easily distinguished by its much stronger double refraction and the monoclinic behavior of all sections not lying in the orthodiagonal zone. Andalusite may be distinguished from sillimanite by determining the value of the vertical axis of elasticity, from zoisite by the cleavage and chemical reaction, from feldspar by the noticeably higher index of refraction, as well as by the chemical reaction and specific gravity.

Sillimanite.

Literature.

E. KALKOWSKY, Die Gneissformation des Eulengebirges. Leipzig. 1878. p. 5 sqq.
A. MICHEL-LÉVY, Sillimanite dans le gneiss du Morvan. Bull. Soc. min. Fr. 1880. III. 30.

Sillimanite as a constituent of rocks always forms long prismatic crystals, which are very thin, and are only recognizable macroscopically when they are grouped in felt-like aggregates. The dimensions of the crystals vary greatly, but the length is always greatly in excess of the breadth ; they sink to such fine needles that they are scarcely transparent even with the strongest magnifying powers. In the prism zone the boundary is given by crystal faces, either by the fundamental prism with the angle 110 : 110 = 111°, or by a combination of this with the prism $\infty P\frac{3}{2}$ (230) with an anterior angle of 88°–89°, or by the latter alone. The pinacoidal faces $\infty P\bar{\infty}$ (010) and $\infty P\bar{\infty}$ (100) are comparatively rare. Terminal faces are not definitely recognizable; the crystals apparently break off or are very finely pointed. The cross-sections are rhombic, octagonal, or apparently quadratic, and then suggest andalusite very strongly; the longitudinal sections are long lath-shaped. The individuals, however, are usually so thin that they lie in the thin sections as whole bodies (Pl. XVII. Fig. 4). Their surface is often striated parallel to the vertical axis, and their cross-sections rounded and notched.

The cleavage parallel to the macropinacoid shows itself in very fine parallel cracks in both longitudinal and cross sections of the larger individuals, but is not noticeable in the very microscopic individuals. All individuals which are not too short exhibit a transverse parting, the segments being sometimes separated by the rock mass (mostly quartz). Sillimanite needles never occur bent or curved, but are frequently broken. H. = 6–7.

Sillimanite in thin section is transparent and colorless; the index of refraction is somewhat higher than for andalusite, $\beta_\rho = 1.660$, according to Des Cloizeaux; the double refraction is considerable, $\gamma - \alpha = 0.020$–0.022 according to Michel-Lévy; and the interference colors in thin section are higher, from the upper half of the 1st order and the lower half of the 2d order. The plane of the optic axes lies in the macropinacoid, the vertical axis is the positive bisectrix, thus $c = \mathfrak{c}$, which is a certain and convenient means of distinguishing it from andalusite. The angle of the optic axes is small, $2E = 44°$. Cross-sections in thin section exhibit a distinct axial figure. Sillimanite from Saybrook, Conn., is strongly pleochroic; cleavage plates parallel to $\infty P\bar{\infty}$ (100) give for $\mathfrak{c}$ dark clove-brown, for $\mathfrak{b}$ light brownish; rock-making sillimanite is not noticeably pleochroic in thin section.

Sp. gr., = 3.23–3.24, is higher than for andalusite. Chemical composition and reaction are the same as for andalusite. Sillimanite is usually perfectly free from inclusions. Decompositions lead to kaolin.

Sillimanite is one of the most characteristic minerals of the crystalline schists, especially of the feldspathic gneisses, in which it is sometimes distributed generally through all the constituents, with the exception of the feldspars, at others it is intimately combined with fibrous quartz in the form of lenticular knots, called fibrolite, bucholzite, wörthite, monrolite, xenolite. It is frequently accompanied by cordierite in cordierite gneisses and kinzigites. It is also found to some extent in rocks exhibiting contact metamorphism.

Sillimanite may be confounded microscopically with andalusite and zoisite. For its distinction from the first, see *Andalusite.* From zoisite it is distinguished by its strong double refraction and its chemical resistance.

Topaz.

As a rock constituent *topaz* has been observed almost always in crystals, very rarely in irregular grains or masses; the crystals have a short prismatic habit, and are bounded principally by the faces ∞P (110) and $2P\breve{\infty}$ (021) with 92° 42', or $4P\breve{\infty}$ (041) with 55° 20'. In addition the pyramids and the prism $\infty P\breve{2}$ (120) are generally very subordinate. The base is usually wanting, or is very slightly developed. The crystals, sometimes blue, sometimes colorless or light yellow, seldom attain macroscopic dimensions, and are usually first recognized microscopically.

The perfect cleavage parallel to oP (001) shows itself by distinct parallel cracks in all sections which are not parallel to the base. H. = 8.

Topaz is always perfectly transparent in thin section; the index of refraction in consequence of its fluorine percentage is lower than is to be expected for a gem; the double refraction is weak, about the same as that of quartz. Hence the relief is not strong, the interference colors quite low; in good thin sections they scarcely exceed yellow of the 1st order. On colorless Brazilian topaz Rudberg determined $\alpha_{na} = 1.612$, $\beta_{na} = 1.614$, $\gamma_{na} = 1.621$; on Schneckenstein topaz, Des Cloizeaux found $\alpha_\rho = 1.614$, $\beta_\rho = 1.616$, $\gamma_\rho = 1.623$.

The axial plane lies in the brachypinacoid, and the vertical axis is the acute bisectrix. The character of the double refraction is positive: thus, $c = \mathfrak{c}$, $a = \mathfrak{a}$, $b = \mathfrak{b}$. The axial angle varies between wide limits, even in plates from the same crystal—according to Des Cloizeaux from about 70° to 120° in air. The dispersion is quite strong, $\rho > \upsilon$. In thin sections the interference figure of both axes is obtained on those sections which show no cleavage—an important criterion for diagnosis. Optical anomalies are frequent, especially in the yellow Brazilian topaz, where they were first observed by Brewster;[*] but in rock-making topaz they are weaker and less frequent. Pleochroism is not noticeable in thin section.

Sp. gr. = 3.52–3.56. Chemical composition = $5Al_2O_3$, SiO_2, + Al_2Fl_6, $SiFl_4$. Acids have no action upon topaz. Through decomposition topaz loses its fluorine, and by taking up water passes into kaolin

---

[*] Trans. Cambridge Phil. Soc. 1822.

(nakrite), or by the addition of water and alkali passes into muscovite. The latter process has been studied by **J. S. Diller** and **F. W. Clarke*** in the topaz of Stoneham, Me.; the formation of mica advances along the cleavage planes and other cracks.

Topaz usually encloses besides plates of hematite and ilmenite abundant fluid inclusions, which sometimes lie in lines and rows, and at others are arranged approximately along concentric faces parallel to

Fig. 64

∞P (110) and $\infty P \tilde{2}$ (120). Their form is very variable and striking: they sometimes consist of water and aqueous solutions; sometimes of liquid carbon dioxide, or of both together. In the fluid inclusions of many topazes are crystalline secretions (Fig. 64), among which colorless cubical crystals are the most frequent. These dissolve in their mother-liquor when sufficiently heated, and crystallize out again upon cooling. Hence they can scarcely be referred to rock-salt. Less frequently there are rhombohedral colorless crystals, long needle-shaped microlites generally crossing one another as if twinned, and reddish-brown pyramidal crystals with a truncated point; these crystallizations are not dissolved in the fluid upon heating. Topaz is peculiar to all granitic rocks which carry tin ore, and is particularly constant in the greisen. It is also sparingly met with in granitic rocks, especially when they bear fluorite or tourmaline, as, for example, many Cornwall occurrences.

Staurolite.

Literature.

A. von Lasaulx, Ueber Staurolith. T. M. M. 1872. II. 173.

K. Peters und R. Maly, Ueber den Staurolith von St. Radegund. S. W. A. 1868. LVII. 646.

Staurolite always appears as single individuals or twins, which occasionally assume the form of grains through the imperfect development of the crystal faces. The absence of elongated forms is characteristic. The forms (Fig. 65) are very constant, $m = \infty P$ (110) with

---

* Amer. Journ. 1885. XXIX. May. 378–384.

$129°\ 20'$, $p = oP$ (001), $r = P\overline{\infty}$ (101), mostly very small, often wanting, $o = \infty P\overline{\infty}$ (010), often very small, and at times absent.   Hence
the cross-sections are acutely rhombic or almost hexagonal, the
longitudinal sections broad rectangles (parallel to $\infty P\overline{\infty}$) or narrow
ones (parallel to $\infty P\infty$).   The twinning parallel to
$\frac{3}{2}P\overline{\infty}$ (032) and $\frac{3}{2}P\frac{3}{2}$ (232) is the same for microscopic crystals as it is for the large individuals.
Often the twinning is not noticeable in the outline
of the crystal, as one individual may be wholly
enclosed in the other, but is recognized optically
by the position of the axial planes or by the pleochroism.   A laminated structure occurs parallel
to oP, producing a parting parallel to this face
which resembles a cleavage.

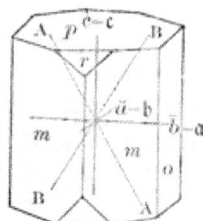

Fig. 65

The cleavage parallel to $\infty P\overline{\infty}$ (010) and ∞P (110) is variable, showing itself at times in sharp cracks, especially in the short diagonal of
the cross-section, and by parallel cracks in the longitudinal sections;
at times it is scarcely noticeable.   H. $= 7$–7.5.

Staurolite becomes transparent and yellowish to reddish brown
according to the thickness of the section and its position.   The index
of refraction is very high—according to Miller $\beta_\rho = 1.7526$; Des
Cloizeaux, $\beta_\rho = 1.749$; therefore the marginal total reflection is very
strong, the relief considerable, the surface very rough.   The double
refraction is strong; the interference colors are brilliant even in very
thin sections.   The axial plane lies in the macropinacoid, the angle for
red rays is about $89°$, the character is positive, the vertical axis is the
first bisectrix; the optical scheme is indicated in Fig. 65.   Cross-sections, even in quite inclined positions, give interference figures in convergent light, which show that the axial plane lies in the longer diagonal of the prismatic cleavage; this is the best means of distinguishing
it from titanite, which often resembles it closely.   The dispersion is
weak, $\rho > v$.   The pleochroism is distinct, though not strong: $\acute{c}$ hya-
cinth-red to blood-red; $\acute{a}$ and $\overline{b}$ yellowish red, often with a tinge of
green.   This pleochroism is often noticeably stronger around interpositions than in the main mass of the mineral, although there is no
difference in the intensity of the coloring in ordinary light.

Sp. gr. $= 3.4$–3.8, varying greatly with the amount of the many
kinds of interpositions; it is higher as the mineral is purer.   Chemical
composition not known with certainty, approximately represented by
the formula $FeO, 2Al_2O_3, 2SiO_2$, in which a small part of the FeO is
replaced by MgO.   Is not acted on by acids including hydrofluoric.

It is therefore easily separated from the rocks by chemical methods, and from the isolated minerals accompanying it by means of its specific gravity and by an electro-magnet.

The larger crystals of staurolite are made very impure by inclusions of the minerals associated with them (tourmaline, rutile, mica, disthene, etc.), but they are especially impregnated with quartz grains carrying rings of dark interpositions of carbonaceous matter. The microscopic individuals, on the other hand, are usually much purer and often completely free from admixtures. Decomposition products are rare, and generally occur only along the cracks in the larger crystals; they appear to be chlorite and a green mica.

Staurolite does not occur in eruptive rocks; it is particularly a mineral of the Archæan rocks, and is very frequently accompanied by disthene. It is very common in gneiss and mica schists, but does not occur in schists rich in amphibole.

The Group of Orthorhombic Pyroxenes.

Literature.

F. Becke, Ueber die Unterscheidung von Augit und Bronzit in Dünnschliffen. T. M. P. M. 1883. V. 527.

J. Blaas, Petrographische Studien an jüngeren Eruptivgesteinen Persiens. T. M P. M. 1880. III. 479 sqq.

H. Bücking, Bronzit vom Ultenthal, Z. X. 1883. VII. 502.

F. Fouqué, Sur l'hypersthène de la ponce de Santorin. Bull. Soc. min. Fr. 1878. III. 46.

J. A. Krenner, Ueber den Szaboit. Z. X. 1884. IX. 255.

H. Rosenbusch, Die Gesteinsarten von Ekersund. Nyt Mag. for Naturvid. XXVII. 1883.

—   Ueber den Sagvandit. N. J. B. 1884. I. 195.

F. Svenonius, Bronzit från Frostvikens socken i Jämtland. Geol. För. i Stockholm Förhdlg. 1883. VI. 204.

G. Tschermak, Mikroskopische Unterscheidung der Mineralien der Augit-, Amphibol- und Biotitgruppe. S. W. A. 1869. LIX. 1. Abthl.

—   Ueber Pyroxen und Amphibol. T. M. M. 1871. I. 17—21.

B. Weigand, Die Serpentine der Vogesen. T. M. M. 1875. 183.

The *orthorhombic pyroxenes* occur in two ways in rocks; they either form short prismatic crystals, perfectly developed, of small or microscopic dimensions in certain porphyritic eruptive rocks; or they appear in lamellar crystalloids and aggregates, often of very considerable dimensions, seldom microscopic, which occur in certain granular eruptive rocks of the oldest geological epochs, as well as in many members of the crystalline schists. The prismatic crystals (Figs. 66

and 67) when referred to the axes $\bar{a}:\bar{b}:\bar{c} = 0.97133 : 1 : 0.57000$ (that is, with the obtuse prism angle in front) show $a = \infty \, P \bar{\infty} \, (100)$, $b = \infty P \bar{\infty} \, (010)$ predominant, and $m = \infty P \, (110)$ subordinate; also $e = P\bar{2} \, (212)$, $i = 2P\bar{2} \, (211)$, $o = P \, (111)$, $k = \frac{1}{2}P\bar{\infty} \, (012)$. The value of the prism angle is about 92°. In sections parallel to $\infty P\bar{\infty} \, (100)$ the terminal edges of $P\bar{2} \, (212)$ intersect at 148° 11′, those of $2P\bar{2} \, (211)$ at 120° 38′; in sections parallel to $\infty P\bar{\infty} \, (010)$ these terminal edges intersect at 119° 11′ and 80° 52′ respectively. Cross-sections through the crystals present rectangles with truncated corners. Sections from the vertical zone present long strips pointed at both ends. Sections of irregular masses naturally exhibit no regular outline and must be oriented by the cleavage. Twinning is quite rare; the pyroxene crystals in the porphyrites and andesites are sometimes intergrown in such a manner that the individuals have the faces $\infty P \bar{\infty} \, (010)$ in common, and appear twinned after a macrodome. From the inclination of

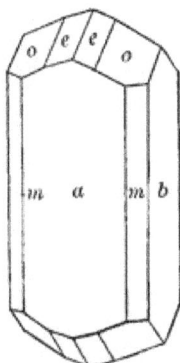

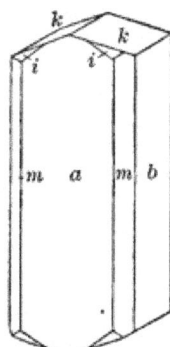

Fig. 66   Fig. 67

the vertical axis, $P\bar{\infty} \, (101)$ may be considered as the probable twinning plane; but there appear to be other faces in the same zone which occasionally act as twinning planes. In massive bronzite of the norites, peridotites, and crystalline schists a twinning parallel to $\frac{1}{4}P\bar{\infty} \, (014)$ is not at all uncommon. This last twinning, however, appears to be secondary, arising from mechanical causes, as is indicated by the "jogging" phenomena in the twinned individuals in the immediate neighborhood of the composition plane.

The cleavage in all orthorhombic pyroxenes lies parallel to the prism of about 92°; the perfection of this cleavage varies, but it is always noticeable in convergent light upon careful observation. In cross-sections of the crystals the cracks corresponding to this cleavage run parallel to the small faces which truncate the pinacoidal edges. In the massive varieties in the older rocks, besides the prismatic cleavage there is always a more perfect one parallel to $\infty P\breve{\infty}$ (010); they also show cracks which indicate an imperfect parting parallel to $\infty P\bar{\infty}$ (100). This last cleavage is seldom found in the crystals occurring in the porphyritic rocks and lavas. Cross-sections of the massive forms, therefore, are traversed by a double system of cleavage cracks apparently intersecting at right angles, and bisecting each other's angles (Pl. X. Fig. 2). In sections from the prism zone all the cleavage cracks run parallel to the vertical axis. It is not unlikely that the pinacoidal partings in some cases correspond to gliding planes, and are the result of mountain pressure, while in others they are brought about by the inclusion of foreign substances. Furthermore, an irregular cracking, approximately perpendicular to the vertical axis, is observed both in the crystals and lamellar masses, which, in spite of its general distribution, does not correspond to a cleavage. The latter cracks play a great rôle in the decomposition of these minerals.

The orthorhombic pyroxenes become transparent in various colors, according to the position of the section and to the iron percentage. Enstatite is almost colorless to grayish or yellowish white; bronzite is yellowish to greenish; hypersthene green, light red, or brownish red. The index of refraction is high, and appears to increase with the iron percentage; hence the marginal total reflection is strong, the surface distinctly rough.

Bronzite from Kupferberg,	$\beta = 1.668$ (Des Cloizeaux).
Hypersthene from Lauterbach,	$\beta = 1.685$ (Des Cloizeaux).
Hypersthene from Soggendal,	$\beta = 1.7125$ (Sanger).
Hypersthene from St. Paul's Island,	$\gamma = 1.7270$ (J. E. Wolff);
	$\alpha = 1.7158$ (J. E. Wolff).

The double refraction is weak for members of the series poor in iron: Michel-Lévy determined on bronzite from Lherz, Pyrenees, $\gamma - \alpha = 0.010$; on pale brown hypersthene from Arvien, $\gamma - \alpha = 0.0115$; from the figures given above for hypersthene from St. Paul's Island, $\gamma - \alpha = 0.0112$. Hence the interference colors are low for enstatite and bronzite, not exceeding yellow of the 1st order; for hypersthene they are noticeably higher, reaching red of the 1st order in sections which are not too thin.

The direction of extinction in the principal zones lies parallel to the pinacoidal cleavages and diagonal to the prismatic. The vertical axis is the axis of least elasticity, the brachydiagonal is that of greatest elasticity; thus, $\overset{\iota}{c} = \mathfrak{c}$, $\breve{a} = \mathfrak{a}$, $\bar{b} = \mathfrak{b}$. Consequently the plane of the optic axis always lies in the brachypinacoid. The angle between the optic axes varies considerably, chiefly with the iron percentage: for enstatites and bronzites the vertical axis is the acute bisectrix—they are optically positive; for hypersthenes $\breve{a}$ is the acute bisectrix—they are negative. On account of the high index of refraction the observations must be made in oil. The following table shows clearly the decrease in the negative axial angle with increasing iron percentage.

In Oil.	FeO+MnO.		Locality.	
133° 8′	2.76%	Enstatite	Mähren	Des Cloizeaux.
123° 38′	5.77	Bronzite	Leiperville	"
114° 15′	11.14	"	Greenland	"
112° 30′	8.42	"	Balsfjord	Rosenbusch.
106° 51′	9.86	"	Kraubat	Tschermak.
101° 30′	10.62	"	Lauterbach	Des Cloizeaux.
98°	13.58	Meteorite	Breiterbach	v. Lang.
98° 22′	15.14	Hypersthene	Farsund	Des Cloizeaux.
85° 39′	22.59	"	Labrador	"
59° 20′	33.6	"	Mont Dore	Krenner.

The dispersion about the negative bisectrix is quite weak for those members of the series poor in iron; stronger, $\rho > v$, for those rich in iron.

The pleochroism changes with the iron percentage. It is scarcely

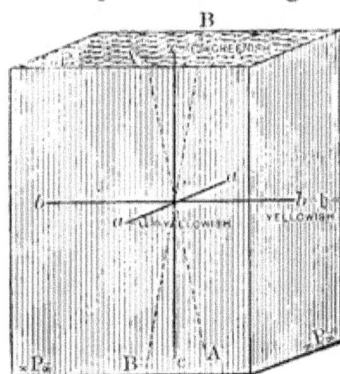
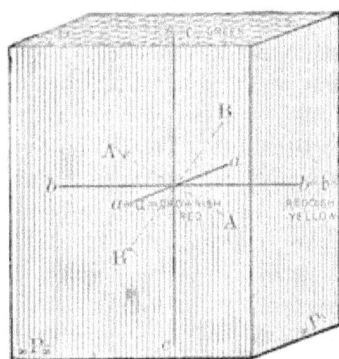

Fig. 68   Fig. 69

or not at all noticeable in the enstatites and bronzites poor in iron; in the more ferruginous bronzites the color of the rays vibrating parallel to $\overset{\iota}{c}$ is pale grayish green; those parallel to $\breve{a}$ and $\bar{b}$ are quite uniformly pale yellow to pale grayish yellow. For the hypersthenes $\breve{a}$ is reddish brown, $\bar{b}$ is reddish yellow, $\overset{\iota}{c}$ is green; the absorption is slightly pronounced. The pleochroism diminishes rapidly as the thickness of the section decreases. The axial plane always lies in the plane of the

green and brownish-red rays ; hence the interference figure is found in
the less pleochroic sections, and the axial bar then lies parallel to the
perfect pinacoidal cleavage.   Figs. 68 and 69 present graphically the
phenomena of cleavage and optical orientation just described.   As a
means of distinction from the monoclinic pyroxenes it is to be noted
that plates parallel to the most perfect pinacoidal cleavage yield no interference figure (bastite gives an acute bisectrix, diallage an axis emerging to one side of the centre of the field of view), and that the interference colors are considerably lower than in most of the monoclinic
pyroxenes.   Cross-sections giving rectangular prismatic cleavage show
in the case of orthorhombic pyroxenes the point of emergence of a
bisectrix, in that of monoclinic pyroxenes an axis.   Sections in the
macrodiagonal zone exhibit an axial bar in convergent light, which is
parallel to the good pinacoidal cleavage in the orthorhombic pyroxenes
and perpendicular to it in diallage.   Isolated crystals and cleavage
pieces show on all faces in the cleavage zone parallel extinction for the
orthorhombic pyroxenes, but partly parallel and partly inclined extinction for the monoclinic pyroxenes.   The hardness is about 5.5 for enstatite and bronzite, and 6 for hypersthene.

The sp. gr. increases with the iron percentage from 3.1 for enstatite
to 3.5 for hypersthene.   The orthorhombic pyroxenes are isomorphous
mixtures of MgO, SiO_2 and FeO, SiO_2, to which may be added inconsiderable amounts of MnO, SiO_2, CaO, SiO_2, and MgO, Al_2O_3, SiO_2.   The
mixtures in which the percentage of FeO does not exceed 5% are called
enstatite ; those with as high as 14% FeO, bronzite ; those higher in iron,
hypersthene.   The limits are entirely arbitrary, the optical character of
hypersthene commences for mixtures with about 10% FeO.   The orthorhombic pyroxenes in general are not attacked by acids, and by hydrofluoric acid with difficulty.   With hydrofluosilicic acid they yield abundant rhombohedral crystals of magnesium fluosilicate and iron fluosilicate, which are strongly refracting and strongly doubly refracting.   The
mixtures poor in iron are very difficultly fusible, hypersthene less so.

The massive *enstatites* and *bronzites*, which seldom exhibit distinct
crystal boundaries, occur in the norites, gabbros, granular peridotites
and the serpentines derived therefrom, and in the olivine aggregations
of basaltic rocks.   They are also found in the olivine-bearing members
of the Archæan and the resulting serpentine rocks.   Occasionally they
form independent rock masses in the Archæan, or are associated
with magnesium carbonates and chromite in peculiar deposits.   A
columnar structure is highly characteristic of enstatite and bronzite,
though not constant; it seems to be due to the growing together of

innumerable thin prisms to form large crystalloids (Pl. XVII. Fig. 5). These minute prisms are not always completely in contact throughout their length, but leave long cylindrical hollows which are often filled with secondary iron ores. In longitudinal sections it is not always easy to distinguish the long cavities from solid inclusions, as they appear dark on account of the high index of refraction of their matrix. Moreover the ferruginous members of the bronzite series carry the same metallic to sub-metallic scales and particles which are so characteristic of massive hypersthene; they also have the same arrangement as in the last-named mineral. Inclusions of magnetite, chromite, picotite, and other older secretions are frequent, but fluid inclusions are quite rare. A lamellar intergrowth of massive enstatite and bronzite with monoclinic pyroxene is very widespread, and is often first recognized in polarized light (Pl. XVII. Fig. 6). · The orthorhombic and monoclinic lamellæ are so placed with reference to one another that the face $\infty P \breve{\infty}$ (010) of the latter coincides with $\infty P \bar{\infty}$ (100) of the former, that is, the acute and obtuse prism angles have the same position in both minerals; $\infty P \breve{\infty}$ (010) or one face of ∞P (110) serves as the composition plane. Since the orthodiagonal zone of the monoclinic lamellæ coincides with the brachydiagonal zone of the orthorhombic ones, the extinction in sections in these zones is the same in both minerals, and hence their intergrowth is not noticeable in parallel polarized light. In other sections in the zone of the vertical axis the extinction is parallel to the cleavage in the orthorhombic lamellæ, and inclined to it in the monoclinic. The intergrowth sometimes extends to complete mutual penetration, and the lamellæ may sink to immeasurable thinness.

Enstatite and bronzite are found in well-developed crystals in many porphyritic rocks, always accompanied by monoclinic pyroxene, less frequently by hornblende and biotite; they also occur in the trachytes and andesites. The microstructure of these occurrences differs entirely from that of the massive forms. The parallel composition of small individuals is wanting, and with it the columnar structure and tubular hollows parallel to the vertical axis; the lamellar penetration with monoclinic augite is also wanting, although the parallel growth of distinct crystals of both kinds occurs, partly as lateral juxtaposition, partly as the surrounding of orthorhombic crystals by monoclinic. Microlitic scales also are absent, but glass inclusions either round or in the form of their host are frequent and characteristic.

The radially fibrous aggregation of bronzite and enstatite in the chondri of many meteorites is wholly unique.

Enstatite and bronzite are comparatively susceptible to the reagents occurring in nature. The massive occurrences alter to parallel fibrous aggregates, a process which sets in from the cross cracks and cleavage; the alteration product being bastite and serpentine, much more rarely talc. The bastite alteration is not uncommon in the crystals of the porphyritic and andesitic rocks (Pl. XVIII. Fig. 1); here the process continues until the silicate is broken up, and there results pseudomorphs of a mixture of carbonates with limonite and quartz.

Hypersthene forms poorly defined masses in the more basic members of the granular eruptive rocks (gabbro, norite), and appears in thin prismatic crystals in the porphyrites, trachytes, andesites, and lavas. It is to be remarked that it is entirely absent from the normal members of the Archæan; where it is met with (in trap granulites, labradorites, and amphibolites) there are generally grounds for considering the rocks as regionally metamorphosed eruptive masses. The massive hypersthene of the gabbros and norites occasionally possess the vertical fibration due to the parallel growth of thin prisms, as well as the lamellar intergrowth and penetration with monoclinic pyroxene. Hypersthene is also intergrown with hornblende lamellæ, so .that the faces $\infty P \bar{\infty}$ (100) of the latter mineral coincide with the faces $\infty \bar{P} \infty$ (010) of the former. Massive hypersthene encloses, besides the older minerals associated with it (magnetite, apatite, zircon, olivine, biotite), tabular microlitic interpositions, which are so frequent as to be almost constant. They lie in three different directions, with their tabular faces parallel to the principal cleavage face, $\infty \bar{P} \infty$ (010), of the hypersthene. The plates are approximately rhombic, almost rectangular or irregular, seldom perfectly straight-edged, grading into short prisms, and appear as very thin, opaque strips or points in all sections which are not parallel to the plane of the most perfect cleavage. Their color according to their thickness is dark brown to opaque, reddish brown, light brown, yellowish, grayish white to almost colorless. The thicker opaque plates have a metallic habit, and even the transparent ones have a submetallic habit in reflected light. These plates and prisms, to which hypersthene as well as bronzite and diallage owe the metallic sheen of their principal cleavage face, lie with their longer axis usually at right angles to the vertical axis, less frequently parallel to it or inclined at about 30° (Pl. VII. Fig. 5). In reflected light these thin plates exhibit the most brilliant Newton colors. Their exact nature is not known: they are considered primary interpositions by some, and referred to titanic iron and brookite; by others they are thought to be secondary infiltration prodncts. It seems probable that they are not always of the same nature, being undoubtedly primary in some instances and secondary in others. Thus Trippke[*] and Kosmann[†] considered them secondary infiltration products, and found them isotropic in diallage; they considered them opal, and referred their shape to that of their host. Judd[‡] has recently made a special study of these and similar inclusions in the pyroxenes and feldspars of the peridotites of Scotland, and has arrived at the conclusion that in these occurrences the orderly arranged inclusions are not definite chemical compounds, but are mixtures of various oxides in a more or less hydrated condition, such as hyalite, opal, göthite, and limonite. These have been deposited in negative crystal cavities, which they may fill completely or only partially: in the first case their boundaries correspond to the crystal form of the enclosing mineral; in the second they are irregular. The cavities have been formed along certain definite planes within the original crystal, which correspond to planes of least resistance to chemical action called *solution planes;* and the solvent has acted under the influence of great pressure, and therefore he concludes that this process of charging a mineral with definitely oriented inclusions, which he has termed *schillerization,* is a secondary process, which only takes place at considerable depths beneath the surface of the earth.

On the other hand, G. H. Williams[§] has called attention to the fact that similar inclusions exist in certain feldspars, hypersthenes, etc., under conditions which clearly indicate that in these particular instances they are primary bodies, which were formed contemporaneously with their hosts, and cannot be considered as the results of subsequent alteration, and that they may be distinguished from those of secondary origin in many cases, but not in all.

The hypersthene prisms of the porphyritic rocks have not the microstructure of the massive occurrences, but possess throughout the relations of the geologically equivalent bronzite crystals: however,

---

[*] P. TRIPPKE, Ueber den Enstatit aus den Olivinknollen des Gröditzberges. N. J. B. 1878. 673–681.

[†] B. KOSMANN, Ueber das Schillern und den Dichroismus des Hypersthens. N. J. B. 1869. 532.

[‡] J. W. JUDD, On the Tertiary and older Peridotites of Scotland. Quart. Journ. Geol. Soc. Aug. 1885.

— On the Relations between the Solution-planes of Crystals and those of Secondary Twinning, etc. Min. Mag. Vol. VII. pp. 81–92. 1886.

[§] G. H. WILLIAMS, Peridotites of the "Cortland Series" on the Hudson River, near Peekskill, N. Y. Am. Journ. Sci. Vol. XXXI. Jan. 1886.

— The Norites of the "Cortland Series," etc. Am. Journ. Sci. Vol. XXXIII. Feb. 1887 and March, 1887.

they sometimes enclose fine needles with a metallic habit, which are not found in the bronzites. Bronzite and hypersthene are widely disseminated in the trachytic and andesitic eruptive rocks.* Hypersthene is rather rare in the Archæan; when it occurs here it is in more or less well-defined crystals.

Hypersthene withstands decomposition much better than bronzite and enstatite do. It very seldom alters to bastite. Its alteration to limonite is not uncommon in the very ferruginous occurrences in andesites. The hypersthene of the gabbro rocks very often alters into amphibole (actinolite and ordinary hornblende). †

Appendix.—With the alteration of enstatite and bronzite to bastite, which may commence with the taking on of water, there appear very rapid changes in the ellipsoid of elasticity of these minerals, and those axes of elasticity coincident with the horizontal axes change places with one another. Hence the optical scheme becomes $\bar{c} = c,\ \bar{a} = b,\ \bar{b} = a$; the axial plane now stands perpendicular to the principal cleavage parallel to the brachypinacoid: it lies in the macropinacoid, and the macrodiagonal is the negative bisectrix. The axial angle is generally quite large, but varies considerably in cleavage plates taken from the same material. Diaclasite represents such a stage in the alteration of enstatite and bronzite to bastite. Cleavage plates of this mineral assume a

---

* J. Niedzwiedzki, Andesit von St. Egidi in Steiermark. T. M. M. 1872. 253.

J. Petersen, Mikroskop. u. chem. Untersuchungen am Enstatitporphyrit aus den Cheviot Hills. Kiel. 1884.

J. J. H. Teall, On the Cheviot Andesites and Porphyrites. Geol. Mag. (2) X. 1883. No. 225. 226. 228.

F. Teller and C. von John, Geologisch-petrographische Beiträge zur Kenntniss der dioritischen Gesteine von Klausen in Süd-Tyrol. Jahrb. d. k. k. geologischen Reichsanstalt. 1882. XXXII. 589-684.

Whitman Cross, On hypersthene andesite. Amer. Journ. 1883. XXV. No. 146. 139-144.

Arnold Hague and J. P. Iddings, Notes on the volcanic rocks of the Great Basin. Amer. Journ. 1884. XXVII. No. 162, and Notes on the volcanoes of Northern California, Oregon, and Washington Territory. ibidem. 1883. XXVI. September. 222-235.

H. H. Reusch, Vulkanische Asche von den letzten Ausbrüchen in der Sundastrasse. N. J. B. 1884. I. 45.

In H. Abich, Geologische Forschungen in den Kaukasusländern. II. Wien. 1882. 329-364.

† F. Becke, In olivine gabbro of Laugenlois, Lower Austria. T. M. P. M. IV. 1882. 355.

G. H. Williams, Preliminary notice of the Gabbros and associated Hornblende rocks in the vicinity of Baltimore. Johns Hopkins University circulars. 1884. No. 30; also Bull. 28, U. S. Geol. Survey. 1886. 42.

Arnold Hague and J. P. Iddings, Notes on the volcanic rocks of the Great Basin. Am. Journ. Sci. Vol. XXVII. June. 1884. 459.

metallic lustre with a brass-yellow color; the hardness and specific gravity decrease with the alteration. While diaclasite may be easily distinguished from enstatite and bronzite by the position of the optic axes, it must be distinguished from bastite by its specific gravity, which for bastite is 2.74, for diaclasite is over 2.8.

Bastite.

Literature.

R. von Drasche, Ueber Serpentine und serpentinähnliche Gesteine. T. M. M. 1871. I. 10–12.

E. Reusch, Ueber das Schillern gewisser Krystalle. Pogg. Ann. 1863. CXX. 115.

G. Tschermak, Ueber die mikroskopische Unterscheidung der Mineralien aus der Augit-, Amphibol- und Biotitgruppe. S. W. A. 1869. LIX. 1. Abthl. 1–12. —Ueber Pyroxen und Amphibol. T. M. M. 1871. I. 20.

C. E. Weiss, Beobachtungen und Untersuchungen über den Schillerspath von Todtmoos. Pogg. Ann. 1863. CXIX. 459.

Bastite or "schiller-spar" is always a pseudomorph after an orthorhombic pyroxene poor in iron. Hence it has no crystal form of its own, but always appears in that of enstatite and bronzite; consequently it is found massive in lamellar crystalloids with distinct vertical fibration in the granular massive rocks and their derivatives, and in prismatic crystals in the porphyritic or andesitic rocks which in a fresh condition bear bronzite and enstatite. The prisms also exhibit a very pronounced vertical fibration, the fibres extending from one transverse crack to another, and not being continuous throughout the entire length of the prism (Pl. XVIII. Fig. 1). In the alteration of massive bronzite to bastite the original microstructure is almost completely retained, but the glass inclusions in the porphyritic crystals are generally destroyed in the process of alteration. Along the transverse cracks, that is, between the different systems of bastite fibres, there are often deposited iron ores (magnetite, limonite), which may be derived from the iron contained in the primary mineral; in other cases these cracks are filled with confusedly fibrous serpentine.

In bastite the cleavage parallel to the brachypinacoid is much more perfect than in the original mineral, but the cleavage parallel to $\infty P'$ (110) is less distinct. The lustre on cleavage faces of the massive varieties is metallic, but more silky for the well-crystallized varieties. The hardness is less than for the original mineral, 3.5–4.

Bastite becomes transparent with a light-yellowish or light-greenish color, and, optically, behaves most uniformly. The extinction in longitudinal sections lies parallel and at right angles to the axis of the fibres. If the fibres do not lie exactly parallel, the optical behavior,

naturally, cannot be uniform.   The axial plane lies in the macropinacoid, and $\bar{b}$ is the negative bisectrix.   The size of the axial angle varies
within wide limits from 20° to nearly 90°; hence the cleavage plates
must be placed in oil to observe the axial figure.   Dispersion $\rho > v$.
The mean index of refraction is lower than for bronzite, about 1.5–1.6.

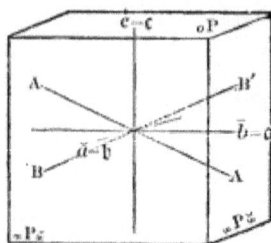

Fig. 70

The double refraction is weak.   The pleochroism is weak ; rays vibrating parallel to
the fibres are the most strongly absorbed,
but it requires rather thick sections to make
out the difference.   Its behavior in convergent light is the surest means of distinguishing it from the original mineral ; for
its diagnosis with reference to diaclasite,
see under the latter mineral.   The optical
scheme of bastite is given in Fig. 70.

Sp. gr. 2.6–2.8, considerably lower than that of enstatite, bronzite,
and diaclasite.   The chemical composition of pure bastite appears to
be the same as that of serpentine, H_2O, $3MgO$, $2SiO_2 +$ aq., with a
variable replacement of MgO by FeO ; small quantities of chromium
are referred to inclusions of chromite or picotite.   Upon being heated
to redness, bastite becomes cloudy and dirty grayish black to brownish.
It gelatinizes with difficulty with hydrochloric acid, easily with sulphuric acid, especially at a high temperature.   Bastite has no independent geological position ; it is always secondary, replacing enstatite
and bronzite.   It is not definitely known whether other pyroxenes, as
diallage, can be altered to bastite, but it seems quite probable.

The Group of Orthorhombic Amphiboles.

Literature.

H. Sjögren, Förekomsten af Gedrit såsom väsendtlig beståndsdel i några norska
och finska bergarter.   Kongl. Vetensk. Akad. Förhandl. Stockholm. 1882.
No. 10. 5–11.

G. Tschermak, Ueber Pyroxen und Amphibol.   T. M. M. 1871. I. 37.

The orthorhombic amphiboles which occur as rock constituents are
anthophyllite and *gedrite*.   Both appear in prismatic and lamellar
aggregations, which never exhibit terminal crystallographic boundaries,
but frequently those in the prism zone, $\infty P\bar{\infty}$ (100) and ∞P (110).
The prism angle lies between 124° and 125°.   Hence the sections
are irregular : those parallel to the prism axis are lath-shaped to broad
and tabular; those at right angles to it are often six-sided, or acutely

rhombic when $\infty P\bar{\infty}$ (100) is wanting. The composition of the larger masses of separate, thin, prismatic individuals produces a fine vertical striation quite analogous to that of bronzite and enstatite. In consequence of the separate individuals not growing together exactly parallel, there arises a more or less divergent arrangement which in some instances becomes almost radial.

The cleavage is very perfect parallel to $\infty P\bar{\infty}$ (100), perfect parallel to ∞P (110), and scattered but sharp cracks indicate an imperfect cleavage parallel to $\infty P\bar{\infty}$ (010). Moreover, there is a transverse parting approximately parallel to oP (001), which is never perfectly plain. In sections in the prism zone all the cleavage cracks are parallel; at right angles to this they form acute rhombs of 55° to 56°, with diagonal cross cracks. The orthorhombic amphiboles, according to their percentage of iron, become transparent and almost colorless to yellowish and reddish brown and yellowish green. The mean index of refraction has been determined by Des Cloizeaux, $\beta_\rho = 1.636$. The double refraction is very strong; hence the interference colors are high in sections which are not perpendicular to an optic axis: even in quite thin sections the colors are red, blue, and green, which distinguish the mineral very well from orthorhombic pyroxenes in longitudinal sections. The extinction in sections in the three principal zones naturally lies parallel and perpendicular or diagonal to the cleavage cracks. The plane of the optic axes lies in the brachypinacoid, and cleavage plates parallel to $\infty P\bar{\infty}$ (100) yield an axial figure in oil, with the bisectrix emerging normally. The optical scheme is $\breve{a} = \mathfrak{a}$, $\bar{b} = \mathfrak{b}$, $\dot{c} = \mathfrak{c}$. In some occurrences (anthophyllite) $\dot{c}$ is the acute bisectrix, and the character of the double refraction is positive, and $2H_a = 81°-82°$; in other occurrences, and especially, as it appears, for the aluminous varieties (gedrite), $\breve{a}$ is the acute bisectrix; the character is negative, and $2H_a$ varies from 47°–82°, even in cleavage pieces from the same occurrence. The dispersion is independent of the character of the double refraction, $\rho < \upsilon$ about $\dot{c}$, $\rho > \upsilon$ about $\breve{a}$; but it is sometimes quite weak, and difficult to determine. The pleochroism is dependent on the depth of the color; the rays vibrating parallel to the axis of the prism are light yellowish brown or greenish to colorless; those at right angles to it are clove-brown. Absorption $\mathfrak{c} < \mathfrak{a} = \mathfrak{b}$. The optical scheme is given in Fig. 71.

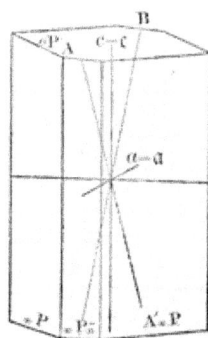

Fig. 71

H. $= 5.5$. Sp. gr. $= 3.15$–3.24. The chemical composition of an-thophyllite is $(Mg, Fe)O, SiO_2$; gedrite also contains an aluminous molecular compound, MgO, Al_2O_3, SiO_2, in variable amount. The percentage of Al_2O_3 rises to 13.5%. Neither mineral is noticeably attacked by acids. Cleavage fragments are very difficultly fusible.

Anthophyllite and gedrite possess no constant microstructure. They are apt to enclose hematite plates, magnetite grains, spinel octahedrons (picotite), and biotite plates. The latter lie with their tabular faces in the cleavage faces of ∞P (110), rarely in those parallel to $\infty P\bar{\infty}$ (100). The same microlitic forms which were described for hypersthene are sometimes found as interpositions in these minerals also. A regular lamellar intergrowth with monoclinic amphibole (actinolite) is not uncommon; both kinds of amphibole then have the axes c and b in common : hence this intergrowth is only noticeable in parallel polar-ized light on sections which do not lie in the zone oP: $\infty P\bar{\infty}$, and in which the lamellæ of monoclinic amphibole extinguish obliquely to the cleavage or to its diagonal. On cleavage plates parallel to $\infty P\bar{\infty}$ (100) the monoclinic lamellæ are recognized by the oblique emergence of an axis, while in the orthorhombic lamellæ a bisectrix emerges perpendicularly.

Anthophyllite and gedrite are Archæan minerals belonging especially to the hornblende gneisses and hornblende schists, in which they are sometimes disseminated as an essential constituent, sometimes are grouped together in radial aggregations. Anthophyllite is often quite abundant in olivine serpentines, generally accompanied by bastite.

Nothing is known concerning the processes of decomposition of anthophyllite.

Olivine.

Literature.

E. Cohen, Ueber Laven von Hawaii und einigen andern Inseln des grossen Oceans nebst einigen Bemerkungen über glasige Gesteine im allgemeinen. N. J. B. 1880. II. 23–62.

R. Hagge, Mikroskopische Untersuchungen über Gabbro und verwandte Gesteine. Kiel. 1871.

E. Kalkowsky, Ueber Olivinzwillinge in Gesteinen. Z. X. 1885. X. 17–24.

F. Kreutz, Ueber Vesuvlaven von 1881 und 1883. T. M. P. M. 1885. VI. 142–148.

H. Rosenbusch, Petrographische Studien an Gesteinen des Kaiserstuhls. N. J. B. 1872. 59 sqq.

G. Tschermak, Beobachtungen über die Verbreitung des Olivins in den Felsarten. S. W. A. 1867. LVI. Juli.

F. Zirkel, Untersuchungen über die mikroskopische Zusammensetzung und Struktur der Basaltgesteine. Bonn. 1870. 55–67.

— Geologische Skizzen von der Westküste Schottlands. Z. D. G. G. 1871. XXIII. 59–95.

Olivine forms either well-developed crystals and incipient forms of growth; or irregularly defined rounded or angular grains; or, finally, granular aggregates. The crystals have the habit of Figs. 72 and 73,

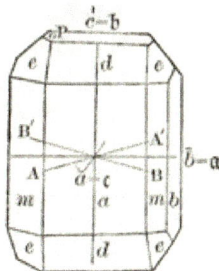

Fig. 72

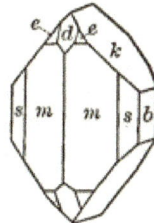

Fig. 73

with the faces $a = \infty P \overline{\infty}$ (100), $b = \infty P \breve{\infty}$ (010), $c = oP$ (001), $m = \infty P$ (110), $s = \infty P \breve{2}$ (120), $d = P \overline{\infty}$ (101), $h = P \breve{\infty}$ (011), $k = 2 P \infty$ (021), $e = P$ (111). The faces oP (001) are usually very small, often entirely wanting; then the sections in the vertical zone are six-sided, otherwise they are octagonal. Individuals of microscopic dimensions sometimes appear monosymmetric on account of a kind of hemimorphism (Pl. XVIII. Fig. 2). The incipient forms of growth are extremely manifold; besides simple forked forms (Pl. III. Fig. 1), there are delicate forms of bisymmetric or hemimorphic monosymmetric character, some of which are represented in Fig. 74. Twins occur, the twinning

Fig. 74

plane being $P \infty$ (011). It is often repeated, the individuals penetrating one another, and the boundaries of the twins being difficultly distinguishable in ordinary light (Pl. XVIII. Fig. 3). The twins are only determined with certainty, optically, in sections parallel to the macropinacoid when the vertical axis of the twinned individuals and their extinctions make an angle of 68° 48′ with one another.

The outline of the crystal sections often exhibits a decided rounding with variously shaped loops, the result of corrosion by the magma out of which they crystallized. The deformations extend to the complete obliteration of the original crystal form; thus arise the isolated olivine grains. In many rocks (lherzolites, olivinite, and olivine schists) in which olivine forms granular aggregations it appears never to have reached the development of a crystal form. The grains are not deformed crystals, but those whose development has been hindered. In meteorites olivine exhibits the chondritic form.

The cleavage of olivine parallel to $\infty P \bar{\infty}$ (010) is shown in thin sections by more or less distinct parallel cracks, which are seldom abundant, and often wedge out in the crystal. Still less frequent and irregular are the cracks corresponding to the cleavage parallel to $\infty P \bar{\infty}$ (100); the cleavage parallel to oP (001) is at least indicated in thin section. It often appears as though the ferruginous varieties (hyalosiderite and fayalite) possessed more perfect cleavage and less of the corrosive deformation of the crystal outline. Besides the cleavages, there is always an irregular fracturing of the crystal, which appears to increase with the alteration into serpentine.

Olivine is transparent and nearly colorless to greenish white, with a high iron percentage; and under certain conditions it is red to reddish brown. The index of refraction is high; hence the relief is considerable and the surface decidedly rough (Pl. XVIII. Fig. 4). The double refraction is very strong, and the interference colors even in quite thin sections are of the 2d and 3d order. Des Cloizeaux determined $\alpha_{na} = 1.661$, $\beta_{na} = 1.678$, $\gamma_{na} = 1.697$; Michel-Lévy found on artificial fayalite $\gamma - \alpha = 0.043$. The plane of the optic axes lies in oP (001) and $\check{a}$ is the positive bisectrix; hence $\check{a} = \mathfrak{c}$, $\bar{b} = \mathfrak{a}$, $\overset{1}{c} = \mathfrak{b}$ (Fig. 72). The axial angle is large, $2 V_{na} = 87° 46'$; the dispersion is weak, $\rho < \upsilon$ about $\check{a}$. In consequence of the large axial angle unsymmetrical sections may show an extinction which is quite oblique to the edges and cleavages, while sections from the principal zones extinguish parallel and perpendicular to the axes of these zones and to the cleavage. The optical behavior may occasionally lead to confusion with colorless pyroxenes of the monoclinic system. It is to be remarked, however, that the latter possess two equivalent cleavages, while those of olivine are unequal; further, that in the three principal zones, when the cleavages intersect, the extinction for olivine lies parallel and perpendicular to them, while for pyroxene it is diagonal. In general there is no appreciable pleochroism; but in the red olivines rays vibrating parallel to c are less absorbed than those parallel to $\check{a}$ and $\bar{b}$.

H. = 6.5-7. Sp. gr. = 3.3-3.45. Chemical composition (Mg, Fe)O, SiO$_2$. The more ferruginous varieties are called hyalosiderite, those with predominating iron percentage are called fayalite. Olivines which are not too poor in iron become permanently red upon being heated to redness, and then exhibit more or less distinct pleochroism. In thin section olivine is attacked but slowly by cold hydrochloric acid, but more rapidly when heated, with the separation of gelatinous silica; ferruginous varieties gelatinize more easily than those poor in iron, and sulphuric acid acts more energetically than hydrochloric. This gelatinization may often be used to distinguish olivine from other minerals of similar habit, by coloring the surface coating as already described. The reaction for Mg and Fe is obtained with hydrofluosilicic acid, when the surface of the mineral becomes covered with etched figures which exhibit acutely rhombic outlines.

Three kinds of olivine may be distinguished: olivine of the granular eruptive rocks, olivine of the porphyritic eruptive rocks, and olivine of the crystalline schists.

The olivine of the granular eruptive rocks, as it occurs in olivine diabases, olivine gabbros, olivine norites and peridoites, and as it is found in the older segregations of the volcanic rocks, exhibits no perfectly regular crystallographic boundary. It is always evident that it is older than the other silicates accompanying it, and from the nature of the decomposition processes it is clear that very ferruginous varieties do not occur. These olivines enclose crystals of magnetite, ilmenite, apatite and chromite; in many gabbro and norite occurrences they are crowded with needles and tabular microscopic interpositions, which are arranged parallel to the three principal sections and possess a metallic habit: they appear to be a titaniferous iron compound. Fluid inclusions are not infrequent.

The olivine of the porphyritic eruptive rocks (melaphyres, basalts, basanites, nepheline and leucite basalts, limburgites) belong to the secretions of the first period of consolidation; their crystallization immediately followed that of the apatites and iron ores, and preceded that of the mica and bisilicates. Hence it appears in crystal forms, which, however, are very often disturbed by subsequent corrosion. Pockets and inclusions of the ground mass are very characteristic of this variety of olivine (Pl. XVIII. Fig. 5), as are also glass inclusions of manifold shapes, irregular inclusions of fluids, among which is liquid carbon dioxide, and interpositions of the older minerals associated with it, especially magnetite, ilmenite, chromite and picotite. Ferruginous varieties are very frequent in these rocks. Here also belong the skeleton crystals and incipient forms of growth which are found in glassy rocks, which only reached incomplete crystalline development; they often enclose remarkably large portions of the rock glass arranged symmetrically (Pl. XVIII. Fig. 2). Indications of a second generation of olivine crystals in the porphyritic eruptive rocks are rarely found. Olivine appears as an accessory constituent even in rocks of the trachytic and andesitic series quite rich in alkalies and poor in bivalent bases, and often stands peculiarly correlated to the orthorhombic pyroxenes.

In the Archæan rocks olivine is sometimes an accessory constituent, sometimes an essential constituent in a series of rocks which occurs in the form of an inclusion, and consists at one end of dolomitic limestone and dolomite, while at the other it is wholly made up of silicate rocks rich in magnesia (olivine rocks), in which no magnesia or iron carbonates are found, or in which there are only traces of them. In this series belong many amphibolites, pyroxenites, and eklogites. The olivine of these rocks has the same habit as that of the granular eruptive rocks, but the ore-like interpositions are generally wanting, and there occur only occasional inclusions of fluids and spinels (chromite, picotite, and pleonaste).

Few minerals exhibit such a variety of alterations as olivine. Three different processes may be distinguished. The proper weathering, brought about by the universal reagents existing in the surface of the earth's crust, water, oxygen, and carbonic acid, leads to the formation of carbonates, silica, and limonite when the olivines are not too ferruginous; mixed with these is generally a variable amount of serpentine, so that serpentinization may possibly be the first stage in this process of alteration. The fact that calcite is always present among the carbonates, even to the exclusion of others, makes it appear as though this alteration was accompanied by a process of impregnation. This process of alteration is comparatively rare. The most common one is the alteration of olivine to serpentine: this always starts from the surface and from cracks and leads to a fibration, at the same time with the separation of the iron in the form of ferric oxide, hydrous oxide, and magnetite. The greenish to yellowish green fibres stand perpendicular to the crystal boundaries and the cracks. This produces a net-like appearance, the strings of serpentine forming the web of the net, the meshes consisting of olivine as yet unaltered (Pl. XVIII. Fig. 6). As the process advances, new cracks form with the increase in volume accompanying the serpentinization, resulting finally in the complete alteration of the olivine. Although the serpentinization of olivine in many cases may be a simple act of weathering, yet in others it is probably due to the action of warm waters. In very ferruginous olivines (hyalosiderites and fayalites) the alteration, which also commences from the surface and from cracks, leads to ferric oxide, which passes secondarily into hydrous oxide of iron (Pl. XIX. Fig. 1). At first this process often imparts to olivine a pleochroism which did not previously exist.

The third process is the alteration of olivine to amphibole; it is only known to take place in the Archæan rocks, where it occurs both in the schistose and in the eruptive rocks. It can generally be shown that this alteration takes place through the mutual influence of the olivine and the adjacent rock constituents. The new formation is first confined to the periphery of the olivine, and advances from here inward. The needles of the amphibole minerals, which may be partly referred to tremolite, partly to actinolite, and partly to anthophyllite, stand at right angles to the boundary of the olivine, and usually group themselves in several zones, differing in color. To the same group of phenomena belong the alteration of olivine to a felt of amphibole needles, with a slight admixture of serpentine or chlorite and magnetite; which has been called *pilite* by Becke (Pl. XIX. Fig. 2). These alterations, together with the quite frequent formation of amphibole (actinolite and brown hornblende) and of biotite from serpentinized olivine in the serpentines of the crystalline schists, may be considered dynamometamorphic.

Cordierite.

Literature.

E. Hussak, Ueber den Cordierit in vulkanischen Auswürflingen. S. W. A. 1. Abth. 1883. LXXXVII. April.

A. von Lasaulx, Ueber Cordieritzwillinge in einem Auswürfling des Laacher Sees. Z. X. 1883. VIII. 76–80.

J. Szabó, Der Granat und Cordierit in den Trachyten Ungarns. N. J. B. B.-B. 1. 1881. 308–320.

A. Wichmann, Die Pseudomorphosen des Cordierits. Z. D. G. G. 1874. XXVI. 675–701.

Rock-making *cordierite* sometimes appears in crystals, often in grains without regular boundaries. The crystals almost always exhibit the simple forms (Fig. 75) ∞P (110), $\infty P\infty$ (010), oP (001). The prism angle is 119° 10′. $P\infty$ (011) and $\frac{1}{2}P$ (112) occasionally occur as narrow truncations of the combination edges of the base with the vertical faces. Hence sections in the prism zone are rectangular, and at right angles to it are hexagonal; or the crystallographic outlines are wanting. Twinning on the whole is rare; when present it follows the law: the twinning plane is the prism face. In cordierite which has been exposed to volcanic influences it leads to penetration trill-ings, in consequence of which in a single individual thin lamellæ are often intercalated in place of the other individuals. Pl. XIX. Fig. 3 shows a basal section of a simple trilling. In the metamorphic rocks there is a development of polysynthetic lamellæ parallel to a single prism face. Re-entrant angles do not occur, but with the twinning just mentioned there is a striation on cleavage faces just as on the principal cleavage face of plagioclase. The cleavage parallel to $\infty P \breve{\infty}$ (010) is very variable; it is occasionally observed in thin sec-tions as distinct, parallel cracks. But an irregular parting is often suggested by crooked and disconnected cracks, especially in occurrences which are not entirely fresh. H. = 7–7.5.

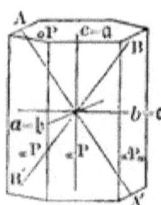

Cordierite generally becomes transparent and colorless, more rarely yellowish, blue, or violet, according to the position of the section. The index of refraction and the double refraction are weak, and strikingly similar to those of quartz, with which mineral cordierite may be easily confounded. Des Cloizeaux determined for yellow light—

On cordierite from Ceylon, $\alpha = 1.537$, $\beta = 1.542$, $\gamma = 1.543$
   "   "   Bodenmais, $\alpha = 1.535$, $\beta = 1.541$, $\gamma = 1.546$
   "   "   Haddam, $\alpha = 1.5523$, $\beta = 1.5615$, $\gamma = 1.5627$

Hence the mean index of refraction is about the same as that of Canada balsam, $\gamma - \alpha = 0.008$–0.009; and the interference colors in thin section seldom exceed yellow of the 1st order, remaining mostly in the gray-blue and white tones as in quartz. The axial plane lies in $\infty P \breve{\infty}$ (100), the character of the double refraction is negative, and c is the acute bisectrix; hence $\mathfrak{c} = \mathfrak{a}$, $\bar{b} = \mathfrak{c}$, $\breve{a} = \mathfrak{b}$ (Fig. 75). The apparent axial angle in air varies within wide limits, from 64° (Haddam) to 150° (Orijärfvi). The dispersion is weak, $\rho < \upsilon$. When twinned, basal sections exhibit the axial planes in convergent light, and the extinction in parallel light inclined 60° to each other in two adjacent individuals. For sections in the prism zone the separate individuals extinguish syn-chronously, whenever the twinning plane lies parallel or perpendicular to the principal section of the polarizer; or else the twinning may be recognized by the fact that the different individuals exhibit different interference colors during a rotation of the section, since their ellipsoids of elasticity are intersected differently. The pleochroism, which is generally very strong in thick plates, is often scarcely noticeable in thin section, yet it is often quite strong in sections from the prism zone. The facial colors, according to Haidinger, are oP (001) blue, $\infty P \bar{\infty}$ (100) bluish white, $\infty P \bar{\infty}$ (010) yellowish white. The following axial colors have been determined:

LOCALITY.	$\check{a}$	$\bar{b}$	$\dot{c}$
Bodenmais...	light Berlin-blue....	dark Berlin-blue..	yellowish white.
Bodenmais...	grayish white.......	milk-white.......	yellowish vinous, yellow-white.
Orijärfvi.....	light Berlin-blue....	dark Berlin-blue..	reddish clove-brown.
Arendal......	plum-blue..........	violet-blue......	reddish clove-brown.
Haddam.....	bluish white........	pale blue.........	yellowish white.
Simiulak....	dark leather-brown..	reddish brown to honey-yellow...	smoke-brown.

The absorption $\bar{b} > \check{a} > \dot{c}$ may even be noticed in colorless sections. Pleochroic halos are very common in cordierite ; the bright yellow halos surround microscopic inclusions of all kinds ; the color is a maximum when the light vibrates parallel to c, and completely disappears when the vibration of the light is parallel $\check{a}$ or $\bar{b}$. Hence basal sections do not exhibit this phenomenon. It is sometimes observed in andalusite, staurolite, augite, muscovite, etc., and is due to the absorption of the blue rays. Hence the yellow halo appears black in blue light, and does not appear at all for red light. Upon being heated to redness, cordierite loses this property which is occasioned by a local accumulation of an organic pigment. It has only been observed in cordierite from the Archæan and from contact zones. The ordinary pleochroism of cordierite becomes more distinct when it is heated to redness, especially in thick sections. It sometimes disappears from thin sections upon very strong heating.

Sp. gr. $= 2.59$–2.66, very near that of quartz, so that it is often difficult to separate the two mechanically. Chemical composition $=$ $2MgO. 2Al_2O_3, 5SiO_2$, in which a variable amount of MgO is replaced by FeO, and to a smaller degree by MnO. It is but slightly acted on by acids, and fuses with difficulty on the edges. The chemical distinction from quartz is furnished most simply by treating the section with hydrofluosilicic acid, in the manner already described. The evaporated solution yields the characteristic prismatic crystals of magnesium fluosilicate (Pl. XII. Fig. 6). The surface of the section becomes covered with etched figures, whose form and distribution vary with the position of the section. In sections in the principal zone these figures have the form of long rectangular depressions. Etched figures are also produced by using hot sulphuric acid, which is a means of distinction from quartz.

Cordierite has no constant microstructure. When it occurs in eruptive rocks it either contains fluid and glass inclusions alone, as in granites and quartz porphyries, or it is almost completely free from interpositions, as in the andesitic rocks. The formation of cordierite in these rocks belongs to an older period of rock development, and appears to antedate that of the feldspars; for many of these occurrences its nature, as a normal constituent, is very doubtful, and its derivation from the Archæan rocks through which the eruptive rocks have passed is quite probable. The real home of cordierite is the gneiss formation, where it is frequently accompanied by garnet, biotite, sillimanite, spinel, pyrrhotite, hematite, and ilmenite. All these minerals, except garnet, occur as inclusions in cordierite. Besides these interpositions fluid inclusions are common, not infrequently with colorless cubes.

Cordierite is present in many schistose hornstones of the granite and diorite contact zones, frequently accompanying andalusite, and possessing the microstructure of the cordierite of the gneiss formation, but is distinguished from this by the frequency of normal crystallographic boundaries.

Cordierite appears to be readily decomposed, altering to more or less fibrous or lamellar aggregates, especially in the gneiss formations. These decomposition products, or mixtures of them with unaltered cordierite substance, are variously termed aspasolite, chlorophyllite, bonsdorffite, esmarkite (in part), pinite, oösite, praseolite, gigantolite, fahlunite, and pyrargillite. Many of these, especially pinite and oösite, consist essentially of potash mica in an irregular intergrowth of lamellæ; in others there occurs, besides dense muscovite, a chlorite or talc. The dirty-brown coloring of these pseudomorphs is due to the admixture of limonite. In still other cases there result yellowish to greenish alteration products, which do not permit of exact determination, and which strikingly suggest the serpentine bands of olivine. The process of decomposition always follows the cleavage cracks and fissures of the mineral.

Zoisite.

Literature.

FR. BECKE, Die Gesteine der Halbinsel Chalcidice. T. M. P. M. 1878. I. 248-250.

O. LUEDECKE, Der Glaukophan und die glaukophanführenden Gesteine der Insel Syra. Z. D. G. G. 1876. XXVIII. 259-260.

A. SAUER, Erläuterungen zur Section Kupferberg der geologischen Specialkarte des Königreichs Sachsen. Leipzig. 1882. 25.

G. TSCHERMAK and L. SIPÖCZ, Beitrag zur Kenntniss des Zoisits. S. W. A. LXXXII. 1880. July.

Zoisite in rocks forms either isolated crystals or prismatic aggregates, consisting of parallel or slightly divergent columns. Crystallographic boundaries only occur in the vertical zone; the faces ∞P (110), with 116° 26', predominate; $\infty P \bar{\infty}$ (010) is seldom wanting: besides these there is often a great number of derived prisms present, among which are $\infty P \overset{\vee}{4}$ (140), $\infty P \overset{\vee}{2}$ (120), $\infty P \bar{2}$ (210), and $\infty P \bar{3}$ (310). Occasionally when terminal faces exist they appear to be P (111) and $2 P \bar{\infty}$ (021). A kind of hemimorphism with respect to the $\bar{b}$ axis is quite common, the prism faces being developed on one side of the crystal, while those in the other side are wanting, being replaced by $\infty P \bar{\infty}$. Sections of the crystals at right angles to the prism axis are rhombic (∞P), or apparently hexagonal ($\infty P . \infty P \bar{\infty}$), or many-sided to round, or finally triangular to trapezoidal; the longitudinal sections are lath-shaped. The length is usually three times the breadth, or more; rarely both dimensions are approximately equal, and the mineral assumes the granular form. Twinning cannot be detected morphologically, but from the optical behavior appears to be quite frequent. The dimensions vary from several centimetres to microscopic proportions.

Zoisite is characterized by a very perfect cleavage parallel to $\infty P \bar{\infty}$ (010), which is shown by numerous sharp cracks parallel to the longitudinal direction in all sections in the prism zone, and in those parallel to the base. A second and much less perfect cleavage runs parallel to $\infty P \bar{\infty}$ (100), and is seen in sections parallel to $\infty P \bar{\infty}$ (010), and parallel to the base. A transverse parting approximately parallel to the base is more noticeable the longer the individuals. The prisms are not infrequently bent. The base appears to be a gliding-plane.

When fresh, zoisite is transparent and colorless, but the larger crystals and aggregates are often clouded peripherally, the smaller ones completely so; they then appear gray to greenish gray. Those varieties colored red by manganese (thulite) are variously colored red, yellow, or almost colorless by transmitted light, according to the position of the section. The mean index of refraction is quite high, $\beta_\rho = 1.69$–1.70; hence the relief is very distinct, and the surface rough; on the other hand, the difference, $\gamma - \alpha$, is small $= 0.0054$–0.0057, consequently the interference colors are very low. In very thin sections not parallel to the axial plane the double refraction is often only recognized by using sensitive tones of color (gypsum or quartz plate). No other orthorhombic mineral has so little double refraction with so high an index of refraction. Even in sections parallel to the axial plane the interference colors in thin section only exceed yellow of the 1st order when the section is quite thick.

The axial plane in some cases lies in the principal cleavage face, in others in the basal plane; in fact the orientation sometimes varies in one and the same crystal. In rock-making zoisite both positions appear to be equally common. In both cases $\breve{a}$ is the acute bisectrix, and the character of the double refraction is positive; in the first instance the optical scheme is $\breve{a} = \mathfrak{c}$, $\bar{b} = \mathfrak{b}$, $\mathfrak{c} = \mathfrak{a}$; in the second, $\breve{a} = \mathfrak{c}$, $\bar{b} = \mathfrak{a}$, $c = \mathfrak{b}$. The optic angle varies between wide limits from almost $0°$ to $100°$ in air; it is usually quite small in rock-making zoisite, and sections parallel to $\infty P\breve{\infty}$ show both axes even in air. The axial figures are not infrequently distorted. The strong dispersion is characteristic; for the basal position of the axial plane the dispersion is always $\rho > \upsilon$, for the brachypinacoidal position $\rho < \upsilon$.

Colorless zoisite exhibits no pleochroism and no absorption; but manganiferous thulite is strongly pleochroic. Rays vibrating parallel to $\mathfrak{c}$ are yellowish, those parallel to $\bar{b}$ rose-red, those parallel to $\breve{a}$ reddish white in very thin sections. The axial plane lies in the plane of the yellow and reddish-white rays, the dispersion is $\upsilon > \rho$; $\breve{a}$ is the positive bisectrix.

H. $= 6$–6.5. Sp. gr. $= 3.25$–3.36. Chemical composition $= \text{H}_2\text{O}$, 4CaO, $3\text{Al}_2\text{O}_3$, 6SiO_2. Acids do not attack zoisite, unless it has been heated to redness. It fuses with intumescence, and then gelatinizes with HCl. With hydrofluosilicic acid the powder gives a strong reaction for calcium; in the solution CsCl produces abundant crystals of cæsium alum.

The small individuals of zoisite are generally free from inclusions; the larger ones frequently contain fluid inclusions, either round or irregularly shaped, and fine tubular canals running parallel to the cleavage, and occasionally curving and branching out in the interior of the crystal. Inclusions of amphibole microlites are not uncommon;

they are usually arranged with their longer axis parallel to the vertical axis of the zoisite.

Zoisite is essentially a mineral of the crystalline schists, especially of the hornblendic members (Pl. XIX. Fig. 4). Its appearance as an essential constituent of the so-called saussurite in gabbros is an altogether different occurrence, in which it must be considered as the product of a dynamo-metamorphic alteration of plagioclase.

Talc.

Rock-making *talc* occurs in the form of plates, usually elongated in one direction like flattened rods; more rarely the plates are developed equally in all horizontal directions, and hence have round to hexagonal (∞P, $\infty P\breve{\infty}$) outlines. The tendency to a rosette-like arrangement is to be noted, leading to more or less complete spherulitic forms; an irregular felting of the plates is very common. The perfect cleavage parallel to oP (001) is just as distinct in all sections, which are not parallel to the base, as in the micas.

The percussion figure (*Schlagfigur*), a six-rayed star, whose rays intersect at 60°, one of them being perpendicular to $\infty P\breve{\infty}$ (010), indicates that there are gliding faces in the prism zone; the plates, however, occur curved and bent in the most irregular manner, without breaks or cracks.

Talc is transparent and colorless, with a low index of refraction, which in fact is not directly determinable, but is calculated at 1.551. The double refraction is very strong, $\gamma - \alpha = $ 0.038–0.043. The interference colors, therefore, are very high, and correspond closely to those of muscovite.

The axial plane is in the macropinacoid, and c is in the negative bisectrix; the axial angle is quite small. A dispersion is not noticeable. Fig. 76 gives the scheme for

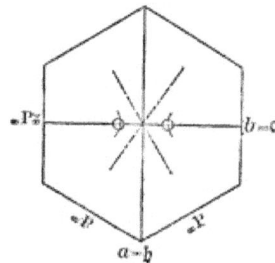

Fig. 76

talc in a plate bounded by the prism and brachypinacoid. The dotted lines indicate the percussion figure, the third ray coinciding with $\bar{b}$. Plates parallel to oP often exhibit a slight division of the field in parallel polarized light, and the interference figure is variously distorted in convergent light. Sections inclined and perpendicular to oP extinguish parallel and perpendicular to the cleavage lines. The rosette-like and spherulitic aggregates exhibit in parallel light between crossed nicols the dark cross parallel to the principal sections of the nicols, with the light quadrants brightly colored. When the sections are not very thin, the colors are whitish red and whitish green of the 4th order.

H. = 1. Sp. gr. = 2.8. Chemical composition = $3MgO, 4SiO_2 +$ H_2O. Almost infusible. Heated with cobalt solution it becomes flesh-red. Acids are almost without action on it.

Talc may be easily confused with brucite and muscovite. It is distinguished from the first by its optical character in convergent light, and by the absence of an alumina reaction in the fusion with alkali carbonates; from the second by the absence of alkalies when treated with hydrofluosilicic action and by the reaction with cobalt solution.

Talc is usually free from inclusions, but the larger lamellæ are often pierced by tremolite needles, and frequently enclose biotite.

Talc is found as an essential rock constituent in the region of the crystalline and dynamo-metamorphic schists. It is often accompanied by rhombohedral carbonates and by quartz. In eruptive rocks it only occurs as pseudomorphs after magnesian silicates, and rarely then. G. H. Williams[*] has observed its occurrence as an alteration product of hornblende in the magnesian rocks from the vicinity of Baltimore, Md.

Natrolite.

Natrolite is never a primary constituent of rocks, but it is extremely common as an alteration product of sodalite, nosean, nepheline and acid plagioclase, partly in actual pseudomorphs after these minerals, partly in cavities and cracks. It almost always forms radial aggregates, which are sometimes parallelly fibrous, sometimes divergent to radially fibrous, not infrequently they form spherulites. The long axis of the individuals corresponds to the vertical crystallographic axis.

Cleavage is seldom observed on the prismatic masses; when present it is parallel to the prism, and in cross-sections appears as a system of cracks intersecting apparently at right angles. It is colorless by transmitted light, if the separate individuals are not too minute, or yellowish to brownish and cloudy for very small transverse dimensions. The coefficient of refraction is small, the double refraction is measurable. $\alpha_\rho = 1.4768$, $\beta_\rho = 1.4797$, $\gamma_\rho = 1.4887$. The axial plane lies in $\infty P\infty$ (010), and the vertical axis is the positive acute bisectrix; hence $\bar{a} = \mathfrak{a}, \bar{b} = \mathfrak{b}. \dot{c} = \mathfrak{c}$. The axial angle in air is over 90°.

* Bulletin 28, U. S. Geol. Survey, 1886, p. 58.

Dispersion $\rho < v$. The extinction in longitudinal sections is parallel to the axis of the prism, in cross-section it is diagonal to the cleavage. The radially columnar aggregates give neat interference crosses in parallel light between crossed nicols, the arms lying parallel to the principal sections of the nicols.

H. = 5–5.5. Sp. gr. = 2.17–2.26. Chemical composition = Na_2O, $Al_2O_3 + 2aq$. Gelatinizes easily with hydrochloric acid. The aggregates usually show very pure substance; they occasionally enclose plates of micaceous iron and interpositions of the parent minerals.

Appendix.—A corresponding lime zeolite, probably scolecite, often occurs as an alteration product of the more basic feldspars. (cf. Kloos, N. J. B. G. H. Williams, Bulletin 28, U. S. Geological Survey.)

Dumortierite.

The translator is indebted to Mr. J. S. Diller for the following notes on the occurrence of dumortierite in the United States.

Dumortierite occurs in columnar to fibrous masses associated with tourmaline in certain Archæan rocks. When crystal form is observed, the planes $\infty P\bar{\infty}$ (100) and ∞P (110) are equally developed, the angle between them being about 152°. Terminal planes rarely visible on imbedded crystals. Cleavage parallel to $\infty P\bar{\infty}$ seldom observed in thin section, but developed by pressure together with one parallel to a prismatic plane. Polysynthetic twinning frequent.

Its index of refraction is higher than that of quartz; double refraction rather strong, interference colors those of the 2d order. Optical character negative; acute bisectrix parallel to the vertical crystallographic axis. It is transparent blue with highly characteristic pleochroism; the extraordinary ray is deep cobalt blue, the ordinary ray is colorless. It loses its color upon being heated to redness.

H. = 7. Sp. gr. = 3.265. Chemical composition not yet definitely determined, a silicate of alumina. Insoluble in acids, including hydrofluoric.

Dumortierite occurs chiefly in quartz, sometimes in hair-like forms, in the pegmatitic portion of a biotite gneiss at Harlem, New York, a rock similar to that near Lyons in which it was originally discovered by Gonnard (Bul. Soc. Min. Fr. Vol. IV. p. 2. 1881). It also occurs in granular quartz near Clip, Arizona.

MINERALS OF THE MONOCLINIC SYSTEM.

Sections of a regularly developed crystal of the monoclinic system and the figures made by the cleavage cracks are only symmetrical when they belong to the orthodiagonal zone. The cleavage is either single (parallel to pinacoids or orthodomes), and lies in the plane of symmetry or stands at right angles to it ; or the cleavage is the same parallel to two faces (prismatic) which make equal angles with the plane of symmetry. A single cleavage gives a single system of parallel cleavage cracks in all sections but those parallel to the cleavage face. Two single cleavages occurring in the same crystal cannot be equal ; they furnish parallel cleavage cracks in all sections in the zone of the cleavages, and intersecting systems of dissimilar cleavage cracks in all other sections.

The ellipsoid of elasticity in monoclinic crystals is triaxial ; one axis coincides with the orthodiagonal or axis of symmetry of the crystal ; the two other axes lie in the plane of symmetry, $\infty P \infty$ (010). If the axis of elasticity which coincides with the orthodiagonal is b, then the optic axes lie in the plane of symmetry (symmetrical axial position), and are dispersed in this plane together with the bisectrices (inclined dispersion) ; if one of the bisectrices coincides with the orthodiagonal, then the optic axes lie in a plane of the orthodiagonal zone (normal symmetrical axial position). The dispersion then is either horizontal or crossed, according as the orthodiagonal is the obtuse or acute bisectrix. Therefore all sections in the orthodiagonal zone during a complete revolution between crossed nicols extinguish light four times, parallel and perpendicular to the single cleavages diagonal to the prismatic cleavage (parallel extinction) ; all other sections which are not perpendicular to an optic axis, during a complete revolution extinguish four times in positions inclined at a certain angle to the principal sections of the nicols (inclined extinction). The inclination of the axes of elasticity lying in the plane of symmetry to the crystal axes is called the *extinction angle*, and is an important means of distinguishing monoclinic minerals. Sections at right angles to an optic axis remain uniformly light during a rotation between crossed nicols. In convergent light sections perpendicular to an optic axis or not much inclined to it, as well as those perpendicular to a bisectrix, exhibit the same interference figures as similarly situated sections in orthorhombic crystals.

But in sections in the first position the axial bar has differently colored borders if the substance has sufficiently strong dispersion and the axial plane is perpendicular to the plane of symmetry; and in sections in the second position the distribution of the colors is not bisymmetrical as in the orthorhombic system, but is monosymmetric with respect to the plane of symmetry if the bisectrix lies in this plane (inclined and horizontal dispersion), and symmetrical with respect to the centre of the interference figure if the orthodiagonal is the bisectrix (crossed dispersion). The distribution of blue and red in the innermost color rings or on the poles of the hyperbolas in the diagonal position determines in this system, as in the orthorhombic system, the relative size of the angles of the optic axes, $\rho > \upsilon$ or $\rho < \upsilon$.

If monoclinic minerals exhibit pleochroism, all sections are dichroic which are not perpendicular to an optic axis; the maximum differences of color lie 90° from one another, and necessarily coincide with the directions of extinction in sections in the orthodiagonal zone; this coincidence generally exists in all other sections, but not necessarily.

Many monoclinic minerals (mica) in their optical characters strikingly approach those of the hexagonal or orthorhombic system, and with the ordinary microscopical investigation are only recognized as monoclinic with great difficulty or not at all.

Gypsum.

Literature.

Fr. Hammerschmidt, Beiträge zur Kenntniss des Gyps- und Anhydritgesteins. T. M. P. M. 1882. V. 245-285.

Rock-making *gypsum* shows no crystallographic boundaries; it appears in irregular granular aggregates. In secondary, possibly primary, veins which traverse the granular masses it is lamellar to fibrous, the fibres standing perpendicular to the walls of the veins. The sections therefore exhibit no characteristic forms.

The perfect cleavage parallel to $\infty P\bar{\infty}$ (010) gives rise to abundant parallel cracks. The fibrous fracture as well as the conchoidal fracture and gliding planes parallel to $P\bar{\infty}$ (101) and $\frac{4}{3}P\bar{\infty}$ (509) are seldom observed microscopically in rocks.

Gypsum becomes transparent and colorless or is gray to grayish blue from carbonaceous matter, and reddish to yellowish from hydrous oxide of iron, or from plates of ferric oxide. The index of refraction is small; the double refraction measurable, $\alpha_{na} = 1.5207$, $\beta_{na} = 1.5228$,

$\gamma_{na} = 1.5305$. The axial plane lies in the plane of symmetry; the inclined dispersion is distinct. $2V = 61° 24'$ becomes rapidly smaller with increase of temperature.   The acute bisectrix lies in the obtuse angle between the vertical and clinodiagonal axes, and is inclined 75° 15' to the former and 23° 42' to the latter.

II. = 2.   Sp. gr. = 2.2–2.4.   Chemical composition = CaO, SO, + 2aq.   Difficultly soluble in water.   Gives off much water in a closed tube, and fused with soda on charcoal gives the reaction for sulphur.

Not infrequently gypsum encloses, besides carbonaceous substances and iron oxides, fluid and gas inclusions of irregular shape or in negative crystals.   Calcite, magnesite, dolomite, and quartz are found in it in crystals or grains.

Gypsum only occurs in the gypsum rock and in the anhydrite of sedimentary formations.

Wollastonite.

Literature.

A. E. TÖRNEBOHM, Nefeliusyenit från Alnö.   Geol. Fören. i Stockholm Förhdl. 1883. VI. No. 82. 542–549.

Wollastonite appears as incompletely bounded prismatic or tabular crystals, which are always elongated parallel to the axis of symmetry (b), or they are in prismatic or fibrous aggregates with a more or less parallel or slightly divergent arrangement of the individuals.   Therefore, sections of the isolated individuals from the orthodiagonal zone are lath-shaped, cross-sections are round to six or eight sided from the faces oP (001), $\infty P\dot{\infty}$ (100), $\frac{1}{2}P\dot{\infty}$ (102), and $-P\dot{\infty}$ (101).   The angles are $001 \wedge 100 = 95° 30'$, $100 \wedge 101 = 44° 27'$, $100 \wedge 10\overline{2} = 69° 56'$.   The face $\infty P\dot{\infty}$ (100) is usually the most broadly developed.   Twinning is quite frequent according to the law: the twinning plane is $\infty P\dot{\infty}$ (100); the faces oP of the individuals making an angle of 169° with one another.

The cleavage is perfect parallel to $o\,P$ (001) and $\infty P\dot{\infty}$ (100), less perfect parallel to $\frac{1}{2}P\infty$ (102) and $-P\overline{\infty}$ (101).   The last-named face is inclined 50° 25' to $\infty P\dot{\infty}$ (100).   In sections from the orthodiagonal zone all the cleavage cracks are parallel to one another; in sections parallel to $\infty P\dot{\infty}$ they form two very distinct systems of cracks intersecting at 84° 30', which are sometimes cut diagonally by two other systems which are neither so distinct nor so numerous.   In the orthodiagonal sections the cleavages run parallel to a longitudinal fibration, usually quite distinct, which is due to the parallel growth of very slender individuals. Larger, irregular cracks stand at right angles to the length of these sections.

Wollastonite becomes transparent and colorless, and possesses a mean index of refraction which is not inconsiderable = 1.635. The interference colors are quite bright parallel to the plane of the optic axes; $\gamma - \alpha = 0.016$. The axial plane lies in the clinopinacoid; $2E_\rho = 70° 40'$; $2E_v = 68° 24'$. The positive acute bisectrix lies in the obtuse angle β, and makes an angle of about $37° 40'$ with the cleavage parallel to $\infty P \check{\check{\infty}}$ (100). The inclined dispersion shows itself by a lively difference of color on the margin of the hyperbola of one axis (red inside, blue outside), while the colors of the second hyperbola are blue inside and outside. Fig. 77 presents the optical scheme for the plane of symmetry. It is seen that an axis stands perpendicular or only slightly inclined to each principal face. Twinning parallel to $\infty P \check{\check{\infty}}$ (100) can be recognized by the fact that in all sections inclined to the twinning plane the lamellæ do not extinguish at the same time; in sections from the orthodiagonal zone, although the lamellæ extinguish together, they can be recognized by their different interference colors, which are due to the fact that the ellipsoid of elasticity is cut differently in each half of the twin and $o - e$ is different in each case. In convergent light sections in the orthodiagonal zone exhibit axial figures, and the points of emergence of bisectrices, and the axial plane always lies perpendicular to the longitudinal direction and the cleavage.

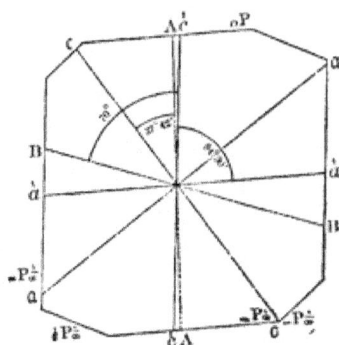

Fig. 77

There is no pleochroism.

II. = 4.5–5.0. Sp. gr. = 2.8–2.9. Chemical composition = CaO, SiO_2. It gelatinizes easily with hot hydrochloric acid; there is an abundant reaction for gypsum upon adding sulphuric acid to the solution: anhydrite forms in a very concentrated solution, the crystals being rhombic with a cubical habit. The powder fuses with great difficulty. Its gelatinization with HCl is an important means of distinguishing it from epidote, with the colorless varieties of which it may be confused, because of the similar prismatic development parallel to $\bar{b}$ and of the same position of the axial plane with reference to the longitudinal axis, in spite of the high index of refraction and higher double refraction of the latter.

Wollastonite has no constant microstructure : it often contains fluid inclusions, grains of calcite and diopside, or other minerals associated with it.

Wollastonite is a frequent guest in granular limestone and in the rocks related to it occurring in the Archæan (garnet rock, epidote rock, etc.), and it not infrequently occurs in feldspathic schists when these are rich in lime. It is also found in contact-metamorphosed limestones and in limestone inclusions in eruptive rocks, where it is usually accompanied by pyroxene and garnet, as in the schists. It is very rarely found in eruptive rocks.

Group of Monoclinic Pyroxenes.

Literature.

P. Mann, Untersuchungen über die chemische Zusammensetzung der Augite aus Phonolithen und verwandten Gesteinen. N. J. B. B.-B. II. 1884. 172–205.

A. Merian, Studien an gesteinsbildenden Pyroxenen. N. J. B. B.-B. III. 1884. 252–315.

A. Michel-Lévy, De l'emploi du microscope polarisant à lumière parallèle pour l'étude des plaques minces de roches éruptives. Ann. des Mines. Paris. 1877. (7). XII. 424–429.

G. Tschermak, Ueber Pyroxen und Amphibol. T. M. M. 1871. 17.

— Mikroskopische Unterscheidung der Mineralien aus der Augit-, Amphibol- und Biotitgruppe. S. W. A. 1. Abth. 1869. LIX. May.

Only those minerals of the monoclinic system commonly referred to the pyroxene family are here grouped as monoclinic pyroxenes, in which the characteristic cleavage parallel to an almost right-angled prism is distinctly noticeable.

The monoclinic pyroxenes belong to the most widely distributed rock-making minerals, both in eruptive rocks and in the crystalline schists; they appear in perfectly developed crystals, in irregularly bounded individuals, or in aggregates. The habit varies with the chemical composition. It may be stated as the rule,—which, however, is not without exceptions,—that pyroxenes of the diopside and acmite series usually form long columnar crystals with highly subordinate prism faces, and columnar masses; pyroxenes of the augite series form short prismatic individuals and grains. The commonest crystal forms based on $\dot{a}:\bar{b}:\dot{c} = 1.0903:1:0.5893$ and $\beta = 74° 11'$ are $m = \infty P$ (110) with 87° 06', $a = \infty P \bar{\infty}$ (100), $b = \infty P \infty$ (010), $s = P$ (111) with $11\bar{1} \wedge 1\bar{1}\bar{1} = 120° 48'$, $u = -P$ (111) with $111 \wedge 1\bar{1}1 = 131° 30'$, $o = 2P$ (22\bar{1}) with $22\bar{1} \wedge 2\bar{2}\bar{1} = 95° 48'$, $p = P\bar{\infty}$ (101), $c = oP$ (001), $n = \frac{1}{2}P\bar{\infty}$ (10\bar{2}) with $10\bar{2} \wedge 100 = 89° 38'$. Fig. 78 shows one of the most frequent forms of rock-making diopside; Fig. 79 such a one of augite. Hence cross-sections more or less perpendicular to *c* exhibit squares with slightly truncated corners, or octagons with sides of almost equal length; sections from the orthodiagonal zone give lath-shaped figures, either pointed quite steeply or cut off straight, some-times almost hexagonal; sections lying more or less parallel to the plane of symmetry are lath-shaped, with one or two-sided terminations or slightly prolonged inclined rhombs. Sections through irregularly bounded individuals may be of almost any shape.

Twinning is extremely common, and usually follows the law: the twinning plane is $\infty P \dot{\infty}$ (100). In the diopsides and acmites the twinning line very frequently runs through the middle of the crystal; hence the outline shows no re-entrant angle and the twinning is only recognized between crossed nicols. Fig. 80 shows the form which

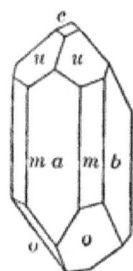

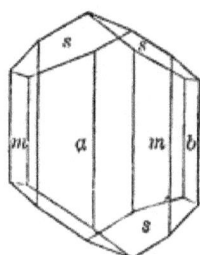

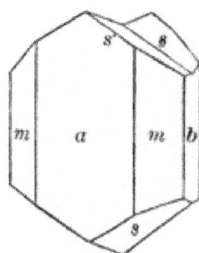

Fig. 78   Fig. 79   Fig. 80

arises in augites; hence the twinning is not noticeable in the outline of sections in the orthodiagonal zone, but it is in that of sections in the prism and clinodiagonal zone. Between the larger halves of twin crystals there often appear a number of smaller twinned lamellæ (Pl. XIX. Fig. 5). A second twinning, occurring especially in the diopsides and diallages, follows the law: the twinning plane is the base. In this case the form of development is usually lamellar, a number of twinned lamellæ being enclosed in a larger individual. This is not noticeable in the outer contours, but is detected on the vertical faces of the crystal as a fine striation at right angles to the axis of the prism; in sections it is generally noticeable even in ordi-nary light, and comes out distinctly between crossed nicols (Pl. XIX. Fig. 6). Both of these kinds of twinning occur in the same crystal in the diallage-like augites of many diabases. The twinnings parallel to $- P \dot{\infty}$ (101) (Fig. 81) and parallel to $P\dot{2}$ (122) (Fig. 82) are rarer, and are principally confined to basaltic augite. They generally appear in the form of complicated intergrowths of several augite individuals (Pl. XX. Fig. 1).

The dimensions of pyroxene crystals vary greatly; in the eruptive rocks especially they sink to microlitic proportions. Here also occur the greatest variety of imperfect crystal forms; not infrequently these incipient forms of growth are found in larger individuals (Pl. XX. Fig. 2). The incomplete development is generally confined to the terminal faces. Skeleton crystals also (Pl. III. Fig. 3), whose arms intersect at pyroxene angles, are not uncommon, besides extremely delicate and capricious forms of growth, at times approaching spherulitic forms (Pl. XX. Fig. 3); these, however, are confined to the glassy eruptive rocks.

Shelly structure is frequent in augite and acmite, and from the variety of chemical composition appears in the form of isomorphous

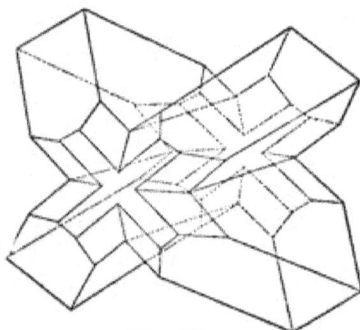

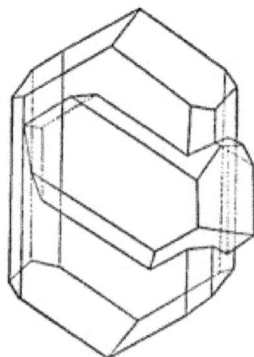

Fig. 81   Fig. 82

layers or as zonal structure. Generally these successive shells are geometrically similar to one another and to the outward form of the crystal, but occasionally the inner shells exhibit a different crystallographic outline from the outermost shell (Pl. XX. Fig. 4). The number of shells varies greatly. Moreover, the outline of the inner shells is not always a crystallographic one : sometimes they merge into one another (Pl. XX. Fig. 2), or it is evident that the kernel was at one time a corroded crystal (Pl. XX. Fig. 4). Quite rarely the shelly structure follows oP (001), when it is accompanied by the twinning and parting parallel to this face. Hour-glass forms (Pl. V. Fig. 6) are produced by the filling up of the gaps of forked crystals by newer pyroxene substance.

Corrosion phenomena are not infrequent, especially on the older pyroxenes of the eruptive rocks; mechanical deformations in the shape of bendings, breakings, shatterings, and tortions occasionally occur in the pyroxenes of all rocks, but are quite rare in those of the Archæan rocks, because the pyroxenes appear to be unable to withstand the mechanical processes which these undergo; they are here converted into amphibole.

The monoclinic pyroxenes cleave with variable perfection parallel to the prism of 87° 06'. The cracks corresponding to this cleavage are almost always distinct and numerous, but they seldom run uninterruptedly and straight through the entire crystal. In sections approximately perpendicular to the prism axis they form two systems, crossing each other nearly at right angles (Pl. X. Fig. 4); in sections in the prism zone they run parallel; in all other sections they make rhombic figures whose angles depend on the position of the section with respect to the crystal (Pl. XX. Fig. 5). Besides the prismatic cleavage there is also a cleavage parallel to one or both vertical pinacoids, especially in diallage and diopside, less frequently in the augites and acmites; it is always quite imperfect, and is only indicated by short or intermittent cracks. Individuals of the diopside and diallage series twinned parallel to oP (001) exhibit quite a perfect parting parallel to this face (Pl. XIX. Fig. 6), which, however, does not represent a proper cohesion minimum, but is due to the twin lamination. In some basaltic rocks there are augites which exhibit neither macroscopic nor microscopic cleavage. All monoclinic pyroxenes when in long prismatic forms exhibit an irregular parting approximately perpendicular to the prism axis.

All monoclinic pyroxenes, even when strongly colored, become perfectly transparent; except the acmites, which are not very transparent. The colors in transmitted light are very different according to the chemical composition, and therefore change in one and the same crystal with the isomorphous layers. The diopsides and diallages are mostly quite colorless to light greenish; the latter are also brown; the augites and acmites are green or brown to violet, in different shades. A deep brownish-red to brownish-violet color seems to indicate a not inconsiderable percentage of titanium. Yellowish augites are rarer, and are almost exclusively confined to certain trachytic and andesitic rocks. A red color only occurs secondarily in augites which have been heated to redness, and may be produced artificially in this way from green augites. The pyroxenes of the acid and alkali rocks are predominantly green; those of basic eruptive rocks and such as are poor in alkali are brown.   The monoclinic pyroxenes of the schists are usually colorless or greenish.

All monoclinic pyroxenes have the optic axes in the plane of symmetry; they all possess a high index of refraction and strong positive double refraction.   But the angle of the optic axes and the inclination of the bisectrices vary considerably, and in the general remarks on the optical orientation acmite must be omitted.

Des Cloizeaux found in the clear diopside from Ala for yellow light $\alpha = 1.6727$, $\beta = 1.6798$, $\gamma = 1.7062$.   Hauser found for the same occurrence, $\beta_{na} = 1.68135$.   Tchihatcheff, in diopside from Zillerthal, $\beta_{na} = 1.67996$.   A. Schmidt, in diopside from Ducktown, Tennessee, $\beta_{na} = 1.6902$.   Tschermak, in coccolite from Arendal, $\beta_p = 1.690$; in dark-green diopside from Nordmarken, $\beta_p = 1.701$; in augite from Borislau, $\beta = 1.70$; and in that from Frascati, $\beta = 1.74$, approximately. These figures explain the strong relief and the rough surface of monoclinic pyroxenes.   The difference $\gamma - \alpha = 0.0335$ determines the bright interference colors; a section parallel to $\infty P \breve{\infty}$ (010) of only 0.02 mm.

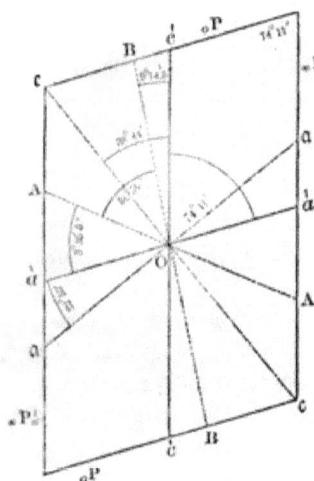
Fig. 83

thickness gives colors of the 2d order; this strong double refraction is an important means of distinguishing them from orthorhombic pyroxenes.   For all monoclinic pyroxenes, except acmite, the positive acute bisectrix lies in the obtuse axial angle, and forms a variable angle with the vertical crystal axis, which, however, is always large.   Hence sections parallel to the plane of symmetry show the maximum of darkness between crossed nicols when the prismatic cleavage is highly inclined to the principal sections of the nicols, and this large extinction angle (between 36° 30' and 54°) is one of the most characteristic properties of the monoclinic pyroxenes.   The scheme (Fig. 83) gives the optical orientation in a diopside poor in iron, and exhibits the position of the optic axes and bisectrices in the plane of symmetry.   The extinction angle varies with the chemical composition of the pyroxene, but in exactly what ratio is not yet definitely known.   It is least in the diopsides and diallages poor in alumina and iron; in these the extinction angle $c \wedge c$ lies between 36° and 40°; it increases with the percentage of iron and alumina; in the augites proper it varies from 41° to 54°, lying mostly between 43° and 48°. It naturally varies in the different isomorphous shells of zonally built crystals.

The behavior of sections in the three principal zones in parallel polarized light between crossed nicols is evident from the foregoing. In sections from the zone $oP : \infty P\bar{\infty}$ (001 : 100) the cleavage cracks form rhombic figures whose anterior angle of 84° 49′ first increases to 87° 06′, then decreases to 0° ; while the side angle decreases from 95° 11′ to 92° 54′, and then increases to 180°. The extinction is always symmetrical to the cleavage cracks ; it bisects their angle as long as they intersect one another, and lies parallel to them when they are parallel to each other. This is the behavior of all monoclinic minerals when the zonal axis coincides with an axis of the ellipsoid of elasticity. Sections from the zone $\infty P\bar{\infty} : \infty P\bar{\infty}$ (010 : 100) are recognized by the fact that the cleavage cracks always run parallel; the extinction angle has its maximum in the plane of symmetry, and decreases steadily to 0° in sections parallel to $\infty P\bar{\infty}$ (100). In sections from the zone $oP : \infty P\bar{\infty}$ (001 : 010) the cleavage cracks form rhombic figures whose anterior angle decreases from 84° 49′ to 0°. In the section parallel to oP (001) the extinction is symmetrical to the cleavage cracks, then rapidly becomes quite inclined, reaches a maximum which is slightly greater than the extinction angle on $\infty P\bar{\infty}$ (010), and then falls slowly to the angle corresponding to this face (Fig. 83).

The angle between the optic axes of monoclinic pyroxenes varies with the chemical composition just as the extinction angle does; in general, it appears to be smaller as the chemical composition approaches that of normal diopside.

Diopside, Zillerthal, Switzerland...$2V_{na} = 54° 43′$ $c \wedge c = 39°$ ....Osann.
Diopside, Ducktown, Tenn........$2V_{na} = 54° 32′$ $c \wedge c = 40° 19′$....A. Schmidt.
Coccolite, Arendal, Norway...... $2V = 58° 38′$ $c \wedge c = 40° 22′$....Tschermak.
Diopside, Ala, Tyrol.............$2V = 58° 59′$ $c \wedge c = 38° 54′$....Des Cloizeaux.
Augite, Bohemia................ $2V = 59° 28′$ $c \wedge c = 46° 40′$....Osann.
Diopside, Nordmarken...........$2V = 60°$ $c \wedge c = 46° 45′$ ...Tschermak.
Augite, Borislau ....$2V = 61°$ $c \wedge c = 45° 30′$.... "
Hedenbergite, Tunaberg, Sweden ..$2V = 62° 32′$ $c \wedge c = 45° 66′$.... "
Augite, Frascati, Tyrol...........$2V = 68°$ $c \wedge c = 54°$ .... "

From Fig. 83 it is evident that all sections of the orthodiagonal zone will show the emergence of axes or bisectrices. Cleavage plates parallel to oP (001) or $\infty P\bar{\infty}$ (100) exhibit an axis somewhat eccentric to the field of view, occasionally almost in its centre. The point of emergence of the acute bisectrix is shown in sections which correspond

to a negative ortho-hemidome, that of the obtuse bisectrix in sec
tions corresponding to a positive ortho-hemidome. One or more
brightly colored axial rings are seen about each axis even in very thin
sections because of the strong double refraction: this is not the case in
the orthorhombic pyroxenes. The inclined ~~extinction~~ is clearly seen
in sections which are perpendicular to the acute bisectrix. In the
diagonal position one hyperbola has brilliant red on the inside and
blue on the outside; on the other hyperbola the colors are reversed,
and are noticeably duller

The pleochroism of monoclinic pyroxenes is usually small, especially
in thin sections, and in general only shows itself as different shades of
the body color, green or brown. It may occasionally, however, be considerable. Thus Tschermak found in the black basaltic augite from
Frascati, c olive-green, b grass-green, a clove-brown. The porphyritic
augites of trachytes, phonolites, and andesites often have b brownish
yellow to reddish, a and c greenish; they thus resemble the orthorhombic pyroxenes, in which, however, a and c show a recognizable
difference of color. In the augites of tephrite b is often green to
greenish yellow, a and c reddish brown; in the titaniferous augites of
basaltic rocks, especially of the nepheline rocks, b is usually violet, a
and c yellowish gray to yellowish. The differences of absorption in
the direction of the principal vibrations are always small—a relation
which is to be noted in contrast to that of the hornblendes.

II. = 5-6. The specific gravity of the rock-making pyroxenes,
when pure, is never lower than 3.3. It is lowest in the diopsides and
diallages poor in iron, rises rapidly with the iron percentage, and reaches
its maximum in those pyroxenes in which the acmite molecule abounds,
when it is 3.55. This high specific gravity is important for its mechanical separation from the amphiboles, whose density is considerably
lower than that of pyroxenes of similar chemical composition.

The chemical composition of the monoclinic pyroxenes is one
which is not yet fully explained. According to Tschermak's conception, they consist of isomorphous mixtures of the molecular combinations $CaMgSi_2O_6$, $CaFeSi_2O_6$, $MgAl_2SiO_6$, $MgFe_2SiO_6$, $FeAl_2SiO_6$,
in which a small amount of manganese can replace iron, and with
which, moreover, may be combined the molecule $NaFeSi_2O_6$, which
preponderates in acmite. The compound $CaMgSiO_6$ is present almost pure in the colorless diopsides, $CaFeSi_2O_6$ in hedenbergite; the
sesquioxide-bearing molecule is not known by itself. The pyroxenes
of the diopside and diallage series consist principally of isomorphous
mixtures of the diopside and hedenbergite molecules, with only sub-

ordinate amounts of the sesquioxide-bearing compound, whose abundant occurrence on the other hand characterizes the members of the augite series. There may also be present in variable amounts, TiO_2, the compound $Na_2Al_2Si_2O_6$, besides $Mg_2Si_2O_6$ and $Fe_2Si_2O_6$.

The pyroxenes generally fuse easily to glasses in which microscopic crystallizations usually take place if the fusion is continued. They are only attacked by hydrochloric acid with difficulty, or not at all. The results in testing for the bases with hydrofluosilicic acid are often only reached upon repeated treatment. The green and yellow varieties become red to brown through the separation of ferric oxide upon being heated to redness on platinum foil.

The processes of alteration of the monoclinic pyroxenes are very different according to their chemical composition, and to the geological moments influencing them. Hence they will be described under the different varieties to which they belong.

Under *malacolite* will be included those rock-making monoclinic pyroxenes which are poor in alumina or free from it, and are not laminated parallel to the orthopinacoid. This variety appears to occur but sparingly in eruptive rocks. It forms well-developed crystals in the augite granitites of Laveline in the Vosges, and also from other localities. The colorless to light-green pyroxenes of many quartz porphyries, and those of kersantite, probably belong to this variety. Besides the perfect cleavage parallel to the prism, they are characterized by traces of cleavage parallel to the vertical pinacoids, and by a well-defined cross parting, as well as by the easy alteration into a greenish fibrous aggregate which belongs to serpentine. The alteration commences at the transverse cracks, the fibres placing themselves parallel to one another, and perpendicular to the walls of the cracks. A crystal then resolves itself into a row of fragments, each passing into a felty aggregate of fibres with which calcite is very often associated.

Malacolite is very widely disseminated in the Archæan rocks, being chiefly confined to the granular limestones, in which it occurs partly as isolated crystals, partly in prismatic or granular aggregates. From the granular limestones it may be traced to those intercalated rocks composed mainly of lime and magnesia silicates (ophiolites), found in the gneisses. In such malacolites Schumacher observed the paramorphic alteration into amphibole. Related to this is the occurrence of a colorless monoclinic pyroxene in prismatic individuals in many amphibolites and gneisses. The lime-silicate hornstones of the granite-schist contact zones contain malacolite quite abundantly, together with garnet and epidote. A confusion with the last-named mineral may be most easily avoided by observation in convergent light. In malacolite the axial plane lies parallel to the longitudinal axis and cleavage cracks, in epidote perpendicular to the same directions. It is distinguished from zoisite by its strong double refraction. The coarse malacolite aggregates from Sala, Sweden; Arendal, Norway; Stambach, Gefrees, Bavaria, etc., often show an alteration into talc scales. Malacolite possesses no constant micro-structure; the inclusions are chiefly fluid and gas interpositions of cylindrical form, which are arranged parallel to the cleavage faces.

Diallage.—The chemical composition of the rock-making diallages is in general the same as that of malacolite, with a slight admixture of the molecular group (Mg, Fe)O, (Al, Fe)$_2$O$_3$, SiO$_2$, with which is associated the acmite molecule, Na$_2$O, Fe$_2$O$_3$, 4SiO$_2$, in rocks rich in alkali (augite syenites). Morphologically they are characterized by the almost complete absence of crystallographic boundary, and the presence of a very distinct parting parallel to $\infty P \dot{\infty}$ (100), in addition to the prismatic cleavage (Pl. XX. Fig. 6). A much less distinct parting parallel to $\infty P \infty$ (010) is occasionally observed, and very rarely one parallel to oP (001). The modes of twinning are the same as those in malacolite; they are mostly developed as polysynthetic lamellæ. Diallage is very frequently filled with lamellæ of an orthorhombic pyroxene (bronzite) (Pl. XVII. Fig. 6); the latter has the prism in common with diallage, and its $\infty P \infty$ (010) coincides with $\infty P \dot{\infty}$ (100) of diallage. Much less frequently prisms of hornblende are found in diallage parallel to the parting along $\infty P \dot{\infty}$, which are probably primary.

Very frequently the same tabular microscopic interpositions occur in diallage which have been described at length under bronzite and hypersthene (p. 206). They lie chiefly in the plane of parting, arranged parallel in such a way that they appear broad and shortened in the direction of the prism axis in sections parallel to the plane of parting, while in sections parallel to $\infty P \infty$ (010) they appear narrow and relatively elongated in the direction of the prism axis. They produce the metallic sheen (schiller) on transverse faces. They also occasionally lie in an inclined face.

All longitudinal sections exhibit more or less distinctly a fibrous to prismatic structure like that of bronzite; here also it is often united with the appearance of cylindrical cavities which are frequently filled with iron ores, carbonates, and other decomposition products. The fibrous structure appears to be the result of prismatic aggregation.

The diallages become transparent with a grayish-green to green color, and in many rocks brown; index of refraction, double refraction, extinction angle ($\dot{c} \wedge c = 39° 41'$), and axial angle are nearly the same as in the malacolites. But the axial angle often falls below that which is characteristic of the diopside series; thus in diallage from Volpersdorf $2V_a = 47° 51'$. The most important optical characteristic of diallage, especially in contrast to the orthorhombic pyroxenes and bastite, is the eccentric point of emergence of an optic axis in cleavage plates parallel to $\infty\, P\,\bar{\dot{\infty}}$ (100) in convergent light.

Pleochroism is seldom observed in the diallages, and then $\mathfrak{b}$ is usually yellowish, $\mathfrak{a}$ and $\mathfrak{c}$ greenish. Differences of absorption are scarcely noticeable in thin section.

Diallage is an essential constituent of gabbro and its derivatives, as well as of many peridotites and serpentines, and forms in these rocks lamellar masses often of considerable size. These enclose the older constituents of these rocks, especially magnetite, titanic iron, chromite, and olivine. It is possible that minute inclusions of these minerals furnish the small quantities of TiO_2 and Cr_2O_3 found in the analyses of diallages. Diallage is frequently surrounded by a parallel or irregular growth of orthorhombic pyroxene and hornblende. Diallage occurs but sparingly in eruptive rocks of basaltic or andesitic habit. In these it forms prismatic crystals in which the parallel fibrous structure, the microlitic interpositions, and the intergrowth with bronzite and hornblende are wanting, but in which, on the other hand, glass inclusions are occasionally present. In the Archæan rocks diallage is seldom met with. It is here confined to the olivine rocks and olivine schists (socalled chrome diopside) and their derivatives, as well as to certain amphibolites, which may be considered as probably dynamo-metamorphic gabbros, and to the rocks of doubtful origin known as trap granulites.

The alteration processes observed in diallage may be divided into two groups. The alteration induced by the taking up of water produces fibrous and flaky aggregates of serpentine or chlorite, which not infrequently preserve the microstructure of the parent mineral, and with which are associated more or less calcite and epidote. In contrast to this process of normal atmospheric weathering, the alteration to amphibole minerals must be referred to mountain-making processes, since it appears to be confined to gabbros in the vicinity of crystalline and phyllitic schists. This alteration usually advances from the periphery toward the centre, so that in the larger individuals the central portion sometimes remains unaltered, and this process is often accompanied by a considerable deformation of the laminated crystalloids to more or less elongated streaks (*Flasergabbro*). The resulting amphiboles apparently belong to common actinolite and also to smaragdite.

Augite.—The aluminous monoclinic pyroxenes are among the commonest constituents of crystalline rocks. In eruptive rocks with porphyritic structure, less frequently in those with granular structure, it occurs in perfectly developed crystals with the form of Fig. 79, and then belongs to the older secretions of the magma. This is the case in certain minettes, more rarely in quartz porphyries, very frequently in the quartzless porphyries, porphyrites, melaphyres, teschenites, rhyolites (liparites), trachytes, phonolites, andesites, and basaltic rocks. It appears in the form of irregular columns and grains in the granular massive rocks, elæolite syenites, augite diorites, diabases, and picrites, and in the porphyritic rocks when it belongs to a second, younger generation of pyroxene as a constituent of the ground mass. Zonal structure resulting from isomorphous lamination is uncommonly wide-spread.

The cleavage parallel to the prism is almost always very distinct; besides this, pinacoidal cleavages occur, especially in the diabases, which give the augite a diallage-like appearance, but never reach the perfection of the parting in the latter mineral. On the other hand, the cleavage parallel to ∞P (110) is sometimes so imperfect in many angites, especially in basaltic and phonolitic ones, that it is not expressed microscopically by cracks, and cannot be produced macroscopically.

The colors by transmitted light are green or brown, more rarely yellow to red or violet; the extinction angle on $\infty P \infty$ (010) almost always exceeds 40°. The axial angle varies greatly, and may occasionally fall below the minimum of diopside; in general, however, it is larger than that of the diopsides and diallages.

The chemical composition varies greatly, especially in the relative amounts of MgO, Al_2O_3, SiO_2 and MgO, Fe_2O_3, SiO_2 as well as in the acmite molecule. It appears as though the latter entered largely into the combination in rocks bearing nepheline and leucite.

The augites of eruptive rocks very frequently enclose, besides the minerals of older origin associated with them (iron-ores, apatite, olivine mica), interpositions of glass, often in great number. The latter are mostly round, egg-shaped or irregularly formed, but also possess the form of their host. Fluid inclusions are less commonly met with, among them liquid carbon dioxide. Gas interpositions also are not uncommon.

Regular intergrowths with amphibole minerals have been frequently observed, both having the vertical axis and plane of symmetry in common (Pl. XXI. Fig. 1); also those with micas of the biotite series, whose basal faces coincide with a prism face of augite, especially in diorites, teschnites, and tephrites, as well as in elæolite syenites. Irregular intergrowths, which may amount to perimorphs, take place particularly with nepheline in tephrites and nepheline rocks, when the development of the augite extends into the period of the nepheline crystallization.

The normal weathering of the augite of eruptive rocks generally leads to the formation of chlorite. It usually commences from the periphery, less frequently from spots within the crystals rich in inclusions, and advances along the cleavage. Sometimes there arise parallel fibrous and parallel flaky or felty, green aggregates (Pl. XXI. Fig. 2), with low double refraction, which gradually replace the entire augite substance. This is often dotted with strongly refracting grains and spines of green epidote, or with small grains of calcite and iron-ores. Further weathering destroys the chlorite, and in its place appears a mixture of carbonates, limonite, clay, and quartz. A decomposition under the influence of stronger acids in a fluid or gaseous condition, that is, a volcanic decomposition, produces pseudomorphs of opal or chalcedony after augite by the removal of all the bases—a process which appears to be limited to the acid rocks of the trachyte and andesite families. The alteration of augite into a hornblende mineral, *uralitization*, is very common. It occurs almost exclusively in augite diorites and diabases, and will be more particularly described under uralite.

Whether the numerous augite gneisses in many gneissic regions with green monoclinic pyroxenes contain a true augite or a deeply colored malacolite, has not yet been determined. The augitic constituent of these gneisses is almost always green, very seldom brown, and usually forms irregular grains or columns elongated parallel to the vertical axis, in which only fluid inclusions are observed, and these only occasionally. The weathering phenomena are most like those of granitic malacolite. The intergrowth with amphibole minerals and the alteration into the latter are frequently observed in this variety also.

Fassaite.—The leek-green and yellowish-green varieties of common augite called fassaite appear to be confined to contact metamorphoses of marly limestone near eruptive rocks.

Omphacite is that variety of light-green common augite occurring in eclogites, which is never crystallographically bounded, and is usually in rounded grains or short columnar aggregates. It shows itself to be pyroxene by its cleavage and high extinction angle. Here it is frequently intergrown with a green hornblende (smaragdite), so that both minerals have the vertical axis and plane of symmetry in common. Many authors employ the term omphacite for pyroxenes which from

16

their cleavage and chemical composition belong to diallage, and especially for the so-called chrome diopsides. Omphacite often encloses
great quantities of rutile crystals and grains, which are so highly
characteristic of the eclogites.

Acmite and *ægirine* are monoclinic pyroxenes rich in soda, which
differ from the ordinary augites in many respects. The crystals are
almost always much elongated prisms, in which $\infty P\breve{\infty}$ (100) and ∞P
(110), with a cleavage angle of 87°, predominate; while $\alpha\check{P}\breve{\infty}$ (010)
is entirely wanting or is very slightly developed. Terminal faces sel-
dom occur, the individuals fraying out, as it were, at the ends. When
crystal boundaries are wanting acmite and ægirine form columns,
scarcely ever grains.

Twinning parallel to the orthopinacoid is common, frequently with
the insertion of several lamellæ between the larger halves. Zonal
structure, with an alternation of brown and green color, is not rare;
parallel growth with augite also occurs. Dark mica plates and amphibole occur intergrown with them in the same manner as with
augite.

The cleavage parallel to the prism of 87° is always distinctly noticeable; pinacoidal cleavage parallel to $\infty P\breve{\infty}$ may reach great perfection.

The color by transmitted light is green, or brown to brownish yel-

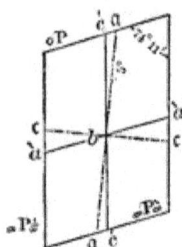

Fig. 84.

low; by incident light the crystals of ægirine are
always blackish green, those of acmite blackish
brown. When both colors occur in the same individual the peripheral portions always appear brown,
the central green. The index of refraction (β_{na} in
ægirine from Laven $= 1.8084$, Sanger) and the
double refraction are very strong. The character
of the double refraction is probably negative; the
axial angle is large. The plane of symmetry is
the axial plane. The orientation of the axes
of elasticity varies considerably from that in the other monoclinic
pyroxenes. The negative bisectrix makes an angle of 4°–5° with
the vertical axis in the acute angle β (Fig. 84); in the zonally built
occurrences, $\alpha \wedge \dot{c}$ is greater in the brown portions than in the green.
The determination of the extinction angle is facilitated by the twinning
parallel to $\infty P\breve{\infty}$ (100). The inclination of the directions of extinction
to one another in two lamellæ does not exceed 10°. The positive
bisectrix stands nearly perpendicular to the orthopinacoid. The extinction angle of rock-making ægirine appears to be somewhat larger.

The pleochroism is strong, suggesting that of hornblende. There has been observed on—

	$\mathfrak{a}$	$\mathfrak{b}$	$\mathfrak{c}$
Acmite, Porsgrund, Norway.	dark brown to greenish brown.	light brown to yellow.	greenish yellow
Acmite, Ditró, Hungary	dark brown........	brownish green.	brownish green (F. Becke)
Ægirine, Låven ...	pure green to blue-green.	olive-green.....	grass-green to yellowish
Ægirine...........	chestnut-brown	olive-green.....	grass-green (Tschermak)
Ægirine, Särna.... Sweden.	blue-green.........	sap-green......	yellowish green (Törnebohm)

The absorption is distinctly $\mathfrak{a} > \mathfrak{b} > \mathfrak{c}$, $\mathfrak{a}$ always being the axis of elasticity lying nearest the prism axis.

Sp. gr. = 3.5–3.6, greater than for the other monoclinic pyroxenes. Chemical composition essentially Na_2O, Fe_2O_3, $4SiO_2$, with variable amounts of the diopside, hedenbergite, and augite molecules. The easy fusibility with a strong coloration of the flame is very characteristic.

Acmite and ægirine appear to be confined entirely to the eruptive rocks, and to develop chiefly in magmas rich in alkalies. Thus they occur in elæolite syenite, phonolites, leucitophyres, and related rocks; also in the phonolitic trachytes of the Azores. The microstructure of these acmitic pyroxenes is the same as that of the geologically equivalent augites.

Jadeite, which is of more interest from an ethnographic than from a petrographic standpoint, forms fibrous columnar aggregates, in which a prismatic cleavage is noticeable. The cleavage according to different authors is from 85° 20′ to 89° 25′, corresponding approximately to the pyroxene prism. Arzruni, however, calls attention to the dissimilarity of the cleavage faces and to the unsymmetrical position of the direction of extinction with respect to the cleavage in cross-section, and places jadeite in the triclinic system, while Krenner refers it to the monoclinic system.

Jadeite is colorless, or almost colorless, with a tinge of greenish or bluish green. The double refraction is great, hence the brilliant interference colors. The axial plane lies in the plane of symmetry, at right angles to which, according to Des Cloizeaux, there is an imperfect cleavage; the extinction angle is large, 31°–45°; the character of the double refraction is positive, according to Krenner, who found for yellow $2H_a = 82°\ 48′$. On the face $\infty P\bar{\infty}$ is the locus of an axis with finely colored rings. Dispersion weak, $\rho < v$. The optical orientation is analogous to that of diopside.

H. $= 7$–7.5. Sp. gr. $= 3.2$–3.4. Chemical composition essentially Na_2O, Al_2O_3, $4SiO_2$; that is, an acmite in which the iron oxide is replaced by alumina. Fusible without difficulty, coloring the flame strongly with sodium.

Arzruni observed its paramorphic alteration into amphibole.

Group of Monoclinic Amphiboles.

Literature.

CH. BARROIS, Mémoire sur les schistes métamorphiques de l'île de Groix (Morbihan). Ann. Soc. géol. du Nord. Lille. 1883. XI. 18–71. cf. also Bull. Soc. min. Fr. 1883. VI. 289 and C. R. 1883. XCVII. 1446.

C. BODEWIG, Ueber den Glaukophan von Zermatt. Pogg. Ann. 1876. CXLVIII. 224.

A. VON LASAULX, Ueber das Vorkommen und die mineralogische Zusammensetzung eines neuen Glaukophangesteins von der Insel Groix. Sitzungsber. niederrhein. Ges. in Bonn. 1884. (3.) XII.

A. MICHEL-LÉVY, De l'emploi du microscope polarisant à lumière parallèle pour l'étude des plaques minces de roches éruptives. Ann. Min. Paris. 1877. (7.) XII. 429–434.

J. STRÜVER, Ueber Gastaldit, ein neues Mineral. Atti R. Accad. Lincei. Roma. (2.) XII.

G. TSCHERMAK, Ueber Pyroxen und Amphibol. T. M. M. 1871. I. 17.

— Mikroskopische Unterscheidung der Mineralien aus der Augit-, Amphibol- und Biotitgruppe. S. W. A. 1. Abthlg. 1869. LIX. Mai.

Next to the monoclinic pyroxenes the monoclinic amphiboles are the most wide-spread and important of the dark-colored ferruginous silicates occurring in rocks. Their forms are here referred to the axial system, $\dot{a} : \bar{b} : \dot{c} = 0.5318 : 1 : 0.2936$, $\beta = 75° 02'$. The rock-making

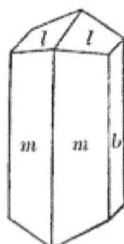

Fig. 85

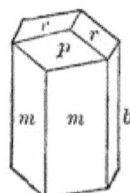

Fig. 86

amphiboles exhibit but few forms; with a constant prismatic habit the completely developed crystals are bounded in the prism zone by

$m = \infty P$ (110) with approximately 124° 30′, $b = \infty P\infty$ (010), rarely $a = \infty P\infty$ (100); in common hornblende (Fig. 85) they are terminated principally by $l = P\infty$ (011) with 148° 16′, occasionally by $p = oP$ (001), in the basaltic hornblendes (Fig. 86), by $r = P$ ($\bar{1}11$) with 148° 30′ and by p. The terminal faces are wanting in the actinolite series and generally in the common hornblendes, and the crystals become jagged and irregular or frayed out, while the basaltic varieties usually appear in well-developed forms. Hence cross-sections are acutely rhombic, with a slight truncation of the acute angles, seldom with both acute and obtuse angles truncated. Longitudinal sections parallel to $\infty P\infty$ (100) are lath-shaped, with an obtuse pair of edges above and below, or are jaggedly terminated; sections parallel to $\infty P\infty$ (010) are also lath-shaped, with inclined terminal edges, or with an obtuse, unsymmetrical termination, or a jagged one. Through the lack of crystallographic boundaries in the prism zone there arise columns, which when much shortened become grains; they are rare, however.

Twinning parallel $\infty P\infty$ (100) is frequent (Fig. 87); the twinning plane is also the composition plane, and generally passes through the centre of the crystal, so that the twinning is not indicated by the outline of the sections. Between the two larger halves of the twin, as in the pyroxenes, one or more twinned lamellæ (Pl. XXI. Fig. 3) are occasionally intercalated.

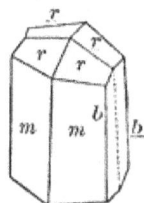

Fig. 87

The amphiboles assume microlitic dimensions less frequently than the pyroxenes do; they take the form of thin needles, more rarely that of plates, and occur as inclusions in accompanying minerals, especially in Archæan rocks, or as alteration products (uralite, pilite). Incipient forms of growth are almost unknown. Shelly or zonal structure of isomorphous layers is not uncommon, and follows the prism (Pl. XXI. Fig. 4). Chemical corrosion is confined to the amphiboles of eruptive rocks; mechanical deformations (bendings, breakings, etc.) are found in massive and schistose rocks.

All amphiboles are characterized by a perfect cleavage parallel to the prism, which generally shows itself in thin sections by systems of sharp cracks crowded closely together. Hence in sections perpendicular to the vertical axis there are two systems of equivalent cracks intersecting at 124°–125° (Pl. X. Fig. 5); in all sections of the prismatic zone the cracks are parallel; in all other sections the cleavage cracks form rhombic figures, whose angles vary with the position of the section. In some amphiboles there are indications of a parting parallel to the plane of symmetry. Cross[*] observed a distinct parting parallel to $P\check{\infty}$ (101) in the actinolite of actinolite schist from Brittany, and in the common green hornblende of a diorite from St. Brieuc.

The rock-making monoclinic amphiboles, with the exception of colorless tremolite and blue glaucophane, become transparent with green or brown colors, the green colors predominating in actinolite and almost exclusively in common hornblende, while brown colors predominate in the basaltic hornblendes. The last named are occasionally red, partly, at least, in consequence of secondary heating. The index of refraction, double refraction, and consequently the relief and interference colors, are high, but always appear, however, to be smaller than for the corresponding varieties of pyroxene. The optic axes always lie in the plane of symmetry; their angle varies considerably. The character of the double refraction is generally negative, less frequently positive. The axis of least elasticity lies in the acute axial angle β, and is in general slightly inclined to the vertical axis; however, the extinction angle varies from 0° to 27°, exceeding this value in exceptional cases. The relation between the optical orientation and the chemical composition has not yet been made out. The variation in the optical characters is shown in the following table:

	Index of Refraction.	Extinction $c \wedge \mathfrak{c}$.	Axial Angle.	Optical Character.	
Tremolite from					
Skutterud	$\beta_{na} = 1.6233$	16°	$2V_{na} = 81° 22'$	—	S. Penfield.
Tremolite	$\beta_{na} = 1.622$	15°	$2V_{na} = 88° 16'$	—	Des Cloizeaux.
Actinolite, St. Gott-					
hard	$\beta_{na} = 1.629$	15°	$2V_{na} = 80° 04'$	—	Des Cloizeaux.
Pargasite	$\beta = 1.64$	18°	$2V = 59°$	+	Tschermak.
Common Horn-					
blende from Vol-					
persdorf	$\beta_{\rho} = 1.643$	19° 53'	$2V = 85°$	+	Tschermak.
Basaltic Hornblende					
from Czernosin	$\beta = 1.710$	1° 40'	$2V = 79° 24'$	—	Haidinger.
Basaltic Hornblende					
from Bilin (?)		1°- 2°	$2H = 92° 37'$	—	Des Cloizeaux.
Basaltic Hornblende					
from Aranyer					
Berg		37° 12'	$2H_{na} = 51° 18'$	+	Franzenau.
Glaucophane, Zer-					
matt		4° 16'	$2H_{na} = 51° 11'$	—	Bodewig.
Glaucophane, Ile de					
Greix		4°	$2V_{ti} = 41° 22'$ [†]	—	v. Lasaulx.
Gastaldite	$\beta = 1.6442$	6°	$2V_{na} = 41° 26'$ [†]	—	Sauger.

Michel-Lévy determined the difference $\gamma - \dot{\alpha} = 0.0265$ for tremo-

---

[*] Studien über bretonische Gesteine. T. M. P. M. 1881. III. 386–400.

[†] Derived from the angle measured in Canada balsam, under the assumption that $n = 1.55$ for Canada balsam.

lite, $= 0.0240$ for pale-brown common hornblende, $= 0.0216$ for glaucophane from Ile de Groix, Brittany. S. Penfield determined on tremoite from Skutterud $\alpha_{na} = 1.6065$; $\gamma_{na} = 1.6340$, hence $\gamma - \alpha = 0.0275$. The dispersion in the monoclinic amphiboles is $\rho < \upsilon$.

The axis of mean elasticity coincides with the axis of symmetry; the axis of least elasticity lies in the acute axial angle β, with variable inclination to the prism axis. The greatest extinction angle is found in common hornblende, but even here it seldom exceeds 20°. It may be assumed as the rule that the angle $c \wedge \mathfrak{c}$ is 15°–18° for actinolite and common hornblende, 4°–6° for glaucophane and arfvedsonite, 0°–10° for the basaltic hornblendes. Fig. 88 presents the orientation of the axes of elasticity and of the optic axes in the clinopinacoid for actinolite and normal common hornblende. All sections from the zone oP: $\infty P\check{\infty}$ (001 : 100) between crossed nicols in parallel light behave like sections from a principal zone of an orthorhombic mineral. The cleavage cracks form rhombs whose obtuse angle of 122° 30′ on oP at first increases to 124° 30′, then decreases to 0°. The extinction lies diagonal to the cleavage cracks, as long as these intersect; it is parallel and normal to these when they are parallel to each other.

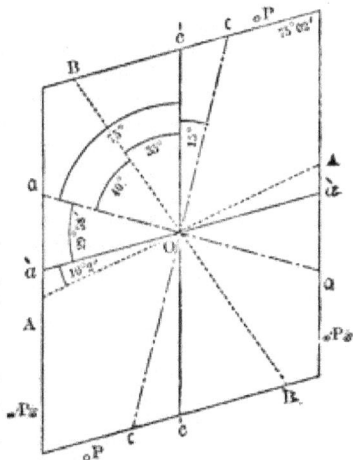

Fig. 88

In the zone $\infty P\check{\infty}$: $\infty P\check{\infty}$ (100 : 010) the cleavage cracks are always parallel to one another; the extinction angle measured from them increases from 0° on $\infty P\check{\infty}$ (100) to a maximum of the angle $c \wedge \mathfrak{c}$ (0° to 20°) on the clinopinacoid. In the zone oP: $\infty P\check{\infty}$ (001 : 010) the acute angle of the cleavage cracks decreases from 57° 30′ on oP (001) to 0° on $\infty P\check{\infty}$ (010). The extinction lies diagonal in the section parallel to oP, and unsymmetrical in all other sections.

All sections in the orthodiagonal zone show the emergence of axes or bisectrices in convergent light; the interference figures in general are not so brilliantly colored nor surrounded by so many rings as those of pyroxene; and the axial plane indicated by the straightened axial bar bisects the field of view and the cleavage cracks, when these intersect one another, and lies parallel to them in sections parallel to $\infty P\check{\infty}$ (100). The dispersion is distinctly noticeable on the hyperbola of the axis emerging from $\infty P\breve{\infty}$ (100), but scarcely or not at all on that emerging from oP (001).

The behavior of sections parallel to the three principal zones in twins parallel to $\infty P\bar{\infty}$ (100) differs according to the value of the angle $c \wedge c$, and may be deduced from what has just been said concerning the action of a single individual in these zones.

All amphiboles, which are not colorless, possess a distinct pleochroism, which increases rapidly with the depth of the color, and is specially strong in the brown varieties. The colors are very different for the different varieties, and will be described under each. Locally stronger pleochroism is sometimes due to a zonal change of coloring, at other times to pigments, which have concentrated themselves about inclusions. Faintly green amphiboles with little pleochroism may often be permanently colored intensely red, and become strongly pleochroic by being heated to redness on platinum foil. The differences of absorption in the direction of the three axes of elasticity are generally very noticeable: the absorption parallel to the negative bisectrix is the weakest; that parallel to the positive bisectrix the strongest; the absorption parallel to the axis of symmetry is sometimes very great. Hence in general $c \gtrless b > a$.

The specific gravity of the amphiboles is always less than that of chemically similar pyroxenes, with the exception of the varieties rich in alkali and iron oxide, which possess nearly the same density in both series. The lightest amphiboles are those free from alumina (sp. gr. $= 2.9$–3.16) and members of the glaucophane series (3.05–3.15); the hornblendes proper have 3.15–3.33, and the density increases with the iron percentage. This smaller density is useful in the mechanical separation of the amphiboles and pyroxenes. In general, the amphiboles are more strongly attracted by an electro-magnet than the pyroxenes of analogous composition.

The chemical composition of the amphiboles is not as well known as that of the pyroxenes. The members of the actinolite series free from alumina are essentially isomorphous mixtures of $3MgO$, CaO, $4SiO_2$, and $3FeO$, CaO, $4SiO_2$, in which the magnesia-lime molecule always predominates. In the arfvedsonites occurs the molecule NaO_2, Fe_2O_3, $4SiO_2$, which corresponds to the acmite molecule of the pyroxenes; in the glaucophanes there is chiefly the jadeite molecule Na_2O, Al_2O_3, $4SiO_2$, with variable amount of the actinolite molecule. It is not yet certain how the alumina and iron-oxide percentages of the hornblendes proper are to be expressed: they are sometimes considered as analogous to the molecule $(Mg, Fe)O(Al_2, Fe_2)O_3SiO_2$ of the pyroxenes, while others consider the compound R_2O, as isomorphous with $\overset{..}{R}SiO_3$. R. Scharizer has undertaken to show that there exists in the hornblendes an isomorphous mixture of the actinolite molecule with a molecule $(\overset{!}{R}_2, \overset{!!}{R})_2 (Al, Fe)_2 Si_3 O_{12}$, which he designates as the syntagmatite molecule. In the latter, according to his conception, the relation between the monoxides is always $(CaO + R_2O) : (MgO + FeO + MnO) = 3 : 4$. The *rôle* of the titanic acid, water, and fluorine given in many analyses of amphiboles is uncertain. Unfortunately, there is only a limited number of investigations of rock-making amphiboles.

Tremolite occurs in columnar and lamellar masses and individuals in the granular limestone of the Archæan, in many silicate hornstones, and with olivine and its alteration products in certain olivine rocks and serpentines. In the latter it sometimes occurs as an original constituent, at other times as a secondary one. The amphibole cleavage and strong double refraction in connection with its colorlessness fully characterize it. A transverse parting is quite common in addition to the cleavage. The individuals often fray out at the ends, and pass over into asbestus-like aggregates. To distinguish it from muscovite, talc, and wollastonite, with which it may be confounded in certain sections, it should be investigated in convergent light. In tremolite the axial plane lies parallel to the cleavage, in the others normal to it. Tremolite appears as an alteration product in the form of a marginal border about olivine in many Scandinavian olivine diabases and olivine gabbros. Tremolite usually alters into talc, the talc scales penetrating the tremolite substance by degrees from the periphery, from fissures and cleavage cracks, until in certain stages of the process it lies in the form of oblong and acutely rhombic meshes within a net of talc scales.

Actinolite also forms prismatic individuals or columnar and fibrous aggregates, on which terminal faces never occur. It is distinguished from tremolite by its more or less green color. Besides the prismatic cleavage, there is occasionally one parallel to $\infty P \overset{\infty}{\infty}$ (010). The separation of the columns at right angles to their axis is common. When but slightly colored and in thin sections the pleochroism is scarcely noticeable; when more strongly colored, the absorption is distinct, $\mathfrak{c} > \mathfrak{b} > \mathfrak{a}$, even when all the rays are green; or there may be a yellowish tone in rays vibrating parallel to $\mathfrak{b}$ and $\mathfrak{a}$, while the color parallel to $\mathfrak{c}$ is green. Actinolite, like tremolite, is free from inclusions of the minerals associated with it. The real home of actinolite is in the Archæan, where it forms, either alone or in combination with pyroxene, epidote, or chlorite, the varied series of actinolite schists; it occurs with quartz and albite in many green schists, and as an accessory in chlorite and talcose schists. It appears as a secondary constituent in diabases and schalsteins and in gabbros, and in these is an alteration product of pyroxene or occasionally of olivine. The emerald-green actinolite, which occurs in the so-called saussurite gabbros, is called smaragdite. It forms very delicate columnar aggregates of a pale greenish white color by transmitted light, which are only transparent when very thin—a consequence of the delicate aggregation. Smaragdite aggregates often appear in the form of diallage, as pseudomorphs or probably as paramorphs after the diallage. Actinolite does not occur as a primary constituent of eruptive rocks. Decomposition processes are seldom observed in actinolite; it passes into fibrous and scaly cryptocrystalline aggregates with a green color which may belong to serpentine. The calcium component is usually secreted as calcite in small grains and rounded masses.

Nephrite or *jade* is a felty, fibrous actinolite with a more or less obscure schistose structure. There are undoubted occurrences of it on Batugol Mountain in Eastern Siberia, in the Kuenluen, in New Zealand. Traube appears to have found an occurrence at Jordansmühl intimately associated with serpentine and granulite.

Common hornblende only forms regularly bounded crystals in those old eruptive rocks with porphyritic structure, for example, in certain granite porphyries, syenite porphyries, and diorite porphyrites; in the granular massive rocks of the older formations and in the Archæan it appears in more or less distinctly prismatic individuals, less frequently in plates or grains. A peculiar variety is the so-called reedy (" schilfige") hornblende, which consists of approximately parallel columnar to fibrous amphibole aggregates with a light-green color and slight pleochroism, and which is usually mixed with epidote, and chlorite, and is common in certain eruptive rocks of the diabase series and in many amphibolites. In many cases it can be shown to have originated from augite, and this is probably true for all its occurrences. It is therefore a uralitic hornblende, and its characters are more closely related to actinolite than to common hornblende. In distinction to these reedy aggregates the hornblendes proper are termed *compact*.

Common hornblende is mostly colored green; it is deep brown to brownish red in tonalite, in many diorite porphyrites, in teschenites; less frequently in the diorites, gabbros, and Archæan rocks. Green hornblende has an extinction angle like that of actinolite, or still higher; in cleavage plates parallel to the prism $c \wedge c = 13°$ or more. The pleochroism is confined to green tones, and only those rays vibrating parallel to a occasionally appear yellow. The green parallel to b often has a tinge of brown, that parallel to c a tinge of blue. Brown hornblende is generally more pleochroic than the green : the colors along c and b are brown in different shades ; a is yellowish or rarely greenish. The angle $c \wedge c$ is smaller; on cleavage plates parallel to ∞P (110) the extinction angle is at most 13°, and may fall almost to 0°.

Common hornblende possesses no constant microstructure ; it generally encloses the ores and apatite, or other minerals associated with it which are older than it is. In many eruptive rocks it carries the interpositions characteristic of hypersthene and diallage, in the Archæan rocks it often contains rutile. Parallel intergrowths with pyroxene are frequent ; the hornblende usually lies peripherally about the pyroxene, having the axes b and c in common. More rarely the pyroxene surrounds the hornblende (in some elæolite syenites) and then appears to have been derived from the hornblende by magmatic processes. Thus in the granular eruptive rocks the formation of pyroxene appears to have preceded that of amphibole. When it is intergrown with biotite, the latter appears to have been the older, and generally lies with its base on the cleavage faces of the hornblende.

The alteration of hornblende to chlorite, with the secretion of epidote or calcite and quartz, is a wide-spread process of weathering ; since the chlorite may further alter into a mixture of carbonates, clay, limonite, and quartz, there arise pseudomorphs of these minerals after amphibole. The hornblende frays out or becomes fibrous during the chloritization, and since the chlorite scales accumulate from the periphery and cleavage faces, such a pseudomorph may closely resemble an aggregate of reedy hornblende. They may be distinguished by treatment with acid, with which chlorite gelatinizes, while hornblende is not attacked, or at most gives up iron to the reagents.

Basaltic hornblende is almost always well crystallized ; when the outward form is wanting it is evident that it has been lost through mechanical processes or magmatic resorption. The cleavage shows a high degree of perfection, and the cleavage faces have a high lustre. It is black by incident light, brown by transmitted light in most every instance, usually in deep tones. A green color occasionally arises from chemical alteration, and produces a decided diminution in the lustre. An isomorphous lamination is not uncommon, in which differently colored zones alternate with one another, usually in shades of brown, rarely brown and green ; they are always few in number, mostly consisting of a kernel and shell. The pleochroism is almost always very strong, and varies from dark brown for c to light yellow for a ; occa-

sionally a is greenish. The absorption, $c \gtrless b > a$, common to all am.
phiboles, reaches its maximum, and at times is as intense as that of
biotite. The extinction angles, with few exceptions (Arany), are
small, and may fall as low as 0°.

Basaltic hornblende is confined to porphyritic eruptive rocks, and
forms crystals in them, which are among the oldest secretions of the
magma. Hence glass inclusions are frequent; besides inclusions of
the ores, apatite, biotite, olivine, and other older constituents. Intergrowths with pyroxene occur, the latter lying peripherally, and being
younger than the hornblende. The alterations produced by the action
of the atmosphere and of thermal waters are the same as those of
common hornblende, and lead to the formation of chlorite, carbonates, limonite, and quartz. Quite different, however, are certain
alterations which can only be explained as the result of resorbing
actions of the magma. The outlines of hornblende crystals in porphyrites, trachytes, phonolites, and andesites, as well as in basalts and tephrites, are variously rounded and melted down; and immediately surrounding the crystal lies a dark zone, which in most cases is formed of
opaque grains of ore and columns or grains of augite: the latter not
infrequently lie parallel to one another and to the hornblende crystal.
That this aggregation of augite and opaque grains is the result of a
magmatic paramorphism of hornblende, is shown by the fact that it may
completely replace the hornblende crystal without changing its form.
This alteration belongs to a period in the development of the rock
when hornblende was no longer capable of existing in the magma, and
became melted and transformed into augite, probably accompanied by
the separating out of an iron oxide. Very rarely the resorption of
basaltic hornblende appears to be followed by a new formation of the
same, which then surrounds the older secretion in the form of microlites.

Arfvedsonite forms columnar individuals in many elæolite syenites,
and in the south Norwegian augite syenites; it forms perfectly developed
crystals in certain phonolites and leucitophyres. Its colors are brown
and green. The pleochroism and absorption are strong, and vary, as in
basaltic hornblende, between deep dark brown and yellow for the brown
varieties, and between deep olive-green to blue-green and muddy yellowish green for green varieties. The extinction angle is rather
higher than for basaltic hornblendes of the same intensity of color. It
is further distinguished from the latter by its higher specific gravity
and strong sodium reaction, together with its very easy fusibility.

Glaucophane always forms prismatic individuals which are bounded
by ∞P (110), occasionally by $\infty P \bar{\infty}$ (010) or $\infty P \bar{\infty}$ (100), and pos-

sess no terminal faces. The perfect cleavage parallel to the prism with the amphibole angle ($124° 25'$–$124° 44'$) and the blue color by incident light make it easily recognizable. It is also characterized by a transverse parting. Its place in the amphibole series corresponds nearly to that of jadeite in the pyroxene series. The extinction angle is very small, $4°$–$6°$, in the plane of symmetry. The pleochroism is very strong and fine: c = sky-blue to ultramarine-blue, seldom blue-green; b = reddish violet to bluish violet; a = almost colorless to yellowish gray. Sp. gr. = 3.0–3.1.

Glaucophane, with which should be classed gastaldite, is almost exclusively confined to the Archæan rocks, occurring in mica schists, eclogite, and phyllitic gneiss. An asbestus-like glaucophane (crocidolite) has been found in contact-metamorphosed limestones of Breuschthal in the Vosges. The paragenesis of glaucophane is the same as that of actinolite and common hornblende; it is associated with diallage, omphacite, garnet, epidote, mica, and rutile. Its occurrence in a minette in the neighborhood of Wachenbach in Breuschthal, Vosges, is exceptional.

Appendix.—*Uralite* is a paramorph of amphibole after pyroxene, having the crystal form of the latter, and the physical characters and usually the cleavage of the former. It appears, though, that in this transformation a part of the lime separates out, for finely divided calcite or epidote often accompanies these paramorphs.

The alteration of augite into hornblende usually proceeds from the periphery toward the centre and from the cracks inward, so that within the uralite there are often remnants of unaltered augite. In this process the vertical axis and the axis of symmetry of the parent mineral remain the same for the new one. The uralite, however, does not form a single compact crystal, but consists of numerous slender columns exactly parallel to one another. Cross-sections exhibit the hornblende cleavage traversing the whole extent of the section (Pl. XXI. Fig. 5), while longitudinal sections appear finely fibrous (Pl. XXI. Fig. 6). If the original augite individual was twinned parallel to $\infty P\breve{\imath}$ (100), then columns of uralite along the original composition plane stand in twinned position to one another.

Uralite is always green, and exhibits the pleochroism of common green hornblende, c and b green, a yellowish green. The specific gravity is that of hornblende.

Uralite is common in diabases, diabase porphyrites, and related rocks when these lie imbedded in faulted schists. It is also found in many augite diorites and augite syenites. It is in general absent from the younger augite rocks, but occurs in these whenever they have been subjected to the same mechanical processes which the palæozoic masses have undergone.   Whether the uralite belongs to common hornblende or to actinolite depends on the original composition of the parent pyroxene mineral.

The Mica Group.

Literature.

M. BAUER, Untersuchungen über den Glimmer und verwandte Mineralien.   Pogg. Ann. 1869. CXXXVIII. 337-370.
— Ueber einige physikalische Verhältnisse des Glimmers.   Z. D. G. G. 1874. XXVIII. 137-186.
E. REUSCH, Ueber die Körnerprobe am zweiaxigen Glimmer.   Pogg. Ann. 1869. CXXXVI. 130 and 632.
G. TSCHERMAK, Mikroskopische Unterscheidung der Mineralien aus der Augit-, Amphibol- und Biotitgruppe.   S. W. A. 1869. LIX. May number.
— Die Glimmergruppe.   S. W. A. 1877. LXXVI. and 1878. LXXVIII ; also Z. X. 1878. II. 14-49 and 1879. III. 122-167.

The micas are distinguished from all other monoclinic minerals by the fact that in the form of their crystals, and in their optical behavior, they approach very closely to hexagonal or orthorhombic substances ; in many instances it is still practically impossible to prove their mono-clinic nature.

Rock-making micas when they exhibit an outward crystal form appear almost exclusively in thin hexagonal plates whose plane angles are exactly 120°.   Less frequently these plates reach a thickness of several millimetres ; in this case it can be shown that only one of the three pairs of vertical faces, namely, $b = \infty P\check{\infty}$ (010), stands at right angles to the basal plane.   It is seldom possible to determine the inclina-tion of the other pairs of faces accurately enough to indicate them crys-tallographically.   The faces most frequently met with on micas of the biotite series are: $c = oP$ (001), $b = \infty P\check{\infty}$ (010), $m = P(\bar{1}11)$, $o = -\frac{1}{2}P$ (112) (Fig. 89).   An orthodome $mP\infty$ (hol) and faces in the zone oP: $\infty P\check{3}$ are very rare.   Occasionally, however, the orthodome and clinopinacoid predominate to such an extent that in many rhyolites, trachytes, and andesites the mica plates appear to be rectangular, as O. Mügge[*] has observed in the hornblende andesites of the Azores. On micas of the phlogopite and muscovite series the faces $M = 2P$ (221) and a clinodome are more commonly observed.   The most im-

---

[*] Petrographische Untersuchung an den Gesteinen der Azoren.   N. J. B. 1883. II. 222.

portant angles are $c \wedge o = 73° 02'$, $c \wedge m = 81° 19'$, $c \wedge M = 85° 38'$, $c \wedge b = 90°$, $m \wedge m = 59° 16'$ for the meroxenes, and but slightly different for the muscovites.

Mica plates from porphyritic rocks often exhibit re-entrant angles on the faces in the nearly vertical zone, which result from a twinning in which the individuals are symmetrical to a left or right prism face. The commonest mode of composition is that in which the twinned individuals join along their basal planes (Fig. 90). It also frequently happens that two or more individuals penetrate one another quite irregularly, so that a thin cleavage plate consists of two or three individuals whose boundaries toward one another are irregular lines. In many rocks, especially the minettes, the mica plates are elongated in

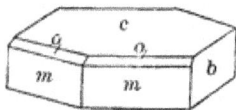

Fig. 89

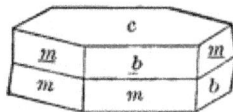

Fig. 90

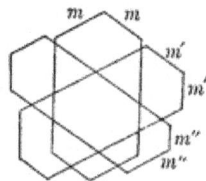

Fig. 91

the direction of a diagonal, and when twinned, the separate individuals project laterally, as indicated in Fig. 91. The composition plane of the twins is very rarely a lateral face. Twins with common terminal faces, in which the individuals are turned 30° to one another, are quite rare, the twinning plane being in the zone $oP : \infty \dot{P}3$ (001 : 130). The cross-sections of mica crystals and twins parallel to the base, therefore, are hexagonal, very rarely lath-shaped; when perpendicular or inclined to it they are more or less narrow lath-shaped.

The dimensions of mica crystals sink to microscopic proportions; incipient forms of growth and skeleton crystals do not occur. However, parallel growths of very small plates forming larger crystals are met with, especially in glassy rocks.

In many rocks the micas possess no crystallographic boundaries except the basal planes; they then form variously notched and jagged plates, or parallel and rosette-like aggregates which may grow to shells and balls. Sections parallel to the face of such plates then are irregular lateral ones always lath-shaped. Zonal structure or isomorphous lamination is not uncommon in the dark micas of the biotite and phlogopite series; from this it is evident that the growth follows the lateral faces, sometimes the base. In the first instance the bands of growth form concentric hexagons; in the second, parallel lines. The intergrowth of different varieties of mica (biotite and muscovite) with one another follows the same directions. Sometimes muscovite surrounds biotite like a mantle, at other times it lies on its upper and lower sides; the biotite is always inside and the muscovite out.

Chemical corrosion occurs, especially on the older secretions of biotite in porphyritic rocks, and is usually in the form of a marginal alteration, which will be described in another place. Mechanical deformations are common to all varieties of mica, and consist of bending, slipping along the gliding plane, curving of the crystals and the rolling out of the same. The first two kinds of deformation are particularly common in the secretions of porphyritic eruptive rocks (Pl. IV. Fig. 5). Micas whose plates have been completely rolled out until they form a row of elongated scales are chiefly met with in granitic rocks of highly faulted mountains and in the Archæan rocks.

All micas cleave very perfectly along the basal plane, and the cleavage plates, when sufficiently thin, are elastic. Hence basal sections show no cleavage cracks, but all others exhibit very sharp and abundant cleavage lines, which are parallel to themselves and to the sides of the lath-shaped sections (Pl. X. Fig. 6). The elasticity of the plates is greatest in the muscovites, decreases almost to brittleness

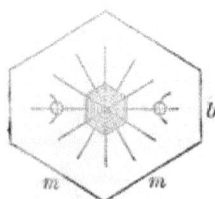 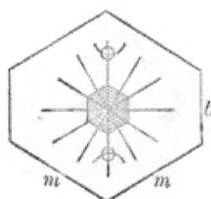

Fig. 92 a   Fig. 92 b

in the phlogopites and biotites, and disappears rapidly upon the alteration of the last-named mica into chlorite aggregates, giving place to ordinary flexibility. This perfect basal cleavage is one of the most important diagnostic characters of the mica minerals, appearing to the same extent and kind only in the chloritoids and chlorites. There are other cohesion minima in mica which are of diagnostic importance. They may be detected by striking the mica plate a quick, elastic blow with a needle point, when there will appear about the point struck a six-rayed star, the rays or cracks intersecting at 60° (Fig. 92) and lying parallel to the edge $c : b$ and to the edges $c : m$. The first ray which lies parallel to the projection of the plane of symmetry on the basal plane is called the characteristic or *leading ray*. This figure is of great importance in the optical determination of the micas, and is known as the *percussion figure*.

If, on the other hand, the mica plate be pressed by a dull-pointed instrument without being pierced, there arises another six-rayed star, or *pressure figure*, whose rays intersect at 60°. The rays of the pressure figure are so placed that they stand at right angles to those of the percussion ·figure, each to each. In Fig. 92 the pressure figure is dotted, its rays lie parallel to the edges $oP : mP\frac{\cdot}{\infty}$ and $oP : \infty P3$; they are not sharp, but fibrous, and generally spread out in tufts. The pressure figure is often incomplete, one or even two of the rays failing to appear. Lines parallel to the pressure figure—that is, normal to the ordinary boundary of the mica plates—faults, and planes of separation parallel to these lines, are very common in rock-making micas (Pl. IV. Fig. 5), and are apparently due to mountain pressures. Moreover, regularly interposed crystals of foreign bodies are usually arranged parallel to the rays of the pressure figure.

The micas become transparent in very different colors according to their chemical composition : the members of the muscovite and phlogopite series are colorless to light yellowish or light greenish, and often exhibit in basal sections beautifully iridescent flakes and circles produced by numerous minute scales loosened in the grinding. The rock-making biotites are deep brown or green, also red to almost opaque. The index of refraction is not large ; the double refraction, however, is very strong : both appear to increase with the iron percentage. Bauer determined on muscovite $\alpha = 1.537$, $\beta = 1.541$, $\gamma = 1.572$; therefore, $\gamma - \alpha = 0.035$. Michel-Lévy found on the muscovite of granite from Montchanin (Saône-et-Loire), $\gamma - \alpha = 0.035$; on meroxene from Somma, $\gamma - \alpha = 0.0404$; on biotite from Pranal, Auvergne, $\gamma - \alpha = 0.060$. Haidinger determined on a Brazilian mica (apparently muscovite) for the ray vibrating at right angles to the cleavage, $n = 1.581$; for that parallel to it, $n = 1.613$. The necessarily brilliant interference colors are important in distinguishing the micas from the faintly doubly refracting chlorites.

All micas are optically negative, and the acute bisectrix $\mathfrak{a}$ is always about normal to the cleavage face oP; generally, its divergence from the normal to oP is scarcely measurable, but in many micas reaches 9°. In some micas (meroxene, lepidomelane, phlogopite, zinnwaldite) the axial plane coincides with the plane of symmetry; its trace on the basal plane is parallel to the leading ray of the percussion figure and is normal to a ray of the pressure figure; these kinds of mica are called "mica of the second order" (Fig. 92b). In muscovite, lepidolite, phengite, paragonite, and anomite the axial plane is normal to the plane of symmetry, and its trace on oP is normal to the leading

ray of the percussion figure; they are called "mica of the first order"
(Fig. 92a). The axial angle varies from almost 0° in lepidomelane,
many biotites and anomites, to 80° in many muscovites.

Since the first bisectrix a is nearly normal to the basal plane,
the second bisectrix c in mica of the first order nearly coincides
with the clinodiagonal a, the axis of mean elasticity exactly with the
axis of symmetry; in mica of the second order b nearly coincides with
a, c exactly with b. From this it follows that in all but basal sections
the extinction between crossed nicols is parallel and normal to the
cleavage. Indeed, an inclined extinction is but seldom observed, and
is most noticeable in sections of twins more or less inclined to the
cleavage face. Basal sections of mica are more distinctly doubly
refracting the larger the axial angle, and approach more closely to the
isotropic behavior in parallel light the nearer the angle 2 V is to 0°.
Hence basal sections of biotite are often not noticeably doubly refracting, but apparently isotropic, and the characteristic feature in the
optical behavior of the micas is that the biotites closely approach
the hexagonal system, the muscovites and phlogopites the orthorhombic.

The phenomena in convergent light are quite analogous to the
foregoing: every cleavage plate furnishes an axial figure—in the dark-
colored biotites a dark cross, which scarcely opens during rotation, often
not noticeably, and whose locus is usually in the centre of the field of
view. In the light-colored micas the interference figure is apparently
that of an orthorhombic mineral cut at right angles to the negative
bisectrix. The size of the axial angle and the dispersion will be given
under the different varieties.

The colored micas are strongly pleochroic, and in all of them
the rays vibrating parallel to the cleavage are far more strongly
absorbed than those normal to it. The absorption and colors are
different for each variety of mica, and will be described under each
variety. Pleochroic halos sometimes occur around microscopic interpositions in all kinds of micas; they always exhibit the minimum of
darkness when the light vibrates perpendicularly to the cleavage.

The specific gravity varies with the composition, between 2.75–3.2,
and is difficult to determine with accuracy on account of the tabular
form of the mica and of the difficulty in moistening the plates with
fluids. Consequently, flakes of mica in a rock powder remain suspended
in a heavy solution of much lower density than that of the mica itself.

The chemical composition of rock-making micas has been only
slightly investigated. Following Tschermak's theory of the constitu-

tion of micas, they consist of the isomorphous molecules $Si_6Al_6K_6O_{24}$ $= K$ and $Si_6Mg_{12}O_{24} = M$, either alone or in combination; with which in some varieties is associated the compound $Si_{12}H_8O_{24} = S$ or $Si_{16}O_8Fl_{24} = S'$. Titanium may enter into the combination to a considerable extent; alumina is replaced by sesquioxide of iron in many varieties of mica (lepidomelane), and magnesia quite generally by protoxide of iron and manganese. The potassium in the compound K is in part replaced by hydrogen, sodium, and lithium. The muscovites and phlogopites are but slightly attacked by acids; biotites are strongly attacked at high temperatures.

Under the name *biotite* are here included those varieties of magnesia mica which Tschermak has termed meroxene and lepidomelane. They are isomorphous mixtures of the molecules K and M, in which the molecule K generally predominates. They contain potash and water, but only a little soda, and scarcely any lithia. They are the heaviest micas, with a specific gravity from 2.8–3.2, which increases with the percentage of iron. All biotites are micas of the 2d order; the axial plane lies in the plane of symmetry; the inclination of the bisectrix $\mathfrak{a}$ to the normal to oP is generally very small, the axial angle extremely variable. In the rock-making biotites the axial angle is mostly very small, so that the biaxial character is scarcely determinable. Nevertheless, even here there are values for $2E$ which exceed the limit of 56°, given by Tschermak. The inclination of the bisectrix to the normal to oP, which is generally less than 1°, occasionally reaches 5° to 8°. Large extinction angles appear to accompany large axial angles. Small axial angles occur both in faintly colored and strongly colored biotites; large axial angles exclusively in strongly colored ones. The pleochroism is always strong; the rays (𝔟 and 𝔠) vibrating parallel to the cleavage are almost completely absorbed in the dark-colored biotites, those parallel to only slightly. Differences between the rays vibrating parallel to 𝔟 and 𝔠 are more noticeable as the angle $2E$ is greater, and sometimes 𝔟, sometimes 𝔠, is more strongly absorbed. The absorption scheme is $\mathfrak{c} \lessgtr \mathfrak{b} > \mathfrak{a}$. Plates parallel to the cleavage are dark brown or dark green to opaque, with slight difference of color when rotated over the polarizer; sections inclined to the cleavage are dark brown or dark green when the cleavage cracks lie parallel to the principal section of the polarizer, light yellow to red or light green when at right angles to it. Similarly strong differences of absorption are only exhibited in basaltic hornblendes, in tourmaline and allanite.

Biotite is equally common in the massive rocks and in the Archæan, and is one of the most characteristic products of contact-metamorphism in certain rocks. In the eruptive rocks it is one of the oldest secretions, being formed immediately after the ores, zircons, and apatites, which minerals are frequently included in the biotite. Fluid inclusions are found in it everywhere, but not constantly. They are not generally observed in the thin sections, as they usually disappear in the process of grinding; on the other hand, they are found in plates split off from the rock more frequently than would be expected, judging from the perfect cleavage of the mineral. Biotite twins occur in all eruptive rocks, which, however, are only recognized by certain distortions of the interference figure in convergent light, and by the pleochroism in sections inclined to the base. When the interference figure is distinctly biaxial, it is often observed that the axial rings are divided into halves, which do not exactly fit one another—a phenomenon which has been imitated by Bauer by placing on one crystal a thin plate turned 60° to the first, or in the position of a twin along $\infty P'$ (110). Since the rays vibrating parallel to b and c are differently absorbed, and the lamellæ of a twinned crystal are cut in different directions by the section, they exhibit different colors in consequence of the pleochroism, and also different interference colors between crossed nicols. Cohen states that the pleochroic halos in biotite are dispelled by being heated to redness, after the mineral has been treated with acids.

The biotites of the older granular massive rocks occasionally enclose great quantities of rutile needles and sagenite webs: in many cases these are undoubtedly primary inclusions; in others probably secondary, since they only occur in the partly altered biotites, and are wanting in the perfectly fresh ones. This indicates the presence of titanium in the fresh biotites. The primary as well as the secondary rutile needles seldom lie irregularly, but are frequently in three systems, which intersect at 60°, and are parallel to the rays of the pressure figure (Pl. XXII. Fig. 1). Occasionally, single needles of rutile are found parallel to the rays of the percussion figure, and then almost always parallel to the leading ray.

In the porphyritic rocks biotite generally occurs only as one of the oldest generations; its crystallization occasionally repeats itself in a second generation as a constituent of the groundmass. If the older biotite has been subjected to magmatic corrosion, it is surrounded, like basaltic hornblende, with a dark border, which consists of a mixture of magnetite and augite. Glass inclusions have not been observed in the biotites.

The biotite of eruptive rocks is found regularly intergrown with

basaltic hornblende and augite, the basal plane of the former coinciding with the cleavage faces of the latter minerals, as in granites, syenites, diorites, trachytes, andesites, etc.   It occurs intergrown with
muscovite in true granites only.   The biotite of Archæan rocks does
not form crystals, but irregularly bounded flakes and plates, usually
elongated in the direction of the schistosity.   Otherwise it possesses
the properties of the biotite in eruptive rocks.   Its frequent intergrowth with muscovite and paragonite is to be noted.   Rutile interpositions also occur as in the granular massive rocks.   Evidences of
chemical corrosion are entirely absent, but those of mechanical deformation are wide-spread.   The color is usually brown in the Archæan
rocks, but green colors are not uncommon, especially when accompanied by green amphibole ; green biotite never occurs in porphyritic
massive rocks, and very rarely in the granular ones.

The biotite, which is one of the most distinctive minerals of the
hornstones in the granite contact zones, forms round or indented
plates of highly characteristic chocolate-brown color; it is green in
only a few localities, and then is generally weakly pleochroic.   The
biotites are comparatively easily decomposed minerals.   At first, under
the action of the natural reagents, the brown color is changed to green
without affecting any other of the optical properties, but the elasticity
of the plates disappears.   In a more advanced stage the green color
fades out, and the mica is completely bleached.   This process, which
appears to be a leeching out of the iron, starts from the periphery and
proceeds along the cleavage, often very irregularly.   In other cases
biotite is altered into green chlorite; the strong double refraction decreases rapidly ; the distinct lamellar structure gives place to scaly
fibrous structure, combined with which there is often a fraying out of
the biotite (Pl. XXII. Fig. 2).   At the same time lenticular masses
of the carbonates are deposited between the lamellæ, together with
quartz and the iron ores ; or, in place of the carbonates and quartz,
epidote may occur under the same conditions (Pl. XXII. Fig. 3).
Upon the further advancement of this process, the biotite may be
completely pseudomorphosed into a mixture of carbonates or epidote
with iron ores and quartz.

Under *anomite* are here included those rock-making micas which
from habit and color belong to the biotite series, but which are distinguished from biotite by the fact that the axial plane is not parallel to
the leading ray of the percussion figure—that is, does not lie in the
plane of symmetry, but is normal to it.   In form and pleochroism
they are quite like the biotites ; the twinning, which is particularly

frequent, is also the same. The inclination of the bisectrix a to the normal to oP is generally greater, as Tschermak found it to be in the mica which he called anomite from Greenwood Furnace, N. J., and from Lake Baikal, Siberia. It reaches 4°, and facilitates the recognition of the twinning in longitudinal sections. The axial angle is small, about 10°, but in many occurrences reaches 25° and over. Tschermak found the dispersion $\rho > v$, while for the rock-making anomites it is oftener $\rho < v$. Evidently there are micas of the second order here called anomite, which cannot be directly united with the anomite of Tschermak. This is in consequence of the lack of chemical investigation upon the dark rock-making micas of the biotite series with normal symmetrical axial position.

F. Becke* found anomite as a constituent of a quartz diorite porphyrite from Steinegg, in Lower Austria, in zonally built crystals; the light greenish brown centre is surrounded by a dark brown shell which behaves uniaxially. According to this, the negative axial angle in anomite decreases with the percentage of iron, which Tschermak found in the occurrences cited. Becke† also found it as a secondary mineral in an altered olivine rock occurring as an intercalated mass in the diorite schist at Dürnstein, Lower Austria. The color is reddish brown, $2E = 18°\,54'$. Eichstadt‡ recognized it in great plates in a mellilite basalt from Alnö, described by Törnebohm; this is also reddish brown, $2E = 8°\text{--}10°$. G. Lattermann found anomite and biotite associated with one another in the same rock: for example, in kersantite from Michaelstein, near Blankenburg, in the Hartz Mts. ($2E = 10°\text{--}22°$); in mica andesite from Repistye, near Schemnitz ($2E = 10°\text{--}40°$); in the nepheline rocks of Katzenbuckel, near Heidelberg ($2E = 40°$ about). The dispersion is that of biotite $\rho < v$; the absorption sometimes $\mathfrak{b} > \mathfrak{c}$, sometimes $\mathfrak{c} > \mathfrak{b}$, always $a < \mathfrak{c}$ and $\mathfrak{b}$. The color of rock-making anomite is always brown or reddish brown, never true green. Moreover, the anomites of rocks are somewhat brittle.

Rubellan is the name applied by Breithaupt to reddish or rust-brown volcanic biotite, more or less impregnated with iron ochre and specular iron, which occur like inclusions in the tuffs and lavas of Lake Laach and of the Bohemian "*Mittelgebirge.*" The name has been subsequently applied to similar biotites of older eruptive rocks which in part are colored by the secretion of iron oxide, and are no longer

* T. M. P. M. 1882. IV. 151.
† T. M. P. M. 1883. V. 332.
‡ Geol. Fören. i. Stockh. Fördhl. 1884. VII. No. 87. 194.

elastic. The inner lamellæ of rubellan often possess' the characteristics of unaltered ferruginous biotites. The rubellans of Lake Laach
and Schima have been investigated microscopically and chemically by
M. U. Hollrung.*

The *phlogopite* series includes phlogopite and zinnwaldite. Both
occur to a limited extent as rock constituents. Both are mica of the
second order; the negative bisectrix is noticeably inclined ($2\frac{1}{2}°$–4°) to
the base; they consist of isomorphous mixtures of the molecules
K, M, S, S'.

Phlogopite is chiefly confined to the granular limestones of the
Archæan, in which it forms crystals and plates. It becomes transparent and colorless to yellowish or greenish, seldom brownish yellow;
its pleochroism is slight, the absorption $c > b > a$. Dispersion $\rho < v$,
as in biotite.

The abundance of inclusions in the large phlogopites of the Canadian occurrences is well known, as well as in those of the United
States. Besides quartz and garnet in quite flat tablets, specular iron
is particularly frequent in opaque or red to yellow and grayish-yellow
transparent plates, elongated and arranged parallel to the rays of the
pressure figure. In other localities it carries tourmaline crystals between its lamellæ, which lie in rows intersecting at 60°, and are so
thin that they glisten with the most brilliant Newton colors. They
give rise to a distinct asterism. Rutile occurs in place of, or associated
with, tourmaline, in the same manner and with the same effect. The
same minerals also occur in three less pronounced systems parallel to
the rays of the percussion figure, giving rise to three more rays of
light, and producing a six-rayed asterism.

Phlogopite alters into fibrous scaly masses, which appear to consist chiefly of talc; rutile, also, not infrequently occurs as a secondary mineral, and indicates that the mica originally contained titanic
acid.

Zinnwaldite or *lithionite* here includes fluorine-bearing lithia-iron
micas, which appear to be isomorphous mixtures of the molecules K,
M, S and S' in very different proportions, and whose color therefore
changes from dark brown by transmitted light to light yellow and grayish white. The axial plane lies in the plane of symmetry, that is, parallel
to the leading ray; the axial angle diminishes from about 60° to 10°,
and becomes smaller as the color becomes darker, that is, as the iron
percentage rises. In the light yellow varieties the inclination of the

* T. M. P. M. 1883. V. 304–331.

negative bisectrix to the normal to oP is distinctly noticeable; it appears to be very small in the dark colored varieties. The dispersion is weak, $\rho > v$. The pleochroism varies between dark brown for c and b, yellowish brown to reddish for a, in the dark varieties; brownish gray for c and b and almost colorless for a in the light varieties. The absorption is always distinct, $c > b > a$. The presence of lithia, recognized in the flame, easily separates this mica from all other micas of the 2d order, and the position of the axial plane distinguishes it from lepidolite.

The specific gravity varies greatly with the iron percentage, and rises to 3.2. Sandberger * describes its occurrence in the tin-bearing granites of the Erzgebirge, Fichtelgebirge, Central France, and Cornwall. Topaz occurs in the same granites, and rutile and cassiterite form microscopic inclusions in these lithionites, with pleochroic halos, not infrequently with zircon and topaz. It also occurs in the pegmatitic secretions of granite and gneiss, when it is usually peachblow-red by incident light and colorless by transmitted light.

The micas of the *muscovite* series, which occur as rock constituents, are always light colored, and do not form regularly bounded crystals, but lamellar individuals and aggregates. They are all micas of the first order, the axial plane lies normal to the plane of symmetry and to the leading ray. The dispersion is $\rho > v$. The specific gravity is 2.83 – 2.9. They are mostly elastic, seldom brittle.

Lepidolite, essentially $3[3(Li,H)_2O, 3Al_2O_3, 6SiO_2] + Si_{16}O_5Fl_{24}$, is generally pale peachblow-red in thick plates by incident light, but becomes transparent and colorless in very thin plates. Pleochroism is not noticeable, yet it can be seen that in longitudinal sections rays vibrating in the cleavage plane are more strongly absorbed than those normal to it. The bisectrix stands apparently normal to oP (001); the axial angle is large, 50°–70°. It is said to accompany muscovite in some granites (Elba, Schaistansk) and in the pegmatitic secretions of many granites and gneisses, and is only distinguished from this with certainty by the reaction for lithia.

Muscovite, $K_2O, 2H_2O, 3Al_2O_3, 6SiO_2$, is wholly foreign to the massive rocks, with the exception of the granites and quite isolated quartz porphyries; it is not a volcanic mineral. On the other hand, potash mica plays a prominent *rôle* in the Archæan rocks and the regionally metamorphosed members of the sedimentary formations. The large tabular occurrences in the granites, gneisses, and mica schists are

---

known as muscovite proper, in distinction to the more microscopic or finely lamellar to scaly and dense sericites of the phyllites, porphyroids, and clay slates. The muscovites are transparent and colorless to light greenish or light yellowish, occasionally colored red by flakes of hematite; they are without actual pleochroism, but with recognizable absorption of the rays vibrating parallel to the cleavage. They are well characterized by strong negative double refraction, by the brightly colored interference figure of cleavage plates, by the large axial angle, 40°–70°, and by the imperceptible inclination of the extinction to the cleavage. The large axial angle distinguishes them from talc. Pleochroic halos are very common. The specific gravity facilitates their mechanical separation from the biotites and zinnwaldites. The freshness of muscovite is very characteristic; it does not appear to suffer from the action of the atmosphere.

Sericite, like muscovite, forms irregularly bounded plates of very small thickness, which are usually drawn out into long narrow stripes. In consequence of their geological position they often appear twisted and bent, or the plates are arranged spirally and like rosettes about a longitudinal axis. All sections through such aggregates show the cleavage cracks slightly curved and not straight, giving the impression that the structure is a fibrous and not a lamellar one; indeed, with strongly crumpled and rolled out sericite-bearing rocks the structure appears to be a felty fibrous one, although it is scaly and lamellar.

The optical behavior is exactly the same as that of muscovite, but strikingly small axial angles (25°–30°) are often observed in the sericites of phyllites. It is probable that substances of different composition are included under sericite. The distinction from talc is only possible through chemical reaction, treatment with cobalt solution, or better, with hydrofluosilicic acid.

Damourite is small-leaved muscovite, which, like sericite, can assume a talc-like habit. The muscovites together with feldspar are the most characteristic minerals of dynamo-metamorphic origin, and arise from true sedimentary rocks (clay slates and grauwacke schists), and from eruptive rocks and their tuffs. They are also formed by the processes which deposit ores along the cracks and fissures of faulted Archaean masses.

The manifold nature of the occurrence of muscovite is shown in the dissemination of this mineral as pseudomorphs after other silicates, such as feldspar, nepheline, leucite, andalusite, cordierite, beryl, etc.

Paragonite, Na_4O, $2H_2O$, $3Al_2O_3$, $6SiO_2$, has not yet been observed in eruptive rocks. It is confined to crystalline schists and phyllites.   Like muscovite, it forms irregularly bounded plates, on which indications of a six-sided outline have only occasionally been noticed, and which yield narrow, lath-shaped longitudinal sections, whose longer sides are parallel to the perfect cleavage.   By transmitted light colorless; axial angle large to very large.   Dispersion and absorption as in muscovite.   Paragonite also sinks to fine scaly aggregates, possessing a dense talc-like appearance, and presenting a sericitic modification. The microscopical distinction from talc lies in the large axial angle.   It can only be distinguished from muscovite, chemically, by treatment with hydrofluosilicic acid, by which process hexagonal crystals of sodium fluosilicate are almost exclusively obtained.   Paragonite schists often contain garnet, staurolite, disthene, tourmaline, rutile, actinolite, magnesite, and dolomite in beautiful crystals.

The Ottrelite Group.

Literature.

C. Barrois, Note sur le chloritoide du Morbihan. Bull. Soc. Min. Fr. 1884. VII. 37–43.

F. Becke, Gesteine der Halbinsel Chalcidice.   T. M. P. M. 1878. I. 269–272.

A. Des Cloizeaux, Sur la forme cristalline et les caractères optiques de la Sismondine.   Bull. Soc. min. Fr. 1884. VII. 80–85.

H. von Foullon, Ueber die petrographische Beschaffenheit der krystallinischen Schiefer der untercarbonischen Schichten und einiger älterer Gesteine aus der Gegend von Kaisersberg bei St. Michael ob Leoben. Jahrb. k. k. geol. Reichsanst. 1883. XXXIII. 220 sqq.

A. von Lasaulx, Ueber Glaukophangesteine der Ile de Groix. Sitzungsber. d. niederrhein. Ges. in Bonn. 3. Dec. 1883.

A. Renard et Ch. de la Vallée-Poussin, Note sur l'Ottrélite.   Annales de la Soc. géol. de Belgique. 1879. VI. 51–68.

G. Tschermak und L. Sipöcz, Die Clintonitgruppe. S. W. A. 1878. LXXVIII. Nov.

Under the *ottrelite* group are here included the very closely related minerals called ottrelite, chloritoid, chlorite spar, masonite, and sismondine.   The name ottrelite is chosen for that of the group because it immediately suggests the geologically characteristic position of these minerals.   The statements of the above-cited authors regarding these minerals differ very widely, and cannot be altogether reconciled.   The following data, which agree with the observations of Tschermak, except in a difference with regard to the pleochroism, are derived from the study of ottrelite from Serravezza, from Ottré and St. Hubert in the Ardennes Mountains, of sismondine from Pregratten. Tyrol, and St. Marcel, Piedmont, of chloritoid from Kossoibrod, Urals, Harvey Hills near Leeds and Inverness in Canada, and of masonite from Natic, Rhode Island.

Rock-making ottrelite forms single crystals, generally with quite incomplete boundaries; also disk-like to lenticular or spindle-shaped grains, or somewhat larger lamellar masses, up to 3 c.m. in diameter; besides sheaf-shaped and tuft-like or irregular aggregates of crystals and crystal grains. They always lie irregularly scattered in the rocks, quite like chiastolites in the hornstones. Whenever a crystal form is observed, it is that of a thin, micaceous hexagonal plate (Pl. XXII. Fig. 4), whose plane angles appear to be 120°. One pair of faces may disappear, and rounding and mechanical deformation produce all the intermediate stages to grains of the most different shape. Hence there arise lateral boundary lines, as with the feldspars of the rhombic porphyries, which do not intersect at 120°, but at any angle whatever. The form will here be given as oP (001), the tabular face, $+ P$ (11$\bar{1}$), and $\infty P\infty$ (010). The base glistens strongly, but is generally scaly, and broken up into small areas; it is also crooked and bent. The lateral faces are dull, and have a resinous lustre, and occasionally are noticeably furrowed parallel to the base. Sections parallel to the base are hexagonal, rhombic, or irregular; all other sections are lath-shaped. Apparently simple crystals almost always show themselves optically as polysynthetic twins, in which the individuals are in contact along their bases (Pl. XXII. Fig. 5), but are so placed that each is turned 120° with respect to the adjacent ones. The twinning law corresponds exactly to that of mica. Less frequently the individuals are in contact along a lateral face, whose projection on the base is parallel to the trace of P (11$\bar{1}$). Hence the twinning is generally not noticeable on the base. More frequently the composition plane is irregular, and the individuals cross one another in hour-glass-like faces (Pl. XXII. Fig. 6).

A very regular zonal structure in concentric hexagons is common, especially in the fine Canadian chloritoids.

The ottrelites possess a good cleavage parallel to the base, which always shows itself in numerous sharp cracks in thin sections, when the plates are thin enough and are well ground. They are wanting in thicker plates, and in those which are so thin that they lie in the section as whole bodies. The cleavage is not so perfect as that of mica and hornblende; it somewhat resembles that of augite. The cleavage plates are extremely brittle, and are only transparent when very thin. Besides the basal cleavage, there is in many ottrelites another parallel to two lateral faces, whose traces on oP (001) intersect at 120°. The corresponding cracks are less numerous, are often interrupted or pass into irregular fractures. The large-leaved ottrelites (St. Hubert, Inverness) exhibit them well; the rounder to spinel-shaped ones, badly or not at all. Finally, there is quite an imperfect parting parallel to the plane of symmetry, whose cracks bisect the obtuse angle of the cleavage just described. All of these lateral cleavages are completely obscured in many occurrences by irregular cracks, which evidently correspond to an internal fracturing due to mountain pressure.

The twinning, fracturing, brittleness, and deep color of the ottrelites place great obstacles in the way of their optical investigation. The sections must be very thin in order to observe simple individuals and not twinned ones. The ottrelites become transparent, and, according to the position of the section, green or blue, or even colorless. Their index of refraction is not inconsiderable; according to Gladstone's law, $n = 1.718$. Hence the rough surface of the sections in Canada balsam. The double refraction is weak even parallel to the axial plane, and the interference colors in sections which are not very thin do not exceed those of the 1st order—a good means of distinction from mica. The extinction in basal sections lies parallel to one edge of the hexagon, or to the diagonals of the cleavage cracks, which intersect at $120°$. In sections inclined to the principal cleavage the extinction sometimes lies parallel to the basal cleavage; the sections are then from the orthodiagonal zone; at other times the extinction takes place at various angles to the principal cleavage. The extinction angle in sections parallel to $\infty P \infty$ (010) is from $12°$–$18°$ in different occurrences. The successive twin lamellæ of these sections almost never show equal extinction angles measured from the trace of the composition plane, which proves that the twinning plane is not normal to the plane of symmetry. The angle between the extinctions in two adjacent twin lamellæ may reach $40°$.

In convergent light a positive bisectrix emerges obliquely from the principal cleavage face. The axial angle must vary considerably, since in many occurrences an axis appears on the edge of the field of view and the dispersion $\rho > v$ can be observed, while in most cases the axes are not visible. The axial plane is parallel to the plane of symmetry; it bisects the obtuse angle of the prismatic cleavage.

All minerals of the ottrelite group, except the Styrian occurrences described by Foullon, are remarkable for a highly characteristic pleochroism, which is of great diagnostic importance. In basal sections the ray vibrating parallel to the axial plane is olive-green, that normal to it plum-blue to indigo-blue; in sections inclined to the cleavage the ray vibrating nearly normal to the cleavage is yellowish green, that almost parallel to it either blue or olive-green. Therefore c = yellowish green, b = plum-blue to indigo-blue, a = olive-green.

H. = 6–7. Sp. gr. = 3.53–3.55. The chemical composition is not definitely known, because of the difficulty of removing the abundant interpositions. Tschermak has given the formula for the purest chlorite spar as probably H_2O, FeO, Al_2O_3, SiO_2. In this a variable portion of FeO is replaced by MgO; in the ottrelite from Ottré a considerable amount is replaced by MnO. Ordinary acids do not attack the minerals of the ottrelite group. Fused with caustic potash, cleavage plates yield etched figures on the basal plane, with apparently triangular or hexagonal outline; they are, in fact, monosymmetric, and their plane of symmetry coincides with the plane of the optic axes (Sanger).

The rock-making ottrelite minerals generally contain a great many different interpositions, among which are quartz grains, ores, carbonaceous particles, rutile needles, and tourmaline columns. The arrangement of these inclusions is irregular.

The ottrelite minerals are almost exclusively confined to phyllitic schists and indicate dynamo-metamorphic processes. Such schists are found in the Ardennes, the Pyrenees, the Apennines, in Styria, and are particularly fine in the Province of Quebec, Canada, and in Rhode Island. Sismondine occurs with glaucophane at Zermatt, Switzerland, in Val Chisone in Piedmont, and on the Ile de Groix in Brittany.

Epidote.

Literature.

C. Klein, Die optischen Eigenschaften des Sulzbacher Epidot. N. J. B. 1874. 1–21.

Rock-making epidote seldom possesses sharp crystal forms, and then most frequently shows the faces $M = oP$ (001), $T = \infty P\check{\infty}$ (100), $r = P\check{\infty}$ (101), in the orthodiagonal zone. The angles at which these faces intersect are $M \wedge T = 115°$ 24′, $M \wedge r = 116°$ 18′, $T \wedge r = 128°$ 18′. The crystals are always more or less elongated parallel to the axis of symmetry. The faces cutting this angle are frequently undeveloped. Therefore sections parallel to the axis b are long lath-shaped, those parallel to the plane of symmetry approximately hexagonal (Fig. 93). Moreover, the face T is generally much smaller than r; it is rarely the reverse. T may also be wanting, and sections parallel to $\infty P\check{\infty}$ (010) are then rhombic. More frequently a crystallographic boundary is entirely wanting, and the epidote forms columns parallel to b, or

irregularly angular individuals and aggregates without regular boundaries to their cross-sections.

Twinning is, on the whole, rare in rock-making epidote; in many
rocks, however, it is very abundant. The twinning and composition
plane is then T (Fig. 94). Between the two larger twins there are
occasionally several delicate twin lamellæ.

The perfect cleavage parallel to M (001) shows itself in sharp
cracks, which, however, are not so numerous as one would expect from
the perfection of the cleavage; the cleavage parallel to T (100) is represented by but few cracks (Pl. XI. Fig. 2). The angle between these

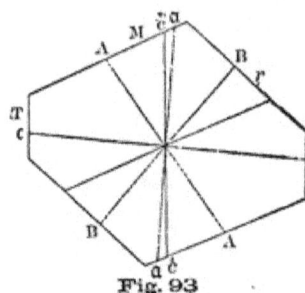

Fig. 93

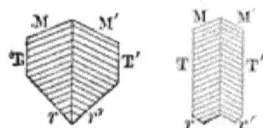

Fig. 94

cleavage cracks varies from $115° 24'$ in sections parallel to $\infty P\infty$ (010)
to $180°$ in those from the orthodiagonal zone.

Rock-making epidote becomes transparent, and almost colorless or
pale yellow, rarely yellowish brown, pale green, very seldom red. The
index of refraction and double refraction are very considerable; hence
the marginal total reflection and the unevenness of the surface are
strongly marked. The height of the interference colors in sections
parallel to the plane of symmetry is greater than for any other silicate,
being only second to those of rutile, anatase, zircon, and the rhombohedral carbonates; even in very thin sections they are of the 3d
order. Klein determined on the epidote of Knappenwand $\alpha_\rho = 1.730$,
$\beta_\rho = 1.754$, $\gamma_\rho = 1.768$. Hence $\gamma - \alpha = 0.038$. Michel-Lévy measured on epidote from the same locality, $\gamma - \alpha = 0.047$, and on that of
the ophite from Lherz, $\gamma - \alpha = 0.0545$; on that from the schists of
Ile de Groix, $\gamma - \alpha = 0.056$.

The plane of the optic axes lies in the clinopinacoid; the axial angle
$2H_\rho = 91° 20'$, $2V_\rho = 73° 40'$; the dispersion is distinctly inclined and
weak, $\rho > v$. The negative first bisectrix is inclined $2°-3°$ to the vertical axis, and lies in the acute angle. Hence the scheme, Fig. 93.
Cleavage plates parallel to M exhibit an axis in the margin of the field
of view, whose hyperbola in the diagonal position is green on the inner

border and red on the outer one; isolated crystals which lie on r show an axis normal to r, the borders of whose hyperbola are red on the inside and green on the outside. All sections from the orthodiagonal zone exhibit in convergent light axial bars, axial figures or the loci of bisectrices, which show that the axial plane is normal to the cleavage— the surest means of distinction from augite, with which mineral epidote may be confounded.

In parallel light sections from the orthodiagonal zone extinguish parallel and normal to the cleavage. In sections from this zone twins cannot be recognized by the extinction in parallel polarized light. In the zone $\infty \check{P} \check{\infty} : \infty P \check{\infty}$ the angle between the extinction and cleavage cracks increases from $0°$ on T to about $28°$ on $\infty P \check{\infty}$ (010). In the zone $o P' : \infty P \check{\infty}$ the extinctions lie between $0°$ and $28°$.

The strong pleochroism of the Sulzbach epidotes disappears entirely in the colorless occurrences in rocks, and is faint in the light colored ones. Thus in the Sulzbach crystals $\mathfrak{a}$ = yellow, $\mathfrak{b}$ = brown, $\mathfrak{c}$ = green, and $\mathfrak{b} > \mathfrak{c} > \mathfrak{a}$; while in the rock-making ones $\mathfrak{a}$ = colorless to light yellowish green, $\mathfrak{b}$ = yellowish green to colorless, $\mathfrak{c}$ = siskin-green to green or light yellowish brown. Absorption $\mathfrak{c} > \mathfrak{b} > \mathfrak{a}$. Though the difference of color is so slight, yet the change from green and siskin-green to colorless or light yellowish in very light colored epidotes is very characteristic; in the uncommon red epidotes the colors change between red, yellow, and colorless.

H. = 6.5. Sp. gr. = 3.3–3.5, increasing with the percentage of iron. Chemical composition, H_2O, $4CaO$, $3(Al_2Fe_2)O_3$, $6SiO_2$. It is not attacked by acids, but is decomposed in HCl, after being heated to redness.

There is no constant microstructure. It is usually free from inclusions; fluid inclusions are more frequent than particles of ore and carbonaceous matter.

Epidote never occurs as a primary constituent in eruptive rocks nor in true Archæan rocks. Still, it is a characteristic constituent of those stratified rocks (garnet rocks and certain amphibolites) which are the equivalent of granular limestones, of paragonite and glaucophane schists, of gneisses in the phyllitic schists, and of metamorphic gneisses, of phyllites, and green schists. It is also one of the commonest formations in the lime-silicate hornstones. As a product of weathering, epidote is the most frequent of all silicates; thus it is formed in the acid and basic rocks from the feldspars under the influence of solutions derived from the micas and bisilicates. Saussurite consists chiefly of epidote. Whenever calcareous iron and magnesia silicates chloritize, epidote is a constant side product, the lime being deposited or removed as a carbonate. Thus it is found accompanying the atmospheric decomposition of pyroxene, amphibole, mica, and garnet.

Allanite.[*]

Literature.

J. P. Iddings and Whitman Cross, Widespread occurrence of Allanite as an accessory constituent of many rocks.   Am. Journ. Sci. Aug. 1885. Vol. XXX. 108.

A. Michel-Lévy and Lacroix, Note sur un gisement français d'allanite.   Bull. Soc. Min. Fr. 1888. Feb. Vol. XI. No. 2. 65.

A. Sjögren, Om Gadolinitens, orthitens, samt med dessa likartade mineraliers förhållande under mikroskopet.   Geol. Fören. i Stockholm Forhandl. 1876. III. No. 37. 258.

A. E. Törnebohm, Under Vega-Expeditionen insamlade bergarter.   Vega-Eped. vetensk. jakttagelser. VI. Stockholm. 1884. 124.

Allanite, which is isomorphous with epidote, occurs as an accessory constituent of many granites, diorites, and other rocks; in the tonalite of Adamello, according to G. vom Rath, [†] it is often so abundant as almost to become an essential ingredient. It frequently forms completely bounded crystals with the faces oP' (001), $\infty P\check{\infty}$ (100) well developed, besides $\infty P'$ (110) and $P\infty$ (011), and sometimes two orthodomes.   (110) $\wedge$ (110) $= 117°$ and (110) $\wedge$ (100) $= 125°$, approximately. The crystals are elongated in the direction of the orthoaxis $\check{b}$, as in epidote, and the sections have similar shapes. It also occurs as irregular grains.   Twinning along the plane $\infty P\check{\infty}$ (100) is frequent.

The cleavages parallel to $\infty P'$ (110), $\infty P\check{\infty}$ (100), $\infty P\check{\infty}$ (010), and also to oP' (001), are occasionally indicated by irregular cracks, but in many occurrences they are entirely absent.

Allanite becomes transparent in thin sections with reddish brown or greenish brown colors.   It usually exhibits a strong pleochroism from light yellowish or greenish brown to dark chestnut-brown.   In the allanite of the granite from Pont-Paul, Finisterre, Michel-Lévy and Lacroix found $\mathfrak{a} =$ greenish brown, $\mathfrak{b} =$ reddish brown, and $\mathfrak{c} =$ yellowish brown.   The mean index of refraction exceeds 1.78. The double refraction is variable: in the allanite from Pont-Paul it is very feeble, but in that from Edenville, N. Y., $\gamma - \alpha = 0.032$, according to Michel-Lévy.   Many allanites are isotropic, without showing any change of form or noticeable signs of decomposition.

---

[*] Expanded by the translator.

[†] Z. D. G. G. 1864. XIV. 255.

The plane of the optic axis lies in the plane of symmetry, $\infty P \overset{.}{\infty}$. The axes of greatest and least elasticity bisect the angles between the vertical axis $\acute{c}$ and the clinoaxis $\acute{a}$; the acute bisectrix is $\mathfrak{a}$. and lies in the obtuse angle between $\acute{c}$ and $\acute{a}$. The optical character, therefore, is negative, $2V = 65°$ to $70°$. The optic axes are nearly normal to the faces oP (001) and $\infty P \overset{+}{\infty}$ (100). There is a large dispersion of the axes of elasticity, which causes confused extinctions in parallel polarized light.

In many occurrences, especially in the granites and gneisses, allanite possesses a marked zonal structure, accompanied by variations in the directions of extinction and in the color. In the porphyrites, porphyries, and volcanic rocks zonal structure is almost entirely wanting, and the color is dark reddish brown.

H. $= 5.5–6$. Sp. gr. $= 3.0–4.2$. Chemical composition similar to that of epidote, except that part of the Ca is replaced by Fe, and the Al is largely replaced by the rare earths, Ce, La, Di, Y, Er. It is decomposed by boiling hydrochloric acid.

Allanite occurs as a primary accessory ingredient of many eruptive rocks. In the granite from Pont-Paul it is one of the oldest constituents, and is enclosed in biotite and surrounded by pleochroic halos. It has a wide distribution through a great variety of rocks in the United States, having been found in gneiss, granite, granite porphyry, quartz porphyry, diorite porphyrite, andesite, dacite, and rhyolite.

It is usually perfectly fresh, without signs of decomposition; occasionally a small zone of the surrounding rock is stained ochre-yellow. In the granite from Ilchester, Md., allanite with pronounced zonal structure occurs at the centre of epidote crystals, the two minerals having parallel crystallographic orientation.[*]

Allanite may be confused with biotite and hornblende in certain instances when they possess the same reddish brown color and do not exhibit their characteristic cleavage or crystal form, but it may be distinguished from basal sections of biotite by its strong pleochroism and larger optic axial angle, and from hornblende by its higher double refraction.

---

[*] Wm. H. Hobbs, On the rocks occurring in the neighborhood of Ilchester, Howard Co., Md.; preliminary notice. The Johns Hopkins University Circulars, No. 65, Apr. 1888.

Titanite.

Titanite is only an accessory constituent of those rocks in which it occurs, but it is at times a very abundant one.   When it occurs in eruptive rocks as a primary component, it is always well crystallized, and belongs to the oldest secretions from the magma.   Less frequently it forms regular crystals in the Archæan rocks.   In both groups of rocks it is common in the form of irregular grains as a secondary product from titaniferous magnetite, from ilmenite and rutile.   The regular boundaries of the primary crystals also are occasionally more or less destroyed through mechanical and chemical processes.   The forms of embedded titanite crystals are less variable than those of attached ones, but there is a certain variableness even in these.   The most predominant type is represented in Fig. 95; besides $n = \frac{3}{4}P\overset{\cdot}{2}$ (123) there appear less prominently $P = oP$ (001) and $y = P\overset{\cdot}{\infty}$ (101), less frequently $x = \frac{1}{2}P\overset{\cdot}{\infty}$ ($\overline{1}02$) and $r = P\infty$ (011).   The combination $l = \infty P$ (110) with n (Fig. 96), besides other subordinate faces, is

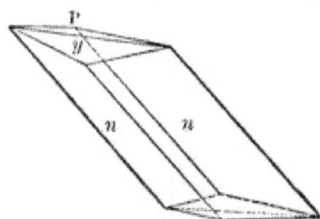

Fig. 95

Fig. 96

especially met with in amphibolites and mica schists.   The combination y, n, r, also, is not uncommon.   The angles most important for cross-sections are $l \wedge l = 133° 52'$, $n \wedge n' = 136° 12'$, $P \wedge y = 60° 17'$, $P \wedge x = 39° 17'$.   The commonest sections are acute rhombs, and long, lath-shaped ones with pointed ends.   Twinning is not infrequent, but is never recognized by the outline, only by the behavior in polarized light. The twinning boundary always bisects the acute angle of the rhombs (Pl. XXIII. Fig. 1).   Hence the base appears to be the twinning plane.   Zonal structure is seldom observed; kernel and shell are then separated from each other by the faces n or l, and spring apart upon being struck.

Titanite only appears in the form of granular or short columnar aggregates when it forms pseudomorphs after one of the above-named minerals.

The cleavage along the prism l only shows itself by occasional rough cracks; since the prism seldom occurs as a predominant form, the cleavage is not parallel to the boundary, which is usually determined by n,—a phenomenon characteristic of titanite (Pl. XI. Fig. 3). Cleavage is rarely observed on secondary grains and aggregates of titanite.

The titanite of rocks becomes transparent and colorless to white, yellowish, or reddish; its transparency, however, is generally small. The index of refraction is very high, $\beta = 1.905$–1.910; the marginal total reflection and the rough character of the surface in Canada balsam are greater than for epidote. The double refraction has not yet been measured, but does not appear to be great; the interference colors are only striking in sections parallel to the axial plane, otherwise they are but slightly noticeable on account of the strong dispersion $\rho > v$. The optic axes lie in the clinopinacoid; thus $b = \mathfrak{b}$, and the positive acute bisectrix is normal to $x = \frac{1}{2}P\overset{.}{\infty}$ (102), from which is derived the scheme Fig. 97. The dispersion of the optic axes for different kinds of light is greater than for any rock-making mineral, and furnishes a positive means of determination. Des Cloizeaux measured $2E_\rho = 53°$, and on another crystal $55°$–$56°$, and $2E_v = 32°\ 27'$ and $34°$. The dispersion of the bisectrices is scarcely noticeable. In convergent light all the acutely rhombic sections from

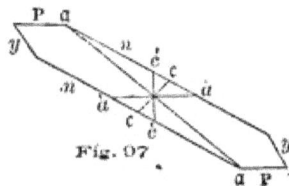

Fig. 97.

the orthodiagonal zone give axial bars, axial figures or loci of bisectrices, from which it can be seen that the axial plane bisects the obtuse angle,—a convenient means of distinction from staurolite, which is otherwise quite similar. Sections lying approximately in the face x show an interference figure, whose hyperbolas are not black in the diagonal position because of the strong dispersion, but are red, green, and blue from the inside outward. By using red and blue glasses the great difference in the axial angle for the two colors can readily be seen.

The extinction angles of the different sections are not characteristic. When the section is considerably inclined to the axial plane, there is no complete extinction in white light, because of the strong axial dispersion.

The pleochroism is scarcely noticeable in very thin sections and for pale coloring: the strongly colored crystals have $\mathfrak{c} =$ red, with a tinge of yellow; $\mathfrak{b} =$ yellow, often with a tinge of greenish; $\mathfrak{a} =$ almost colorless.

H. $= 5$–5.5. Sp. gr. $= 3.4$–3.6. Chemical composition $=$ CaO,

SiO_2, TiO_2. Not attacked by hydrochloric acid. Decomposed by sulphuric acid; the solution becomes orange yellow upon the addition of hydrogen superoxide. On account of its density it falls with the ferruginous constituents in the heavy solutions, and can generally be easily separated from these by the electro-magnet.

Upon decomposition titanite bleaches and loses its lustre; at the same time carbonate of lime separates out. The dull secondary substance has not been investigated. In other instances of decomposition an opaque iron-ore, probably ilmenite, is deposited on the cleavage cracks. Its alteration into rutile has been observed by P. Mann* in elæolite syenites; a decomposition of titanite with the production of anatase was observed by J. S. Diller† in the amphibole granitites of the Troad, Greece.

The titanite of eruptive rocks encloses the older constituents associated with it, as the ores, apatite and zircon, rarely glass and fluid inclusions; in the Archæan rocks it is generally free from inclusions.

Its distribution is considerable: it occurs in the acid rocks which are not too poor in magnesia and iron, as granitites, amphibole granites, syenites, diorites, trachytes, and abundantly in the elæolite syenites and phonolites; it is rarer in the corresponding porphyritic rocks. It is absent from the basic eruptive rocks rich in ilmenite and titaniferous magnetite. Among the Archæan rocks also it occurs to a notable extent in rocks rich in MgO and FeO, that is, in the biotite and amphibole-bearing gneisses and schists. As a secondary product it is found in all rocks bearing ilmenite and rutile.

Monoclinic Feldspars.

Literature.

A. Des Cloizeaux, Observations sur les modifications permanentes et temporaires que l'action de la chaleur apporte à quelques propriétés optiques de plusieurs corps cristallisés. Ann. des Mines. 1862. II.
— Nouvelles recherches sur les propriétés optiques des cristaux naturels ou artificiels, et sur les variations que ces propriétés éprouvent sous l'influence de la chaleur. Mém. Sav. étrangers. Paris. 1867. XVIII.
— Examen microscopique de l'orthose et des divers feldspaths tricliniques. C. R. 1876. LXXXII. 1017–1022.
A. Michel-Lévy, De l'emploi du microscope polarisant à lumière parallèle pour la détermination des espèces minérales en plaques minces. Ann. des mines. 1877 (7.) XII. 392–471.
G. Tschermak, Die Feldspathgruppe. S. W. A. 1864. December. L.
Ch. E. Weiss, Beiträge zur Kenntniss der Feldspathbildung. Haarlem. 1866.

---

* N. J. B. 1882. II. 290.
† N. J. B. 1883. I. 187.

The monoclinic feldspars are classed as orthoclase or sanidine, according to whether they occur in the older massive rocks and Archæan rocks or in younger volcanic rocks. With this difference in geological position are connected certain peculiarities in habit and in physical behavior. For simplicity of expression, the term orthoclase will be here used for all monoclinic potash feldspars, including sanidine, while the name sanidine will be confined to the latter variety of feldspar.

Orthoclase always appears in rocks with more or less complete crystallographic boundaries, whenever they possess a distinctly porphyritic structure; the outward form disappears more and more as the structure becomes more distinctly granular. In the schistose rocks of the Archæan the orthoclase is generally not crystallographically bounded. But a distinct crystal form is also developed here whenever a porphyritic structure occurs.

The crystals of embedded orthoclase always show the faces $P = oP'$ (001), $M = \infty P\infty$ (010), predominant; $l = \infty P'$ (110), $x = P'\infty$ (101), $y = 2P'\infty$ (201), more subordinate; rarely $n = 2P'\infty$ (021), $o = P'$ ($\bar{1}11$), and in the zone $l : M$, $z = \infty P'3$ (130). The angles important for cross-sections are $l \wedge l = 118°$ 48′, $l \wedge M = 119°$ 36′, $P' \wedge x = 129°$ 40′, $P' \wedge y = 99°$ 37′, $P \wedge M = 90°$. The faces P and x are almost equally inclined to the vertical axis. The habit of the crystals is either more

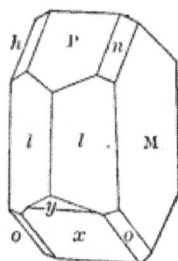

Fig. 98

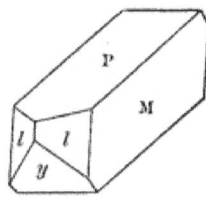

Fig. 99

or less tabular parallel to M (Fig. 98), or prismatic parallel to the axis a (Fig. 99). The shape of sections in different directions is evident from the figures.

The commonest variety of twinning is that according to the *Carlsbad* law. The twinning axis is the vertical axis, and the twinned individuals either join along the plane of symmetry or penetrate each other irregularly. The characteristics of these twins is that the basal faces slope in different directions. In the orthoclases of many rocks (granite of Elba) the faces P and x lie apparently parallel (Fig. 100). In cross-sections the twinning boundary either lies parallel to the intersection of M and the cutting plane, or it is an irregularly bent or jagged line. The twin character is often not recognizable in the outline of these sections when the crystal is a contact twin ; but it is shown by the cleavage and optical behavior (Pl. XXIII. Fig. 2).

The *Baveno* law, by which the normal to n is the twinning axis, is far more rare in rocks, and is always sporadic. The twinning plane n is also the composition plane (Fig. 101), and since the faces inclined to the axis $\dot{a}$ are seldom well developed, and $n \wedge n$ is almost 90°, it happens that the outline of the sections give no indication of the twinning. The basal faces stand at right angles to one another, and since M is also a cleavage face the twinning cannot be detected by the cleavage, but is found through the optical behavior. Twins of this

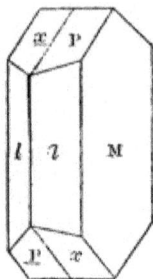

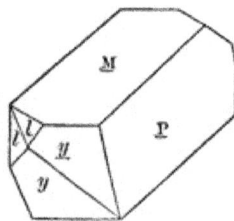

Fig. 100   Fig. 101

kind are always prismatic in the direction of $\dot{a}$. Hence the cross-sections are mostly square or rhombic, with the trace of the twinning plane running diagonally through them (Pl. XXIII. Fig. 3). Through the repetition of this law along one or two more faces of n, there arise trillings and fourlings which are only recognized optically.

The rarest twinning is according to the *Manebacher* law ; twinning axis is the normal to P. P is also composition plane, and the twinning is rarely recognized by the outline. The cleavage is the same in both parts of the twin, and the optical behavior alone reveals the twinning. The trace of the twinning plane is parallel to that of P in the section. This law only occurs in a few rocks, but then quite frequently. The quartz porphyries are the principal rocks which show it.

The combination of Carlsbad and Baveno laws is not infrequent. Confused intergrowths, which are probably brought about by twinning, have been observed quite frequently, but can seldom be made out from the cross-sections. Such exceptional twinnings have been described by Klockmann* and Haushofer.†

The dimensions of the crystals vary extraordinarily, and the crystallizations of the second and third generation in porphyritic eruptive rocks are often extremely small, even for microscopical examination. Nevertheless, true incipient forms of growth or skeleton crystals of orthoclase have not yet been observed with certainty. The habit is always that of the larger crystals. Still, in the older porphyries the tabular forms predominate, while in the trachytes and phonolites it is prismatic; in the rhyolites both occur, but almost never together. The minute prisms often group themselves together in radially columnar aggregates. They form spherulites, which either lie free in the rock or attach themselves in tufts to the older orthoclase crystals. Such orthoclase microlites, like all nearly trichitic forms, often exhibit a fraying out into diverging curved processes.

Zonal structure is very common, indicating the original crystal form when this has been destroyed by subsequent changes. If the crystals are perfectly fresh it is often unnoticeable; it then first appears when the condenser is lowered in order to produce strongly divergent rays, or when the crystal is observed between crossed nicols in a semi-dark position. It becomes very distinctly marked through interpositions, especially glass inclusions in sanidines, and by the first stages of decomposition (Pl. XXIII. Fig. 4).

The crystal form of orthoclase is often completely rounded through chemical corrosion by the magma, brought about by changes in its composition or physical condition; in this way occasionally they become spherical grains (as in many quartz porphyries). Mechanical deformations are very common in the porphyritic eruptive rocks; the crystals are broken in consequence of movements of the magma enclosing them, and become angular and sharp-edged grains (Pl. XXIII. Fig. 5). In the Archæan rocks the mechanical processes which have been active in mountain-making have often produced a marginal rubbing and crushing of the orthoclase crystals (Pl. IV. Fig. 2), which may lead to their complete destruction, so that an originally simple individual is converted into a confused aggregate. In other instances these processes only lead to small molecular displacements and strains, which are first recognized by optical phenomena, such as the

---

* Die Zwillingsverwachsungen des Orthoklases auz dem Granit des Riesengebirges. Z. X. 1882. VI. 493–510.
† Orthoklaszwillinge von Fichtelberg. Z. X. 1879. III. 601.

gradually changing orientation of the axes of elasticity and the resulting undulatory extinction (Pl. IV. Fig. 2).

The cleavage of the feldspars parallel to the faces P and M is one
of the most important factors in their microscopical diagnosis. The
cleavage parallel to P is the most perfect and easiest; that parallel to
M varies somewhat, and occasionally reaches the perfection of that
parallel to P. But both cleavages are not complete enough to become
noticeable in thick sections. When sufficiently thin, both cleavages
show themselves in quite sharp and straight cracks (Pl. XI. Fig. 1),
which are somewhat more numerous and continuous parallel to P than
parallel to M. In many sanidines and orthoclases, however, P and M
can scarcely be distinguished from one another by their cleavages. The
cleavage cracks lie parallel to one another in all sections from the zone
P and M; they intersect at right angles in all sections from the zone
$oP : \infty P\bar{\infty}$ (001 : 100). In sections from the zone $\infty P\bar{\infty} : \infty P\breve{\infty}$
(100 : 010) they intersect at angles which vary from $90°$ to $63° 53'$
($\beta = 63° 53'$).

In the sanidines there is frequently a rude parting approximately
parallel to $\infty P\bar{\infty}$ (100). The corresponding cracks are never straight
nor strictly parallel (Pl. XXIII. Fig. 6); they, however, appear considerably sooner in thin sections than the cleavage cracks parallel to P
and M, and are the only ones noticeable in the thicker sections.

The orthoclases become transparent and colorless. Their index of
refraction is small, very rarely the same as that of Canada balsam;
their double refraction is weak to very weak—weaker than that of
quartz and the lime-soda feldspars. Des Cloizeaux determined:

On adular from St. Gotthard............ $\alpha_{na} = 1.5190$ $\beta_{na} = 1.5237$ $\gamma_{na} = 1.5260$
On sanidine from Wehr, with normal
symmetrical axial position............ $\alpha_\rho = 1.5170$ $\beta_\rho = 1.5239$ $\gamma_\rho = 1.5240$
The same, with symmetrical axial posi-
tion.... .... $\alpha_\upsilon = 1.5256$ $\beta_\upsilon = 1.5355$ $\gamma_\upsilon = 1.5356$

$\gamma - \alpha$ under normal conditions varies between 0.007–0.005, and the
interference colors do not exceed the 1st order even in thick sections;
in good sections they reach yellow of the 1st order at most.

All orthotomic feldspars are optically negative; in general the
plane of the optic axes is normal to the plane of symmetry, and forms
an angle of $3°-7°$, with the plane of the crystallographic axes $\dot{a}$ and $\bar{b}$ in
the obtuse angle β. In exceptional instances, apparently when the
percentage of soda is high, this inclination increases to $10°-12°$. The
horizontal dispersion is very noticeable, and $\rho > \upsilon$. The axial angle
varies within wide limits; it is always large for orthoclases proper,

$2E' = 119°-125°$; and for sanidines it varies in crystal from the same rock, even in plates of the same crystal, but is always smaller, between 50° and 0° in air. The scheme Fig. 102 illustrates these relations; it is evident that the obtuse positive bisectrix emerges from the face M, while plates parallel to the parting face, which is approximately parallel to the orthopinacoid, lie nearly at right angles to the acute negative bisectrix, and give an interference figure. In the same way all sections from the prism zone show axial bars or the loci of axes slightly eccentric to the field of view. The trace of the axial plane lies nearly parallel to the most perfect cleavage.

In all sections from the orthodiagonal zone the extinction is parallel and normal to the perfect cleavage; in the zone $P : M$ it makes small angles with the parallel cleavages along P and M which increase from 0° on P to 3°-7°, seldom to 12° on M. In the zone $\infty P\check{\infty} : \infty P\check{\infty}$ it is better to measure the extinction angle from the second cleavage

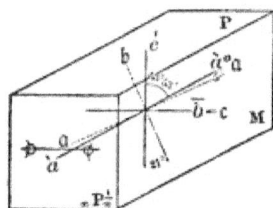

Fig. 102

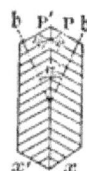

Fig. 103

along M; when near this face it is 21°, increases slowly until a section is reached, which is inclined 45° to M, then rapidly to 90°.

In Carlsbad twins the traces of the axial planes of the twinned individuals are parallel. In parallel light both individuals extinguish simultaneously in the zone $oP' : \infty P\check{\infty}$ (001 : 100), and the cleavage cracks lie parallel to the directions of extinction. In the zone $\infty P\check{\infty} : \infty P\check{\infty}$ (100 : 010) the cleavage cracks and extinctions lie parallel and normal to the twinning line on the first-named face; on the second face the cleavage along M is parallel to the twinning line; that along P makes an angle of $127° 46' = 2 <\!) \beta$, and is bisected by the twinning line (Fig. 103). The directions of extinction of the two individuals make an angle of $2 \times 21° = 42°$, which is also bisected by the twinning line. In sections from this zone with varying inclination to M the angle between the cleavages increases from 127° 46' to 180°; the angle between the directions of extinction increases from 42° to 180°, at first slowly, then very rapidly. The twinning line always bisects the cleavages and extinction angles (Pl. XXIII. Fig. 2). The zone $P : M$ is not common

to both individuals of a Carlsbad twin ; the zone $P : M$ of one individual
is approximately the zone $x : m$ of the other.   In the first individual
all sections have parallel cleavage cracks, with which the directions of
extinction make angles of 0° on P to 3°–7° or rarely 12° on M; in the
second individual, since a basal section of the first one is approximately
parallel to x of the second, it will show rectangular cleavages and the
extinction parallel to them.   With increasing inclination to M the
cleavage angle decreases to 54°.   The extinction angle measured from
the twinning line increases from 0° to 74° on the same side as in the
first individual.

In Baveno twins only the cross-sections and the zone $P : M$ are of
consequence.   In cross-sections the twinning line is diagonal to the
cleavage (Pl. XXIII. Fig. 3) and to the outline, and in convergent
light there are two interference figures standing at right angles to one

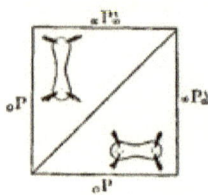

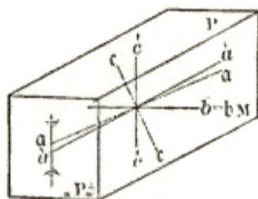

Fig. 104   Fig. 105

another (Fig. 104).   In sections from the zone $P : M$ the cleavages of
both individuals are parallel to one another and to the twinning line;
the extinction angle in one individual increases from 0° to 3°–7°, while
in the other it decreases within the same limits.   The maximum extinction angle in one individual coincides with the minimum in the
other.

The optical behavior of the Manebacher twins is easily understood from what has been said.   They have the zones $P : M$ and
$oP : \infty P\check{\infty}$ in common.   The zone $M : \alpha.P\check{\infty}$ of one corresponds to
a zone $M : m P\check{\infty}$ of the other, which is not characteristic.

Occasionally in the sanidines of lavas, more frequently in those
sanidines which have been thrown out loose, and in those of lapilli, the
axial plane lies in the plane of symmetry (Fig. 105), that is, normal to
the most perfect cleavage.   The dispersion is then $\rho < v$.   The axial
angle is always small, $2E = 40°–0°$; indeed, instances occur in which
the axial plane for red light is normal to M, while that for blue is
parallel to M.   The orientation of the axes of elasticity is nearly the
same in all cases.   The inclined dispersion which accompanies the
symmetrical axial position is not usually great.

The orthoclases exhibit no pleochroism nor noticeable difference of absorption.

Des Cloizeaux has shown that by raising the temperature the axial angle of feldspars diminishes so long as the position of the axial plane is normal-symmetrical, and increases when it coincides with the plane of symmetry. With sufficient heating the axial angle of normal-symmetrical axes decreases gradually to 0° for all colors commencing with blue; the axes pass into the plane of symmetry without noticeably changing the position of the obtuse bisectrix, and gradually open as the temperature increases. Upon cooling, the axes return to their original position if the temperature has not exceeded 500° C. If the temperature is kept at from 600°–1000° C. for some time, the resulting changes remain fixed, and do not alter upon cooling.

The specific gravity of sanidine and orthoclase, when unaltered, is the same, 2.54–2.56. This permits a mechanical separation from the lime-soda feldspars without difficulty. Chemical composition, K_2O, Al_2O_3, $6SiO_2$; but this is always isomorphously mixed with a variable amount of a similarly constituted soda molecule. Since mechanical intergrowths with a soda feldspar are also quite common, it cannot be seen from the analyses to what extent soda has replaced potash. Orthoclase is not noticeably attacked by hydrochloric acid even when heated, but it is very readily decomposed by hydrofluoric acid.

Sanidine occurs in rocks either as older secretions or as a later crystallization of the groundmass. In the first case it has exactly the form of the macroscopic crystals, quite thinly tabular parallel to M or slender prismatic parallel to $\check{a}$. Its crystallization has followed that of the ferruginous constituents, of the haüyne minerals, of nepheline, and to some extent that of the plagioclases; it preceded that of quartz. These secretions are occasionally free from inclusions, and then they are pellucid. More frequently they enclose the associated minerals, and especially gas and glass inclusions, the latter often more or less devitrified. The shape of these inclusions is either irregular or is borrowed from their host, and then shows the combination P, M, y, l. Fluid inclusions are rare. The arrangement of these inclusions is seldom irregular; they generally lie in concentric zones, or are crowded together centrally or peripherally. Occasionally (Drachenfels) they are distributed in layers parallel to M, less frequently parallel to P.

Regular intergrowths of sanidine crystals with triclinic feldspars are very common, and though apparently very diversified, always follow the law that both feldspars have M and the edge $M:l$ in common. This intergrowth may amount to a complete envelopment (Pl. XXIV.

Fig. 1), in which the sanidine is almost always on the outside, very seldom on the inside; or the feldspars may join one another only along one side, or they may penetrate each other with irregular boundaries, so that in thin section they mutually enclose one another in irregularly shaped patches. Sanidine very rarely exhibits a microline-like structure such as Mügge[*] described in the olivine-bearing trachytes from Fayal.

When sanidine occurs in a second generation in the groundmass it is usually free from inclusions.

In general, the sanidines exhibit no signs of decomposition; an alteration into zeolitic aggregates is quite common in phonolites (Pl. XXIV. Fig. 2). The red color occurring in some sanidines arises from infiltrations of iron oxide.

Orthoclase, even when perfectly fresh, does not have the glassy habit of sanidine, or the parting along a face approximately parallel to the orthopinacoid. The perfectly fresh examples resemble adular. It is convenient to separate the orthoclase of porphyritic rocks from that of granular rocks; with the latter is closely related the orthoclase of Archæan rocks.

The orthoclase of porphyritic rocks resembles sanidine in its forms, when it occurs as porphyritic crystals. But inclusions are much less abundant, and glass inclusions can seldom be recognized as such on account of the state of preservation of the rocks. The orthoclase of later generation is free from inclusions, and is more equally developed in all directions than the sanidine. The intergrowths with triclinic feldspars are analogous to those of sanidine; mutual penetrations with quartz are very frequent, and are known as granophyric intergrowths (Pl. VIII. Fig. 3). Intergrowths with microcline only occur in those porphyritic rocks which, like granite porphyry, are very closely related to granular rocks.

The orthoclase of granular rocks and of Archæan rocks shows but imperfect crystallographic boundaries or none at all; glass inclusions never occur. On the other hand, fluid inclusions are very common in fresh orthoclases, but disappear in the processes of alteration. Besides the older associated minerals, orthoclase occasionally encloses scales of specular iron and microlitic interpositions. But this is always a local or individual phenomenon, not a general one. The arrangement of the inclusions in orthoclase also is generally regular, zonal, central, or peripheral. The tendency of orthoclase to form an intergrowth with

---

* N. J. B. 1883. II. 204.

triclinic feldspar is quite extraordinary. As with sanidine, it is either
an envelopment of one by another—the rarest case—or a simple juxtaposition; or finally a complete penetration, the last being the commonest case. The combined feldspars always have the second cleavage
face *M* and the edge *M* : *l* in common. These intergrowths are generally only perceptible in polarized light because of the great similarity
in the form of all the feldspars; in many cases, however, they can be
recognized microscopically by dull places on the principal cleavage
face, or by a banded appearance on the second cleavage face.

Microcline, albite, and oligoclase are known to take part in such intergrowths with orthoclase. Pl. XXIV. Fig. 3 gives an example of the
penetration of orthoclase and plagioclase. In sections parallel to *P* the
orthoclase is recognized by its extinction parallel to the cleavage along *M*,
while the plagioclases and microcline extinguish more or less obliquely
to this cleavage. The cleavage along *M* passes uninterruptedly through
the different feldspars. In sections parallel to *M* the cleavage along *P*
runs only approximately parallel through orthoclase and microcline on
one side and albite and oligoclase on the other. In such sections orthoclase and microcline are distinguished from one another with difficulty,
while albite and oligoclase are easily determined by their different extinction angles. In chance sections the intergrowth is recognized by
the different extinction angles in the different feldspars, in part also
by the local abundance of the twin lamellæ of plagioclase, and by the
differences in the interference colors. But the determination of the component individuals can seldom be made with certainty in such sections.

The lamellar intergrowth of orthoclase (with or without microcline) and albite, like that which exists macroscopically in perthite, is
particularly common. The albite lamellæ are often so extremely fine
that they are not perceptible as such with low magnifying powers.
They appear to lie parallel to the prism or orthopinacoidal faces in orthoclase, which then assumes a striated appearance in sections from the
prism zone (Pl. XXIV. Fig. 4). When the lamellæ are still smaller
these sections appear finely fibrous, and exhibit very different degrees
of brightness, according to whether the light travels parallel or perpendicular to the longer direction of the lamellæ—as, for example, in the
feldspars of many Saxon granulites. Finally, these albite lamellæ
reach such minuteness that they are only recognized as such by very
high magnifying powers; then the orthoclase occasionally exhibits a
beautiful blue lustre on the orthopinacoid and on faces lying near it, as
in many adulars, the moonstone of Ceylon, and the schillerizing orthoclases of Frederiksvärn. It is possible that these submicroscopic

albite lamellæ explain the high extinction angle on M in such feld-spars, which is nearly the mean of the values for orthoclase and albite. This microscopic lamellar intergrowth is called *microperthite.* Pl. XXIV. Fig. 5 shows such microscopic mixtures of orthoclase and albite in sections in different directions. In approximately basal sections it is seen that the albite forms thin rods; when of larger dimensions they become small lamellæ and spindle-shaped bodies. An acid lime-soda feldspar also forms microperthitic intergrowths.

The dull and cloudy appearance of orthoclase is due to a more or less advanced alteration into muscovite or kaolin. The two processes, which are so closely related chemically, and arise from a partial or total removal of the potash by water, together with the separation of $4SiO_2$, exhibit the greatest similarity morphologically, and are scarcely deter-minable microscopically. In both cases there form along the cleavage cracks aggregates of a perfectly uniform substance, which is colorless and is strongly doubly refracting. The feldspar appears to be dis-tended, and is the more opaque and earthy the finer the scaly structure of the secondary product. The process often commences in the centre of the orthoclase crystal, especially when there were many central in-clusions, so that the attackable surface was as great as possible.

In the alteration to kaolin the dimensions of the secondary prod-ucts are always smaller than in that to muscovite. They can be dis-tinguished by the fact that an alteration to muscovite raises the specific gravity of the orthoclase, while that to kaolin lowers it. Pl. XXIV. Fig. 6 represents an orthoclase completely altered to muscovite (pini-toid). Quartz is almost always mixed with these pseudomorphs in vari-able amounts. Moreover, the mass becomes penetrated by solutions carrying iron, manganese, and lime, from which are deposited limonite, pyrolusite, and calcite. Under the influence of accessory solutions the epidote is produced which is so often present in decomposed ortho-clase.

In the so-called pseudomorphs of cassiterite after orthoclase from Huel Coates in St. Agnes parish, Cornwall, tourmaline and quartz, besides cassiterite, form a principal part of the muscovite.* The alteration of granite to greisen must be ascribed to the same processes which give rise to these pseudomorphs.

---

* J. Arthur Phillips, On the structure and composition of certain pseudo-morphic crystals having the form of orthoclase. Journ. of the Chem. Soc., Aug. 1875.

MINERALS OF THE TRICLINIC SYSTEM.

The minerals of the triclinic or asymmetric system are chiefly distinguished by negative characteristics. Sections of all such minerals are unsymmetrical in all zones; the same is true of all figures produced by intersecting cleavages. Each cleavage is parallel to only one face; hence there are no equivalent cleavage cracks which intersect one another. Cleavage cracks which intersect always belong to crystallographically dissimilar faces. In general, those faces parallel to which there is cleavage are made the pinacoids.

The triclinic minerals are optically biaxial; their ellipsoid of elasticity is triaxial, but is different for each wave-length. Hence there is dispersion of the optic axes and of all three axes of elasticity, although these dispersions are generally small, and practically may be neglected in most instances. From the absence of all symmetry, there is no definite relation between the position of the axes of elasticity and the arbitrary co-ordinates, which are chosen as crystallographic axes. In general, no axis of elasticity coincides with a crystal axis; when this is approximately the case (oligoclase), the optical behavior resembles that of a monoclinic crystal, as far as concerns the extinction angles on certain faces. In parallel polarized light all sections which are not cut at right angles to an optic axis are doubly refracting, and between crossed nicols exhibit the quadruple alternation of darkness and light. The direction of extinction is, in general, inclined to the crystal outline, to the cleavage, and to the diagonals of these forms. Sections at right angles to an optic axis remain uniformly light in all positions between crossed nicols, and exhibit an axial figure in convergent light, whose appearance is analogous to that of an orthorhombic or monoclinic mineral.

Sections at right angles to a bisectrix give an interference figure in convergent white light which is distinguished from that of an orthorhombic or monoclinic crystal by the fact that the distribution of the colors is unsymmetrical, both with respect to the trace of the axial plane and to one normal to it, as well as unsymmetrical to the centre of the axial figure. Thus several dispersions occur together which would be distinguished in monoclinic crystals as inclined, horizontal, and crossed. The optical character is designated as positive or negative in this system also, according to whether the axis of least or greatest elasticity bisects the acute angle between the optic axes.

In triclinic minerals, which exhibit pleochroism, all sections are pleochroic which do not lie at right angles to an optic axis. The maximum differences of color are 90° apart, and generally coincide with the directions of extinction, though not necessarily.

Microcline.

Literature.

A. DES CLOIZEAUX, Mémoire sur l'existence, les propriétés optiques et cristallographiques, et la composition chimique du microcline, nouvelle espèce de feldspath triclinique à base de potasse, suivi de remarques sur l'examen microscopique de l'orthose et des divers feldspaths tricliniques.   Ann. de Chim. et de Phys. (5). IX. 1876.—Also C. R. 1876. LXXXII. 885–891.

Microcline is so closely related to orthoclase in habit and angles that often the two cannot be distinguished crystallographically. The angle $P \wedge M$, which for orthoclase is 90°, for microcline is 90° 16′–90° 25′.   As a rock constituent microcline never forms regularly bounded crystals, but irregular grains, which, however, are partly bounded by crystal faces when they project into cavities of the rock (as in many granites).   The form is then that of the orthoclase represented in Fig. 98.   These crystals and grains are scarcely ever simple individuals, but are polysynthetic masses, composed of lamellæ and stripes arranged according to two laws of twinning, the *albite* and *pericline*.   Moreover in these crystals and grains the microcline is more or less intergrown with orthoclase and albite.   The dimensions of the microcline lamellæ as well as of the intercalated orthoclase and albite masses are almost always microscopic.   On the faces P and x (the faces bear the same notation as for orthoclase with the modifications necessitated by the triclinic system) the double twin lamination shows itself in two systems of very fine striations, one of which is parallel to the edge $P : M$, the other is normal to it, or, rather, is not noticeably inclined to it.   The albite lamellæ are intergrown with microcline in the same manner as with orthoclase, and often give the face M a distinctly striated appearance.   Furthermore, the apparently simple microcline crystal, which in reality is polysynthetic, forms twins according to the Carlsbad and Baveno laws.

Microcline cleaves along P and M exactly as orthoclase; there is an imperfect cleavage parallel to the left-hand prism ∞ / P (110) indicated by occasional cracks.   The position and inclination of the cleavage cracks in the different zones is exactly the same as in orthoclase, since the slight difference in the angle of the cleavage faces is scarcely or not at all noticeable. In microcline, also, the cleavage cracks are only perceptible in very thin sections.

The specific gravity $= 2.56$. The chemical composition and chemical reactions are the same as for orthoclase. Hence the distinction betweer microcline and orthoclase lies essentially in their optical behavior.

Microcline becomes transparent and colorless; the index of refraction and strength of double refraction have not been measured, but so far as the polarization phenomena can be relied upon, they appear to correspond exactly to those of orthoclase. The position of the axial plane is analogous to that in orthoclase, but is not absolutely normal to M, making with this face an angle of $82°-83°$; its trace on M is inclined $5°-6°$ to the edge $P:M$ in the direction of a positive orthodome. The acute axial angle is $88°-90°$ in oil. The obtuse positive bisectrix is not normal to M as in orthoclase, but varies $15°\ 30'$ from this normal. The dispersion about this bisectrix is $\rho < v$. Therefore cleavage plates or sections parallel to P and M in polarized light behave as follows: a simple cleavage plate parallel to P, which is bounded by M and K (100), as in the left-hand half of Fig. 106, and which is in the conventional crystallographic position, that is, has the acute edge $P:M$ above on the left, becomes dark between crossed nicols when the directions $\mathfrak{a}$ and $\mathfrak{c}$, the bisectrices of the angle of the optic axes, are parallel to the principal sections of the nicols. In other words, the direction of extinction is inclined $15°\ 30'$ to the trace of the cleavage parallel to M, or the extinction angle on P is positive (cf. plagioclase), that is, it is so that the axis of elasticity $\mathfrak{a}$ passes from the left front to the right back, when the crystal is properly placed above the upper basal plane. If, now, a second plate parallel to P be placed in twin position according to the albite law, the twinning axis normal to M, it will have the position of the right-hand half of Fig. 106, and its direction of extinction will be inclined $15°\ 30'$ to the trace of M, but on the opposite side. The sum of the extinction angles in the two halves of the twin will therefore be $31°$. This extinction angle of $15°\ 30'$ on P is the most characteristic, surest, and simplest means of distinguishing microcline from orthoclase. Since all apparently simple microcline crystals generally consist of many slender lamellæ twinned after the albite law, a cleavage plate parallel to P exhibits a great number of differently colored stripes between crossed nicols, the alternating stripes having the same color when of the same thickness. Each system of these stripes becomes dark when their longer direction

19

(parallel to M) is inclined 15° 30′ to a principal section of the nicols. But nearly all microclines are also twinned polysynthetically according to the pericline law; the twinning axis is b. Lamellæ arranged according to this law are bounded in sections along P by lines running parallel to the edge $P : K$ (001 : 100) (Fig. 107). Since the angle $M : K$

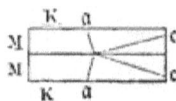
Fig. 107

(010 : 100) is almost exactly a right angle in microcline, the boundary lines of the lamellæ twinned according to the pericline law are normal to the edge $P : M$, also normal to the boundaries of the lamellæ twinned after the albite law. Hence both systems of lamellæ intersect at right angles. The twinning axis of the albite law, normal to M, and that of the pericline law, b, do not diverge perceptibly from one another. Consequently, the extinction angles in each system of lamellæ coincide with one another. Between crossed nicols sections along P exhibit a colored rectangular grating or plaid (Pl. XXV. Fig. 1), in which there are always two sets of bars perpendicular to one another which become dark at the same time, with an extinction angle of 15°–16°. This striking phenomenon, which, except for a change of angles, is the same in all sections which are not parallel to M, immediately distinguishes microcline from all other feldspars.

Both systems of lamellæ often reach such microscopic dimensions that it is no longer possible to determine the extinction of the different lamellæ even with the highest magnifying-powers. The eye then only receives a general impression of this rectangular grating. The different lamellæ very rarely attain the breadth they possess in ordinary lime-soda feldspars; occasionally also one or both systems of lamellæ is wanting. In these cases the characteristic extinction angle on P (15°–16°) is always the means of distinction from other feldspars.

Usually such sections parallel to P when between crossed nicols exhibit irregularly bounded flakes, which are dark when the twinning boundaries run parallel to a principal section of the nicols. They have straight or parallel extinction, and belong to orthoclase. In the same way there are bands which run nearly or exactly parallel to an edge $P : K$, less frequently to an edge $P : T$ or $P : l$, and exhibit a stronger double refraction than microcline and orthoclase, and show themselves finely twinned parallel to the face M, but which possess an extinction angle of about 4°. They belong to albite (Pl. XXV. Fig. 1).

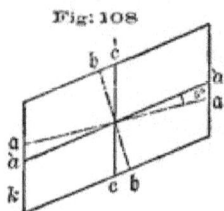
Fig. 108

Cleavage plates parallel to M would have the axes of elasticity a

and b in the position shown in Fig. 108 ; the directions of extinction, then, are the same as in orthoclase, and microcline plates parallel to this face would be dark between crossed nicols when the cleavage along *l'* makes an angle of 5° with a principal section of the nicols. The inclination of the axis of elasticity a to the crystal axis lies, as in orthoclase, in the sense of a positive hemidome ; it is positive (cf. plagioclase). Hence microcline sections parallel to M cannot be distinguished from similar sections of orthoclase in parallel polarized light, and inclusions of the latter in microcline cannot be recognized in such sections in this way. Albite stripes in 'microcline in sections along M run nearly parallel to the vertical axis; they stand out because of their stronger double refraction, and consequently higher interference colors, and exhibit a different extinction ($+18°$ to 20°). They are shown in Pl. XXV. Fig. 2. Occasionally, there is another system of bands which are inclined about 16°–18° to the vertical axis, and whose extinction lies between that of microcline and normal orthoclase and that of albite, and is inclined about 12° to a. They belong to another feldspar, which possesses the optical orientation of the schillerizing orthoclase of Frederiksvärn.

In convergent light plates of microcline along M do not exhibit the emergence of a perpendicular bisectrix, as in orthoclase, but of a rather oblique one. On the margin of the field of view the rings belonging to one axis are noticeable, the axis itself being situated outside of the field.

According to E. Mallard * and A. Michel-Lévy,† it seems highly probable that orthoclase and microcline are not dimorphous, but identical, since they proved that the optical behavior of orthoclase would be a necessary consequence of an intimate multiple twinning of microcline lamellæ after the albite and pericline law. This theory is strongly supported by the fact that in these bodies the relative cohesion and the specific gravity are the same in each, while these properties are generally different in heteromorphous bodies.

The alteration processes of microcline are exactly the same as those of orthoclase.

Microcline occurs with orthoclase, often almost completely replacing it, in granites, syenites, elæolite syenites, and gneisses. The feldspar of so-called graphic granite is almost always microcline. Microcline appears less frequently in quartz porphyries and other porphyries, and

* Explication des phénomènes optiques anomaux. Paris, 1877. 103.
† Bull. Soc. Min. Fr. 1879. II. 135.

still more rarely as the groundmass of these rocks becomes microfelsitic or glassy. In the younger eruptive rocks the sanidine very rarely exhibits a structure which entitles it to be placed under microcline (cf. Sanidine).

The Group of Plagioclases.

Literature.

A. Des Cloizeaux, Mémoire sur les qualités optiques biréfringentes caractéristiques des quatre principaux feldspaths tricliniques et sur un procédé pour les distinguer immédiatement les uns des autres. Ann. de Chim. et de Phys. 1875. (5). IV. and C. R. 1875. LXXX. 364–371.

— Examen microscopique de l'orthose et des divers feldspaths tricliniques. C. R. 1876. LXXXII. 1017–1022.

— Nouvelles recherches sur l'écartement des axes optiques, l'orientation de leur plan et de leurs bissectrices et leurs divers genres de dispersion, dans l'albite et l'oligoclase. Bull. Soc. min. Fr. 1883. VI. 89–121.

— Oligoclases et andésines. Ibidem. 1884. VII. 249–336.

E. Mallard, Sur l'isomorphisme des feldspaths tricliniques. Bull. Soc. min. Fr. 1881. IV. 103.

G. vom Rath, Die Zwillingsverwachsung der triklinen Feldspathe nach dem sog. Periklingesetz und über eine darauf gegründete Unterscheidung derselben. B. M. 1876. Febr. and N. J. B. 1876. 689–714.

M. Schuster, Ueber die optische Orientirung der Plagioklase. T. M. P. M. 1880. III. 117–284.

— Bemerkungen zu E. Mallard's Abhandlung "Sur l'isomorphisme des feldspaths tricliniques." Nachtrag zur optischen Orientirung der Plagioklase. Ibidem. 1882. V. 189–194.

G. Tschermak, Die Feldspathgruppe. S. W. A. 1864. December. L.

Under *plagioclases* are here included the lime-soda feldspars, that is, albite and anorthite, and their isomorphous mixtures from the albite, oligoclase, andesine, labradorite, bytownite, and anorthite series. The chemical composition of the theoretical albite is $Na_2O, Al_2O_3, 6SiO_2 =$ $Na_2, Al_2, Si_6O_{16} = Ab$; that of anorthite, $2CaO, 2Al_2O_3, 4SiO_2 = Ca_2, Al_2,$ $Al_2, Si_4O_{16} = An$. All other lime-soda feldspars, then, are isomorphous mixtures of albite and anorthite $= Ab_n, An_m$. Of the many possible mixtures certain ones occur more frequently, and have received particular names. If these be enlarged by the addition of those compounds closely connected with them, then, following Tschermak, the lime-soda feldspars or plagioclases may be brought into the following table:

Albite series embraces the compounds	Ab_1, An_0	—	Ab_9, An_1	
Oligoclase series "	"	"	Ab_6, An_1	— Ab_3, An_1
Andesine series "	"	"	Ab_3, An_2	— Ab_1, An_1
Labradorite series "	"	"	Ab_1, An_1	— Ab_1, An_2
Bytownite series "	"	"	Ab_1, An_3	— Ab_1, An_6
Anorthite series "	"	"	Ab_1, An_9	— Ab_0, An_1

In petrography, where so sharp a determination of the proportions of the mixtures in many cases is not possible, it becomes necessary to unite the andesine series with the oligoclase series, and the bytownite series with the labradorite series, and to speak of albite, oligoclase, labradorite, and anorthite as the plagioclases, since the name of the feldspar is also used for that of the series. There has been also included under the term plagioclase in petrography a number of feldspars which have been but slightly investigated, and which, by their small percentage of CaO and high percentage of K_2O, present a separate series of compounds, if they do not resolve themselves into very intimate mechanical mixtures. The following statements relate exclusively to plagioclases proper, or lime-soda feldspars :

The crystal forms of the plagioclases exhibit great similarity of habit and angle measurements among themselves, and also with those

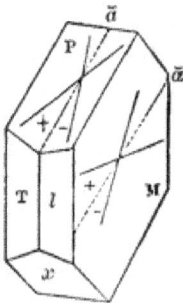

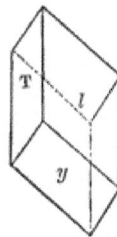

Fig. 109   Fig. 110

of orthoclase and microcline. The most essential difference rests in the fact that the angle $P \wedge M$ is not 90°, but lies between 93° 36′ for albite and 94° 10′ for anorthite ; there are also certain differences of angle in the inclinations of the other faces. The rock-making plagioclases do not always exhibit crystal boundaries, but are very often massive. Well-developed crystals only occur in rocks possessing a clearly marked porphyritic structure. They are then bounded principally by the faces $P = oP$ (001), $M = \infty P \breve{\infty}$ (010), $T = \infty_{,}'P$ (1̄10), $l = \infty P_{,}'$ (110), $x = {}_{,}P_{,}\breve{\infty}$ (1̄01), $y = 2{}_{,}P_{,}\breve{\infty}$ (2̄01), which are accompanied, as in orthoclase, by the subordinate faces $n = 2'P_{,}\breve{\infty}$ (02̄1), $O = P_{,}$ (11̄1), $v = {}_{,}P$ (11̄1), and others. The habit of the simple crystals is sometimes tabular parallel to M (Fig. 109), sometimes slender prisms parallel to a, like Fig. 99 for orthoclase ; it is also peculiarly rhombic (Fig. 110) in certain rocks because the faces P and M are wanting, or because the latter is but slightly developed. The angles of most importance in determining the cross-sections, whose forms are readily derived from the figures just given, are the following for albite, and but slightly different for the other plagioclases : $P \wedge M = 93° 36'$, $P \wedge T = 110° 50'$, $P \wedge l = 114° 42'$, $P \wedge x = 52° 17'$, $P \wedge y = 97° 54'$, $T \wedge l = 120° 47'$, $T \wedge M = 119° 40'$, $l \wedge M = 119° 33'$.

Simple crystals are comparatively rare, and the polysynthetic twinning, which is the most important outward character of the plagioclases when considered macroscopically, plays just as important a *rôle* microscopicallly. The commonest law of polysynthetic twinning in plagioclase is the albite law; the twinning axis the normal to M, composition plane, M. Fig. 111 represents a simple twin of this kind having a prismatic habit,, Fig. 112 represents such a one with tabular habit and with very small prism faces. In this kind of twinning the P faces of the two individuals make a reëntrant angle of 172° 48' with one another, their x faces one of 172° 42', and in the prism zone similar prism faces adjoin one another. Rock-making plagioclases are char-

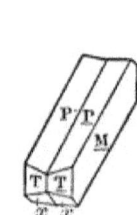
Fig. 111

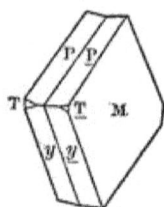
Fig. 112

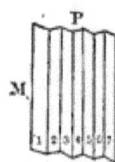
Fig. 113

Fig. 114

acterized by the frequent repetition of this twinning, so that a crystal consists of a great number of thin plates parallel to M. The reëntrant angles between the P faces then give rise to the well-known twin striation parallel to the edge $P : M$ on the basal plane of such crystals. A section parallel to K through such a multiple twin would have the form represented in Fig. 113; in a section which is parallel or inclined to the base the reëntrant angles would be cut off, but the twinning planes are often seen quite distinctly by transmitted light, especially when the section is inclined to P and the boundaries of the lamellæ are illuminated obliquely (Fig. 114). The twinning must be visible in all sections which are not parallel to M.

Much more rarely the twinning is according to the pericline law; the twinning axis is $\bar{b}$, the composition plane parallel to the rhombic section. By this method, when it is repeated polysynthetically, there must be a striation on the face M. In albite, according to G. vom Rath, this is inclined forward $13° - 22°$ less than the edge

edge $P : M$, to which the cleavage is parallel (Fig. 115). In oligoclase the angle between this striation and the edge $P : M$ is only 4° in the same direction, in andesine 0°, in labradorite 2°–9° in the opposite direction, that is, the striations are inclined more steeply forward than the edge $P : M$; for anorthite 18° in the last-named direction. The polysynthetic twinning after the pericline law not infrequently occurs in combination with that after the albite law; twin striation is then present on P and M. Fig. 116 represents a crystal bounded by P, M, and K (100) with albite and pericline lamellæ. On the basal plane the two systems of lamellæ intersect nearly at right angles; the crystallographic axial angle γ, which has different values for different plagioclases, is never more than 1° from a right angle. Fig. 116 shows that all sections through such a polysynthetic crystal must exhibit intersecting systems of lamellæ whose inclination to one another is

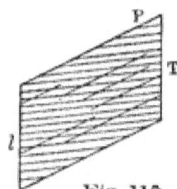

Fig. 115

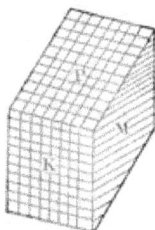

Fig. 116

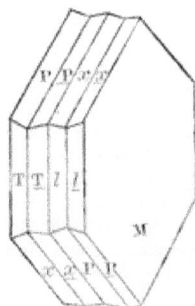

Fig. 117

dependent on the position of the section. The lamination is only single on the face M.

Such polysynthetic individuals, after the albite or pericline law, or after both together, often grow together according to laws corresponding to the Carlsbad, Baveno, and Manebacher laws in orthoclase. Fig. 117 presents a Carlsbad twin of two twins after the albite law, which is a very frequent occurrence. It is evident that the lamellæ on a basal section cannot all belong to the P faces, but partly to P and partly to x faces, which is important in considering their optical behavior. The great variety which is introduced into the twinning of the plagioclases by the combination of these laws is still further increased by the fact that the lamellæ are by no means formed with theoretical regularity. They often wedge out in the middle of the crystal, change their breadth, fork and branch, throng in one part of the crystal and fail in another; they do not always run parallel to the twinning plane, but show by their boundaries that the composition faces may be quite irregular. Their breadth bears no relation to the size of the compound individual, varying quite irregularly. But it appears that quite broad lamellæ in the embedded and rock-making plagioclases are chiefly confined to the more basic series.

The dimensions of plagioclase crystals vary between the widest limits. In general, they seldom reach the upper limits of the ortho-clases; they sink to microlitic dimensions, and then usually form very thin prisms parallel to the edge $P : M$ (Pl. XXV. Fig. 3), the so-called lath-shaped plagioclases or plagioclase microlites. The more acid plagioclases particularly tend to the prismatic development parallel to the edge $P : M$. In other cases the plagioclase microlites assume a tabular form parallel to M; they are then occasionally of scarcely measurable thickness, and sometimes have a rhombic outline formed by P and x or by P and y (like the face M in Fig. 112), sometimes an appropriately hexagonal one like the M face in Fig. 109, or an irregularly six-sided one from P, x, and y. This tabular form appears to be particularly characteristic of the microlites of basic plagioclases.

Actual incipient forms of growth and skeleton crystals are not definitely known.

Anomalies of crystallization are extremely common among the plagioclases. Thus ruin-like, indented terminations are very frequent in the larger individuals, as shown in Pl. XXV. Fig. 4; it almost appears as though small completed crystals had grouped themselves together to form a compound individual. Through chemical corrosion originally sharp-edged crystals have become more or less rounded to grains, whose original form can only be surmised from the zonal structure or the arrangement of interpositions. In other cases there arise "bays" or pockets of greater or less depth, which may amount to a hollowing out of the crystal, or in the other extreme may simply consist of a slight etching of the crystal face. Besides these chemical deformations, which are chiefly confined to the porphyritic eruptive rocks, there are in eruptive and schistose rocks the same fracturings (Pl. XXIII. Fig. 5) as those described for orthoclase, and the same marginal fissurings and crushings (Pl. IV. Figs. 3 and 4); further, a bending of the twin lamellæ (Pl. IV. Fig. 6), or a dislocation of the same through broken and faulted individuals. According to L. van Werveke,[*] it is very probable that a twin lamination may arise in

---

* N. J. B. 1883. II. 97.

plagioclases through the forces which brought about these mechanical deformations (movement in the magma and mountain pressure). Such mechanical twin lamellæ are chiefly characterized by the fact that their extent and course appear to depend on fracture lines in the crystal.

Zonal structure is extremely frequent in the plagioclases of all rocks, excepting in those of later generation in the groundmass of porphyritic rocks. It is in very many cases simply a consequence of repeated interruptions in growth. There is then no physical difference noticeable in the behavior of the kernel and of the different shells. In other cases, however, a zonal structure is first noticeable between crossed nicols by the fact that the extinction does not take place at the same time in the kernel and in the different shells, but the kernel and shells extinguish light in azimuths, sometimes differing by a number of degrees. So far as experience goes, the extinction angles are always so related to each other as to indicate that the character of the kernel is more basic than that of the shells. This phenomenon is explained by the assumption that there exists an isomorphous lamination, in which an original, basic, central crystal is surrounded by shells of other plagioclases, which gradually become more and more acid.* Another explanation of this phenomenon, which is shown in Pl. XXV. Fig. 5, is given by A. Michel-Lévy.† He considers it the result of a submicroscopic twin lamination after the albite and pericline laws.

The rock-making lime-soda feldspars, like the monoclinic potash feldspars, appear in two kinds of habit: In the granular and porphyritic, older, massive rocks and in the schistose rocks they have the dull, cloudy appearance which characterizes orthoclase; in the younger eruptive rocks they appear glassy and colorless, like sanidine. The latter appearance is called the *microtine* habit.

The plagioclases cleave along the faces P and M, the more perfect cleavage being that parallel to P. Both cleavages show themselves in sufficiently thin sections by cracks, which resemble those of orthoclase, except for their inclination. They do not generally show themselves in thicker sections. Cleavages parallel to the faces T and l are but rarely indicated by distinct cracks. The parting parallel to an oblique face, which is so characteristic of sanidine, seldom occurs in the plagioclases. The diagnostic importance of the cleavage in the plagioclases

---

* C. Höpfner, Über das Gestein des Mte. Tajumbina in Peru. N. J. B. 1881. II. 164–192.

† C. R. 1882. XCIV. 93 and 178.

is not so great as in the orthoclases, since the twin lamination takes its place to a certain extent as a means of optical orientation.

All plagioclases become transparent and colorless. Their indices of refraction are nearly equal to that of Canada balsam, and somewhat larger for anorthite than for albite. There is no direct determination; but from the axial angle Des Cloizeaux determined $\beta_\rho = 1.537$ for albite, which corresponds to the indices of refraction calculated by Gladstone's law, which for anorthite would be 1.573. The double refraction is not large, but is always greater than for the orthoclases, as shown by the interference colors, and apparently decreases with the percentage of lime. A. Michel-Lévy determined on anorthite $\gamma - \alpha = 0.013$. Little is known concerning the true position of the optical constants, with the exception of albite. However, the numerous investigations of Des Cloizeaux, and especially of M. Schuster, have completely determined the behavior of cleavage plates and sections parallel to the faces P and M in parallel and convergent polarized light, and have rendered it the most important, surest, and quickest means of distinguishing these minerals.

Since in triclinic minerals there is no regular relation between the position of the optic axial plane and the crystal form, the extinction on a crystal face between crossed nicols in parallel light will not generally take place parallel to a crystal edge or to the trace of a cleavage face, but will make an angle with it. If, now, a simple plagioclase crystal (Fig. 109) stands in the conventional crystallographic position, so that the end face is inclined toward the observer and slopes from left to right, the acute edge $P : M$ will be above to the left, the obtuse edge below to the right. On a plate parallel to P, the direction of extinction nearest to the edge $P : M$ can either deviate from this line so that its trace on P runs in the direction of the edge $P : l$ or in the direction of the edge $P : T$. The deviation in the first direction is called positive, that in the second negative. In the same manner, the direction of extinction on the right-hand face M can either deviate from the edge $M : P$, so that it runs in the direction of the edge $M : x$, or in the reverse direction; the first deviation is called positive, the second negative. All statements concerning the directions of extinction and other optical constants made in the following pages relate to the upper face P and the right-hand face M, in the position of Fig. 109.

For pure albite, the extinction angles on P are between $+ 4°$ and $+ 5°$, on M about $+ 19°$; for an oligoclase, with the composition Ab_2An_1, on $P + 10° 4'$, on $M + 4° 36'$; for an andesine, Ab_3An_2, on $P —$

2° 12′, on M — 7° 58′; for labradorite, Ab_3An_2, on P — 5° 10′, on M — 16°; for bytownite, Ab_1An_3, on P — 17° 40′, on M — 29° 28′; for pure anorthite, on P — 37°, on M — 36°. These relations are represented in Figs. 118 and 119 (s is always the position of the direction of extinction), and it is evident that the extinction angle on both faces assumes greater negative values with increasing percentage of lime. The transition from positive to negative extinction takes place on both the faces P and M on the borders of the oligoclase and andesine series. Thus,

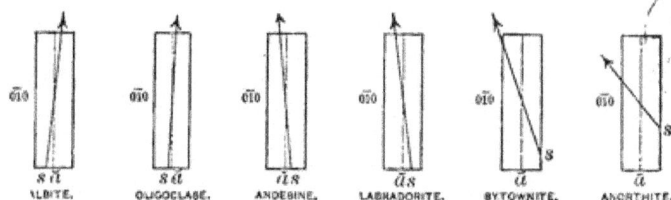

Fig. 118

there is a particular orientation of the directions of extinction on both cleavage faces, corresponding to every variety of composition. These relations have been carefully investigated experimentally by M. Schuster, and mathematically, from a theoretical standpoint, by E. Mallard, and the striking correspondence between their results leaves no doubt about the correctness of Schuster's law—that for every combination of albite and anorthite there exists a certain extinction angle on the faces

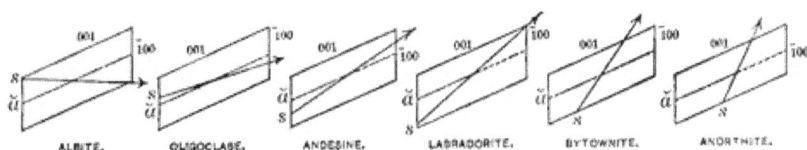

Fig. 119

P and M, which is dependent on the amount of these substances in the compound. The table on page 300 presents the relations between the extinction angles and the compounds, from which, when either the composition or the extinction angle is given, the other may be found.

In a basal section of a plagioclase between crossed nicols the lamellæ twinned after the albite law must in general be differently colored, since the section cuts them in different directions with respect to their ellipsoids of elasticity. But since in Figs. 113 and 114 the lamellæ marked with even numbers have the same

Formula of Mixture.	Extinction angle on $P = \infty P \bar{\infty}$ (001).	Extinction angle on $M = \infty P \infty$ (010).	Inclination of the rhombic section to the edge $P : M$.	On M in convergent light.	Mean specific gravity.
Ab	+ 4° 59'	+ 19°		Positive bisectrix slightly inclined.	2.62
Ab₁₃An₁	+ 3° 58'	+ 15° 35'	+ 2° to + 30°		
Ab₆An₁	+ 3° 12'	+ 13° 49'			
Ab₅An₁	+ 2° 45'	+ 11° 59'	+ 9° to + 7°	Positive bisectrix emerges almost normal, with a small inclination upward.	2.64
Ab₃An₁	+ 2° 25'	+ 10° 34'			
Ab₂An₁	+ 1° 55'	+ 8° 17'	+ 6° 42' to + 3° 20'		
Ab₃An₂	+ 1° 04'	+ 4° 56'			
Ab₂An₂	+ 0° 35'	+ 2° 15'	+ 1°		2.65
Ab₃An₂	− 0° 12'	− 2° 58'	+ 0°		
Ab₄An₃	− 2° 58'	− 10° 56'			2.69
Ab₃An₃	− 5° 19'	− 16°	− 1° to − 3°	Positive bisectrix emerges quite obliquely. One axial bar visible; the axis itself not in the field.	
Ab₂An₃	− 6° 50'	− 19° 12'			
Ab₂An₄	− 7° 33'	− 29° 32'			
Ab₁An₂	− 13° 58'	− 35°			2.71
Ab₁An₃	− 17° 40'	− 29° 28'	− 9° to 14°		
Ab₁An₄	− 21° 05'	− 31° 10'			
Ab₂An₄	− 23° 37'	− 32° 10'			
Ab₁An₄	− 27° 33'	− 33° 29'			
Ab₁An₄	− 28° 04'	− 33° 40'		One axis emerges in the margin of the field.	2.73
Ab₁An₁₃	− 30° 23'	− 31° 19'	− 15° to 18° 48'		
An	− 37°	− 36°			
Microcline	+ 15° 39'	+ 5°		Positive bisectrix inclined forward.	2.57
Orthoclase	0°	+ 5° to + 7°		Positive bisectrix normal.	2.58–2.59
Soda orthoclase	0°	+ 9° to + 12°	− 4° to − 8°	Positive bisectrix normal.	2.58–2.60
Anorthoclase	+ 1° 39' to + 5° 45'	+ 6° to + 9° 48'		Positive bisectrix slightly inclined.	

position throughout, and those marked with odd numbers have another position, then all the even lamellæ and all the odd lamellæ will exhibit two sets of interference colors. The result is a colored lamination, which is extremely characteristic of the plagioclases (Plate XXV. Fig. 6). If the section be rotated between crossed nicols, then one set of lamellæ will become dark for a particular inclination of the boundary lines between the lamellæ (trace of *M*) to the left or to the right of a principal section of the nicols, which inclination varies with the chemical composition of the feldspar. If the section be now rotated through the same angle to the right or left of the twinning line, the second set of lamellæ would become dark, if they were cut parallel to the face *P*. But this is not the case, and therefore the extinction angle of the second set of laminæ is not exactly the same as that of the first. But the difference is always small, and in general it is more convenient and sufficiently exact to determine the extinction angle on *P*, by rotating the section between the points of maximum darkness for each set of lamellæ, and halving the angle so obtained. If the extinction angles should be the same, right and left, for both, it would show that the section was not parallel to *P*, but normal to the twinning plane *M*. In many basal sections or cleavage plates of plagioclase there are lamellæ which do not exhibit the same interference colors or extinction angles as the two sets of lamellæ, although they appear to be inserted according to the same twinning law. Such lamellæ belong to a set twinned according to the Carlsbad law, which, as Fig. 117 shows, are not cut parallel to *P*, but to *x*. There will also be two sets of these latter lamellæ, arranged according to the albite law, which in turn extinguish almost symmetrically on both sides of the twinning plane. Plate XXV. Figs. 4 and 6 exhibit these relations in the brightness of the different lamellæ. Lamellæ twinned after the pericline law would cross the albite lamellæ nearly at right angles, and would furnish two sets of lamellæ, each of which would extinguish the light at approximately the same time as the sets of albite lamellæ, since both twinning axes very nearly coincide (compare microcline). In all other sections not parallel to *M*, the different sets of lamellæ will always be differently colored, and will extinguish in different azimuths, which are unsymmetrical to the twinning plane. Only in sections lying in a zone at right angles to *M* will the extinctions in both sets of lamellæ be symmetrical to the twinning plane. When pericline and albite lamellæ occur together, the angle between the lamellar systems in irregular sections varies with the position of the section; otherwise, the relations remain the same

(Plate XXVI. Fig. 1). The presence of Baveno twins in a plagioclase shows itself as in orthoclase, through the occurrence of a twinning
boundary, running diagonal to P and M, in sections parallel or inclined
to the cross-section (querfläche) (Pl. XXVI. Fig. 2).

Sections of a plagioclase parallel to M will only exhibit twin lamination when there are lamellæ according to the pericline law. Their
boundary will be inclined to the cleavage along P, according to the
position of the rhombic section for the particular composition of the
feldspar (column 4 of the table just given), or, if the composition
plane is the base (the rarer case), it will be parallel to this cleavage.
The extinction is to be measured from the cleavage along P.

If the section is very much inclined to the twinning plane, and the
lamellæ are very thin, it may happen that a complete extinction does
not take place. It is due to the fact that within the thickness of the
section two wedge-shaped lamellæ are
superimposed. The conditions, then,
are those described on page 62.

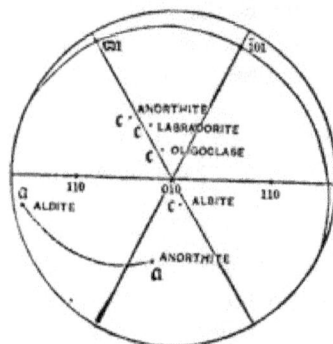

All plagioclases from albite to anorthite in convergent polarized light show
a positive bisectrix more or less inclined
to the face M. The axial angles about
this positive bisectrix vary in the neighborhood of 90°; they are acute for albite
with $\rho < v$, obtuse for oligoclase with
$\rho < v$, acute for labradorite with $\rho > v$
and obtuse for anorthite with the same
dispersion. The size of the axial angle
changes for different light, and diminishes in a complicated ratio with
the percentage of anorthite.

The approximate position of the axial plane is best understood
from the projection on M (010), Fig. 120, taken from Schuster's work.
The positive bisectrix for all plagioclases lies very nearly in the plane
of the zone $P : M$ (001 : 010), but is inclined on the right-hand M face
toward the acute edge $P : M$ for albite, rights itself with increasing
anorthite percentage so that in the normal oligoclases it is slightly
inclined toward the obtuse edge $P : M$, and this inclination increases
more and more with the labradorites, bytownites, and anorthites. In
certain oligoclase-albites the positive bisectrix very nearly coincides
with the normal to M. With this rising up of the point of emergence
of the positive bisectrix there is combined a rotation of the axial plane
so that the negative bisectrix, which in albite emerges from the macro-

Fig. 120

pinacoid, in anorthite appears to be turned about 70°, and leaves the crystal in the neighborhood of the right lower front corner. The inclination of the positive bisectrix downward from the normal to M (010) is about 18° in albite, the inclination upward for anorthite about 42°.

Cleavage plates and sections of albite parallel to M in convergent polarized light show the emergence of a positive bisectrix to one side of the field of view (for the proper position of the right-hand M face downward). The axes themselves do not come within the field, but their outer rings are equally distinct on both sides when the convergence of the light is sufficient and the plate is not too thin. The dispersion is inclined and slightly horizontal. There is no axial figure on P.

For oligoclase a bisectrix emerges nearly normal to M, the inclination being toward the obtuse edge $P : M$. The dispersion is very slightly inclined and slightly crossed. There is no axial figure on P.

Labradorite and bytownite show curves and an axial bar on the right-hand M face, which indicates that an axis emerges outside of the field of view below to the left. Plates parallel to P show the same phenomenon, but for proper crystallographic position the axis emerges outside of the field above to the right. The dispersion is distinctly crossed and slightly inclined.

Anorthite plates on the right-hand M face show an axis within the field not far from the margin and below, and on the upper P face an axial figure within the field and back. There is no distinct dispersion of the bisectrices.

From the foregoing it is clear that it is possible to determine the proportions of a mixture within certain limits which depend on the perfection of the material, its freshness, and not too complicated twinning structure. The difficulty lies in the determination of the character of the extinction angle in the cleavage plate or thin section under investigation. They are diminished by combining certain extinction angles on P with those on M. Small extinction angles on both faces indicate oligoclase or andesine, and it is generally impossible to distinguish between these unless the crystals in question are measurable. Large extinction angles on both faces characterize bytownite and anorthite. Medium extinction angles on P and M occur in albite and labradorite. In order to distinguish between the last-named varieties plates parallel to M are used. If cleavage cracks parallel to the prism are present in ordinary light, the character of the extinction is easily told. It is negative when it lies in the acute angle between the cleavage cracks parallel to the prism and base, positive when in the obtuse angle between these cleavages. If the cleavage is wanting, convergent light is used. In albite a positive bisectrix emerges almost normal to *M*, but in labradorite *M* shows no recognizable bisectrix and no axis. As between anorthite and bytownite, the former shows an axial figure on *M* in convergent light, which is situated in the margin of the field of view; in bytownite the axis is no longer in the field.

For the correct determination of cleavage plates they should be bounded by two plain, smooth cleavage faces. If they have but one such face, they should be secured by this one and a second face ground parallel to it. If the material is too fine-grained to furnish such cleavage plates, unstriated sections should be sought out in the thin section, whose outlines, if possible, are evidently those of *M*, and these tested in parallel and convergent light. It is possible in this way to arrive at a conclusion as to the approximate basicity of the plagioclase, and in particularly good cases to determine it accurately. If the plagioclase is greatly twinned after the pericline law, the *M* faces are recognized with less certainty, and the determination is made more difficult if not impossible. In such cases, when good cleavage pieces cannot be had, a sort of statistical process may be employed which has been specially elaborated by A. Michel-Lévy.[*] It is evident that sections of a plagioclase at right angles to the twinning plane *M* can always be recognized by the fact that the extinctions in alternate lamellæ are symmetrical to the twinning plane *M*. These extinction angles when measured have very different values, but for each plagioclase a maximum. For microcline it is 18°; for albite, 15° 45′; for oligoclase, 18° 30′; for labradorite under certain suppositions, 31° 15′; for anorthite, over 37° 21′. Thus it is evident that, for example, the occurrence of symmetrical extinction angles of 25° would indicate that the feldspar was not albite nor oligoclase, but a distinction between labradorite and anorthite would not be possible. In general, this process is not applicable unless it is certain that the maximum extinction has been observed. When this cannot be assumed, such a determination should be employed with the greatest caution.

For the determination of plagioclase microlites A. Michel-Lévy proposed to employ the zone *P* : *M*, in which they are developed prismatically, so that their longitudinal axis corresponds to the axis of this zone. In this zone the extinction angles of a lamella measured from the zonal axis vary in microcline from 0° to 16°; in albite, from 0° to 19°; in oligoclase, from 0° to 2°; in labradorite from 0° to 17°, or from

0° to 27°, according to the size of the axial angle 2 V; in anorthite, from 0° to 37°. From this it is seen that oligoclase microlites are well characterized by the fact that they extinguish light almost parallel to their length. The frequent recurrence of extinction angles over 27° would show the presence of anorthite. Further than this these data cannot be used.

In cases where the optical determination of the plagioclases is impracticable, their specific gravity may be used to advantage. It may be determined on small grains by suspending them in a heavy solution whose density has been determined by one of the methods described on page 104, or by taking it during the mechanical separation of the mineral constituents immediately before and during the settling of the plagioclase powder. Tschermak first showed that the specific gravity of the plagioclases increases with the percentage of anorthite in such a manner that it can be calculated for a particular plagioclase from its relative composition. For that purpose it was assumed to be 2.624 for pure albite, and 2.758 for pure anorthite. V. Goldschmidt* made a large series of determinations on feldspars, which average somewhat lower than the values given by Tschermak, although the differences are not great. Bärwald determined the specific gravity on the ideally pure albite of Kasbek at 2.618. In the following table the values given by Tschermak and Goldschmidt are correlated:

	Sp. gr. according to Tschermak.	Sp. gr. according to Goldschmidt.	Typical average value.
Orthoclase } Microcline }	......2.56–2.57	2 50–2.59	2.57
Albite............	..2.62–2.64	2.61–2.63	2.62
Oligoclase.........	2.64–2.66	2.62–2.65	2.64
Andesine	2.66–2.69	2 65	2.65
Labradorite.......	2.69–2.71	2.68–2.70	2.69
Bytownite.........	2.71–2.74	2.70–2.72	2.71
Anorthite.........	2.74–2.76	2.73–2.75	2.75

From the great exactness with which the density of a heavy solution can be regulated, this determination of the feldspars is very reliable, so long as the material is pure and fresh. This certainty is considerably lessened by the presence of interpositions as well as by alteration processes in the feldspars whose specific gravity is to be determined. It is to be remembered that the commonest inclusions of the feldspars in porphyritic rocks (gas and glassy parts of the magma) diminish the specific gravity, while the individualized inclusions of the feldspars

* N. J. B. B.-B. I. 1880.

in granular rocks tend to increase it. The influence of alteration processes is less easily foreseen, because the alteration products are difficult to recognize with certainty. A kaolinization and zeolitization would diminish the density; the development of carbonates, the formation of mica and saussurite, must increase it.

The chemical investigation of the feldspars in cleavage plates or in powder after its isolation should be used to control the results found in other ways. Without regard to the fact that anorthite and bytownite are decomposed by boiling hydrochloric acid with the formation of gelatinous silica, while the other feldspars are not attacked at all or only very slightly, they are distinguished with the greatest certainty by Bořický's method from the amount of potassium, sodium, and calcium fluosilicates. The advantages of spectrum analysis also should be borne in mind.

The alteration processes of the plagioclases are partly the same as those of orthoclase; sometimes kaolin, sometimes muscovite, or perhaps paragonite, is formed. From the nature of things, calcite is more common besides quartz as side products in these processes. Zeolitization is particularly frequent when the plagioclases are associated with nepheline or minerals of the sodalite group in younger eruptive rocks, but also occurs in certain rocks of the diabase and gabbro families. The so-called saussurite alteration of the basic plagioclases (labradorite, bytownite, and anorthite) is chiefly confined to dynamo-metamorphic regions; they are converted into an aggregate, which consists principally of epidote or zoisite, with which scapolite is occasionally associated, while the soda gives rise to the formation of albite. The alteration of feldspars to pseudophitic substances has rather the character of a local process; it has been observed in granular limestones. The lime and alkalies of the feldspars must have been replaced by magnesia and protoxide of iron from associated minerals.

Albite has a greater distribution in rocks than was formerly supposed. In the massive non-glassy condition it is a constituent of certain granitic rocks, its occurrence in which up to the present time has been investigated but little. The crystalloids of albite exhibit the normal polysynthetic twinning of the granitic plagioclases. In the form of microperthitic intergrowths with orthoclase and microcline, albite is quite generally present in granites and gneisses, especially in those with high percentage of silica. It is very probable that albite is sometimes very abundant in the microcrystalline groundmass of porphyries and porphyrites, and is confounded with orthoclase on account of the lack of twinning. Its presence is rendered quite certain by the chemical composition of the groundmass of such rocks, but it has not been directly proven as yet. As long prismatic microlites of microtine habit it occurs in the groundmass of acid trachytic and andesitic eruptive rocks, and probably it is not infrequent among the porphyritic secretions. Here also the evidence has been derived mainly from the chemical composition. Albite has a distribution in the Archæan rocks which was formerly quite overlooked, especially in those whose crystalline condition has been brought about by dynamo-metamorphic processes. Thus it has been described by A. Böhm * in distinctly polysynthetic grains from gneiss in the Wechselgebirge, the north-eastern extension of the central range of the Alps. In the sericite gneisses, phyllite gneisses, feldspar phyllites and porphyroids, it sometimes forms more or less distinct crystals, at other times grains, which occur like porphyritic sections; sometimes intimately associated with quartz and muscovite, it forms more or less fine-grained aggregates. When very fresh it is white, and often cloudy to dark gray from abundant inclusions of carbonaceous matter, rutile needles, minute fluid and gas interpositions; it is also reddish from infiltrations of hydroxide of iron. Not infrequently the twinning is entirely absent, or there are simple twinned halves in whose separate individuals very small lamellæ are occasionally inserted in twinned position. The twinning boundary is often a very irregular face. Finally, albite occurs with very similar habit, but generally in much smaller grains in the adinoles of diabase contact zones and in many so-called green schists, as well as in quartz nodules and veins in phyllites and clay-slates.

Oligoclase in massive grains and crystals is one of the most frequent feldspars in granites, syenites, diorites, and their porphyritic equivalents, and particularly accompanies orthoclase. The inclusions and structure are exactly the same, and have the same arrangement as in the orthoclase of the same rock, with which it is frequently intergrown. When a form can be made out it has the more equidimensional to tabular habit of Figs. 109 and 117. The twin lamination is seldom if ever absent, and the lamellæ are not very broad. Oligoclase occurs in the same form and with the same microstructure in gneisses. The weathering leads to the formation of kaolin and light-colored mica, with an accessory secretion of calcite and epidote. In the diabase rocks and their porphyritic varieties the habit of the oligoclase is generally lath-shaped, with P and M equally developed. In this group of rocks, even when granular, the oligoclase occasionally contains glass inclusions. The pe-

---

* T. M. P. M. 1883. 5. 202.

culiar crystals bounded chiefly by T, l, y (Fig. 110), occurring in the so-called rhombic porphyry, belong to oligoclase, according to O. Mügge.*

Oligoclase with microtine habit forms one of the principal constituents of trachytic and andesitic rocks, and when occurring as porphyritic secretions has chiefly a tabular form ; as a constituent of the ground-mass, it has a lath-shaped form.   It is particularly characterized, like all the plagioclases of these rocks, by an abundance of glass inclusions, which are often scattered through it like a net (Pl. XXVI. Fig 3), are often arranged zonally, peripherally, or centrally, and are rarely isolated or irregularly arranged.   In the basaltic rocks the oligoclases are mostly lath-shaped.

Sunstone is the name applied to certain oligoclases, which have a beautiful red sheen from the interposition of lamellæ of specular iron. The familiar occurrence at Twedestrand has been investigated micro-scopically by Th. Scheerer.†   The lamellæ of specular iron lie chiefly along the faces P', M, and a prism, in part also parallel to a face $2P'$, $(22\bar{1})$.

Andesine has exactly the same geognostic position and development of forms as oligoclase in the older and younger eruptive rocks and in the gneisses.

Labradorite appears to be confined to the more basic eruptive rocks and to certain Archæan rocks rich in amphibole and pyroxene; it always appears to avoid the proximity of orthoclase and quartz.   The massive labradorite of the older granular massive rocks of the diorite family possesses the same habit as oligoclase and andesine ; the same is true in general of the lath-shaped labradorite of diabase and ophite. On the other hand, the spathic labradorites of the gabbro and norite series are often distinguished by peculiar gray or grayish brown to reddish brown colors, which arise from interpositions, which, in spite of all differences of form. appear to belong essentially to iron-ores and titaniferous iron-ores. Long opaque or brownish translucent plates of hexagonal, rhombic, or irregular outline are particularly characteristic, and are probably lim-onite and specular iron.   Moreover, there are also acicular microlites which are mostly straight, but are also curved and bent or separated into points.   In many labradorites of the gabbros and norites, partly also in the ophites, these interpositions sink to the finest dust-like forms,

---

* N. J. B. 1881. II. 107 sqq.
† Pogg. Ann. 1845. LXIV. (153.) cf. Isaac Lea, Proc. Acad. Nat. Sci. Phil. 1866 110.

not resolvable even with the highest powers. Moreover, these labradorites often contain microlites of pyroxene and hornblende, crystals and grains of associated minerals, and quite frequently fluid inclusions. Labradorite occurs with the same habit in many amphibolites of the Archæan which are evidently dynamo-metamorphic gabbros.

To this group of labradorites belong those from St. Paul's Island, Ojamo, and the neighborhood of Kiew, which are well known on account of their beautiful play of colors and their broad cleavage, and whose interpositions and microstructure have been carefully investigated. Vogelsang was the first to refer the iridescence of these labradorites to their orderly arranged interpositions. These interpositions differ from those of ordinary gabbro labradorites only in the beauty of their development and in their usually very regular arrangement parallel to the vertical and brachydiagonal axes.

The tendency of these gabbro labradorites to the simultaneous development of albite and pericline twin structure is to be noted, as well as the rarity of the formation of carbonates in the processes of decomposition, which almost always lead to the formation of saussurite. The description of the alteration processes which take place with the aid of solutions arising from the associated minerals (pyroxene, olivine, ilmenite) belongs to the petrographical part of this work.

The labradorites of the older porphyritic eruptive rocks (porphyrite, augite porphyrite, melaphyre, etc.) as well as of the younger volcanic rocks (trachyte, andesite, basalt, and tephrite) exhibit exactly the same development of forms and microstructure as the more acid feldspars of the same rocks.

Bytownite possesses the same geological position, the same development of forms, and the same microstructure as labradorite in the older and younger granular and porphyritic rocks. F. Zirkel * has shown that the occurrence which gave the name to this series of plagioclases, lying between labradorite and anorthite, is a mixture.

Anorthite occurs in granular individuals or broad tabular aggregates in a few diorites, in lath-shaped forms in occasional diabases and in the teschenites, in large spathic masses in gabbro and norite, especially in the olivine-bearing varieties, and here possesses the structure of labradorite. It forms tabular crystals in the most basic porphyrites. The microtine form of anorthite is found in many andesites and basalts, especially in the older granular segregations in these rocks, which occasionally reach the surface as bombs, or lie like inclusions in

---

* T. M. M. 1871. 61.

the lava-flows. In the Archæan rocks labradorite, bytownite, and anorthite are only found in amphibolites, which were probably once gabbros.

Fischer[*] has investigated the basic plagioclases belonging to anorthite and bytownite which have been named amphodelite, latrobite, indianite, rosellan, polyargite and pyrholite, and the pseudophitic alteration of the same. Des Cloizeaux (l. c.) carried through their optical investigation. V. Lasaulx[†] and Liebisch[‡] described the feldspathic mixture saccharite, which forms nests and seams in serpentine.

1. *Appendix.*—Besides the true plagioclase group, including albite and anorthite with their isomorphous mixtures with the general formula Ab_nAn_m, and whose members never possess any considerable proportion of the compound $Or = K_2Al_2Si_6O_{16}$, there appears to be series of triclinic potash-soda feldspars whose percentage of $An = Ca_2Al_2Al_2Si_4O_{16}$ never exceeds a certain limit. This group has not been definitely known until recently, and its occurrence and properties have been but little studied. Feldspars belonging to this group from the island of Pantelleria were first described by H. Förstner[§] as soda orthoclases, and considered monoclinic. C. Klein[‖] recognized their triclinic nature, as well as that of a similar feldspar from Hohenhagen, and placed them near oligoclase. W. C. Brögger[¶] then found in the augite syenites, whose intergrown orthoclase and albite have already been mentioned, feldspars which showed no mechanical mixture of orthoclase and albite, and behaved optically, in part monoclinic, in part triclinic. Their chemical composition, which was the same in both cases, indicated that they were isomorphous mixtures of potash and soda feldspars, with an inconsiderable percentage of lime feldspar. More recently, H. Förstner[**] has investigated the feldspars of Pantelleria anew, and has determined a considerable number of such triclinic potash-soda feldspars with small percentage of lime, chemically, crystallographically, and optically.

---

[*] Kritische, mikroskopisch-mineralogische Studien. 1. Fortsetzung. Freiburg. i. Br. 1871. 40. sqq.

[†] N. J. B. 1878. 623.

[‡] Z. D. G. G. 1877. XXIX. 735.

[§] Ueber Natronorthoklas von Pantelleria. Z. X. 1877. I. 547.

[‖] Ueber den Feldspath im Basalt vom Hohen Hagen bei Göttingen und seine Beziehung zu dem Feldspath vom Mte. Gibele. Göttinger Nachrichten. 1878. No. 14, and N. J. B. 1879. 518.

[¶] Die silurischen Etagen 2 und 3 im Christiania-Gebiet. Christiania. 1882. 260 sqq. and 293 sqq.

[**] Ueber die Feldspathe von Pantelleria. Z. X. 1883. VIII. 125.

The series of triclinic potash-soda feldspars, one of whose most important properties must be that they possess an apparent cleavage angle $P \wedge M$, which varies scarcely any from a right angle, and yet must do so, is to be designated as the series of *anorthoclases* in distinction to the plagioclases which plainly cleave obliquely.

The *anorthoclases* are isomorphous mixtures of Ab and Or in the ratio of $2:1$ to $4.5:1$; that is, Ab_2Or, to $Ab_{4.5}Or$, to which is added An in varying amount. The ratio $An : Ab + Or$ varies from $1:3$ to $1:22$. The habit is like that of the other feldspars, but there is occasionally a type in which the crystals are developed in prisms parallel to c. T and l predominate; M sinks to almost nothing. Of the macrodomes, y is the only one which occurs. The triclinic character is very obscure. Twinning according to the Carlsbad, Baveno, and Manebacher law is very common; the separate individuals (halves of the twin) are multiple twins after the albite and pericline law. The twin lamellæ are almost always of the most extreme fineness, so that P and M are apparently plane faces, and appear to intersect at right angles. A third law of lamellar arrangement occurs locally: twinning axis the normal to y ($\overline{2}01$).

The lamellæ twinned according to the pericline law, that is, parallel to the rhombic section, are inclined $4°-6°$, rarely $8°$, to the cleavage parallel to P on the M face in the negative sense; hence in the opposite direction to those of albite, to which anorthoclase stands nearest chemically. The cleavage is parallel to P and M, as in all feldspars. The specific gravity lies between that of orthoclase and albite, $2.57-2.60$, and rises with the percentage of albite. It is exactly the same as for the perthites.

Index of refraction low; $\beta_{na} = 1.504-1.531$, according to Förstner, not determined directly, but calculated from the axial angle. Double refraction somewhat stronger than for orthoclase. The extinction on P is positive, and varies between $5° 45'$ and $1° 30'$; it is also positive on M, and lies between $6°$ and $9° 48'$, so that, if the percentage of anorthite be overlooked, it appears to grow less on P with the albite percentage, and to increase on M. The twin lamination is often only visible on basal sections when they are the thinnest possible, because of the minuteness of the lamellæ; anorthoclase is best distinguished from orthoclase in sections at right angles to P and M. In these sections, also, highly twinned areas pass into others free from twinning without there being any visible boundary between them. The positive bisectrix emerges from M, as in all feldspars, and with slight inclination; it bisects the obtuse axial angle; the negative acute bisectrix emerges approximately normal to y. About this the dispersion is distinctly horizontal, and $\rho > v$. $2E_{na}$ varies from $71° 40'$ to $88° 27'$.

Anorthoclases are known to occur in the augite syenites of Southern Norway, and perhaps in their porphyritic equivalents, as well as in siliceous varieties of the amphibole and augite andesites. From a consideration of the rock analyses, it is probable that they will be found in trachytes and rhyolites, as well as in dacites.

2. *Appendix.*—Fouqué[*] described a very remarkable feldspar from Quatro Ribeiras, on Terceira, one of the Azores. It has the composition of a CaO- and K_2O-bearing albite, but possesses optical properties which approach those of microcline very closely; it has fine lamellar twinning, and specific gravity $= 2.593$. The extinction angle on P is $1° 30'$, on M it is $+ 9°$ to $+ 9° 30'$. Almost normal to M stands a positive bisectrix, which bisects the obtuse axial angle. About the negative bisectrix, which is almost normal to y, the dispersion is distinctly horizontal, and $\rho > v$. $2E = 65° 40'$ to $75°$. The indices of refraction are $\alpha_{na} = 1.5234$, $\beta_{na} = 1.5294$, $\gamma_{na} = 1.5305$. Changes of temperature up to $200°$ C. are without effect on it.

Disthene.

Literature.

M. BAUER, Beiträge zur Kenntniss der krystallographischen Verhältnisse des Cyanits. Z. D. G. G. 1878. XXX. 283–326 ; 1879. XXXI. 244–254. 1880. XXXII. 717–728.

F. BECKE, Die Gneissformation des niederösterreichischen Waldviertels. T. M. P. M. 1882. IV. 225–231.

E. COHEN, Ueber einen Eklogit, welcher als Einschluss in den Diamantgruben von Jagersfontein, Orange-Freistaat, Süd Afrika, vorkommt. N. J. B. 1879. 864–870.

G. VOM RATH, Ein Beitrag zur Kenntniss der Krystallisation des Cyanit. Z. X. 1879. III. 1–12 ; 1881. V. 17–23.

E. R. RIES, Untersuchungen über die Zusammensetzung des Eklogits. T. M. P. M. 1878. I. 195–198.

Disthene occurs in rocks as crystals or as columnar crystalloids, and also in parallel columnar aggregates, less frequently in twisted ones. The crystals are only well crystallized in the prism zone, and are elongated parallel to the prism axis; terminal faces are not

---

[*] Feldspath triclinique de Quatro Ribeiras (Ile de Terceira). Bull. Soc. min. Fr. 1884. VI. 197.

so rare, but are usually so uneven and bent that they furnish no measurable angles. Hence sections parallel to the prism zone are lath-shaped, with round, jagged, or quite irregular ends; at right angles to this zone they are six-sided, with one large and two small edges, or are rounded. By the suppression of one pair of faces the basal sections become obliquely rhombic. The predominant faces are $M = \infty\,P\bar{\infty}$ (100), $T = \infty P\bar{\infty}$ (010), $l = \infty P_{,}'$ (110), $o = \infty_{,}'P$ (1$\bar{1}$0), $P = oP$ (001), $k = \infty P$, $\bar{2}$ (210). The most important angles are $M \wedge T = 106°\ 4'$, $M \wedge l = 145°\ 13'$, $M \wedge o = 131°\ 42'$, $P \wedge M = 101°\ 30'$, $P \wedge T = 105°\ 4'$, according to G. vom Rath's calculation. Twinning is very frequent, and takes place after the following laws: (1) Twinning axis normal to M. The faces P and T form protruding and re-entrant angles; this is the most common law, and is often repeated polysyn-thetically. (2) Twinning axis normal to the edge $M : T$ lying in the face M, composition plane M. The faces T form re-entrant angles. (3) Twinning axis the edge $M : T$, composition plane M. The faces P form re-entrant angles. (4) Twinning axis normal to P, generally repeated a number of times, and, as Bauer has shown, it is a pressure twinning. Crossed twins like staurolite, twinned parallel to a face ($\bar{2}1\bar{2}$), are not uncommon in the smaller crystals of paragonite schists; their vertical axes intersect at about 60°.

The cleavage parallel to M is very perfect, and gives rise to sharp cracks, which, however, do not traverse the whole section when in rather thick plates. The cleavage parallel to T is less distinctly notice-able microscopically; its cracks are shorter and rougher, end abruptly, and are less numerous. The parting parallel to P corresponds to a gliding plane, as Bauer has shown. Hence, longitudinal sections ex-hibit more or less sharp cracks parallel to the length of the section, and fissure-like cracks at right angles to it; cross-sections show distinct cleavage cracks parallel to the longest edge, sometimes with another set parallel to one of the shorter edges.

Cyanite becomes transparent and colorless; many varieties, how-ever, are blue or greenish blue. The pigment is generally dissemi-nated quite irregularly. The crystals may become almost opaque from carbonaceous matter. The index of refraction is high (Des Cloizeaux determined $\beta_{g} = 1.720$), therefore the surface is quite rough, the mar-ginal total reflection strong. The double refraction must be not in-considerable because of the height of the interference colors, which is greater than that of andalusite, but less than that of sillimanite. The axial plane stands almost normal to M, and its trace cuts this face like the diagonal of the acute plane angle on M from the edge $P : M$, with an

inclination of about 30° to the edge $M : T$ (Fig. 121). The character
of the double refraction is negative ; the dispersion about a weak, $v < \rho$.
The axial angle is generally quite large, so that the axes often are only
visible on M in oil ; $2 V = 82°-83°$; but smaller axial angles occasionally occur (Litchfield). The extinction on M from the edge $M : T$ is
30°–31°, on T from the same edge 7°–8°; on P it is approximately parallel and normal to the cleavage parallel to M, and
in convergent light an inclined positive bisectrix
emerges from P. The trace of the axial plane
runs obliquely to the cleavage. Twins, according
to the 1st law, are not recognizable in polarized
light, since the axial plane is the same in both
individuals. Those following the other laws in
which M is the composition plane give extinctions symmetrical to the twinning boundary,
whose difference may reach 60°. The twinning
boundary lies parallel to the cleavage parallel to
M in sections of the zone $M : T$ and $M : P$, but
intersects these cleavage cracks at considerable angles in oblique sections.

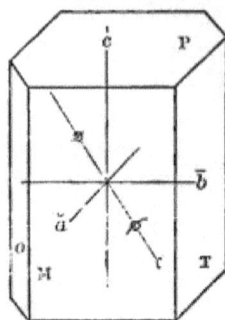
Fig. 121

Pleochroism is only noticeable in distinctly colored varieties ; the
colors vary between bluish and colorless, and are strongest in sections
parallel to T.

Disthene is generally free from inclusions : inclusions of biotite plates
or those of colorless mica, and of specular iron, quartz grains, tourmaline and rutile needles, are rare ; fluid inclusions more so. Their
arrangement is without order ; but the mica plates usually lie along
the principal cleavage, M.

Specific gravity, 3.5–3.7. Unattacked by acids. Hence it is easily
isolated from the rocks. Chemical composition and reactions like those
of andalusite. Proper decomposition is very seldom observed ; a finely
laminated aggregate of mica appears to be developed from disthene,
which is then clouded by limonite.

Disthene is highly characteristic of crystalline schists, gneisses,
granulites, paragonite schist, muscovite schist, and eclogite, and appears to occur especially where these rocks are of metamorphic origin.
It is almost universally accompanied by garnet.

Axinite.

When *axinite* occurs in rocks it usually forms irregular grains, which are rarely bounded crystallographically by the hatchet-like forms. The cleavages which truncate the sharp edges of the faces P (1$\bar{1}$0) and u (110), and P (1$\bar{1}$0) and r (1$\bar{1}$1), are but imperfectly expressed by cracks in cross-sections.

By transmitted light it is colorless to very light yellowish, pale grayish brown, or violet. Index of refraction high, and double refraction strong. Des Cloizeaux determined $\alpha_\rho = 1.6720$, $\beta_\rho = 1.6779$, $\gamma_\rho = 1.6810$, $\alpha_\nu = 1.6850$, $\beta_\nu = 1.6918$, $\gamma_\nu = 1.6954$. The bisectrix is normal to x (111). The axial angle is large, so that the axes do not emerge from sections at right angles to the bisectrix in air. $2\,V_\rho = 71°\,38'$ to $74°\,17'$, $2\,V_\nu = 71°\,49'$ to $74°\,39'$. Character of the double refraction negative; inclined and horizontal dispersion very distinct.

The pleochroism, $\mathfrak{a} =$ pale olive-green to colorless, $\mathfrak{b} =$ dark violet-blue, $\mathfrak{c} =$ cinnamon-brown, is scarcely noticeable in thin sections.

Specific gravity $= 3.3$. Chemical composition not known exactly, a boron-bearing lime-alumina silicate. Not attacked by acids.

Bears no interpositions besides the associated minerals, chiefly tremolite and chlorite, and occasional fluid inclusions.

Axinite occasionally occurs on the borders of diabases and granites, and among their contact products. Zirkel * described a mixture of axinite, light-greenish augite, dark-green hornblende, quartz, calcite, titanite and iron-ores, called limurite, from the valley of Lesponne, in the Pyrenees.

Cossyrite.

Literature.

H. FÖRSTNER, Ueber Cossyrit, ein Mineral aus den Liparitlaven von Pantelleria. Z. X. 1881. V. 348–362.

The crystal forms of *cossyrite*, which has not yet been completely investigated, are very similar to those of hornblende, and exhibit almost monoclinic symmetry. Crystals only 1.5 mm. long and 0.5 mm. broad show the prism and both vertical pinacoids in the prism zone. The prism exhibits the most noticeable difference of angle as compared with hornblende; $\infty'P : \infty P_{\prime} = 134°\,09'$. It is placed in the triclinic system chiefly from the fact that the crystals are almost always twins parallel to $\infty P\bar{\infty}$ (010).

Cossyrite cleaves very readily parallel to both prism faces.

---

* Limurite aus der Vallée de Lesponne. N. J. B. 1879. 379.

It only becomes transparent occasionally in very thin sections. Microlites of cossyrite appear coffee-brown to rust-brown by transmitted light. Extinction angle on $\infty P \bar{\infty}$ (100) from the cleavage parallel to (110) $= 3°$, on the longitudinal face (010) from the same cleavage, 39°. The rays most inclined to the prism axis appear to be the most strongly absorbed.

Sp. gr. $= 3.74$–3.75. Chemical composition approaching that of a hornblende rich in iron and soda. Fuses readily to a brownish-black glass, and is strongly attacked by boiling hydrochloric acid. It forms a constituent of the acid dacitic lavas of the island of Pantelleria.

HOMOGENEOUS AGGREGATES.

THE doubly refracting aggregates, when of sufficiently fine grain, are characterized under the microscope by the fact that their thin sections do not become dark in any position between crossed nicols. For the different substances composing the aggregate lie beside, through, or over one another, in such a way that their principal optical sections never coincide. Consequently the phenomena produced are those described on page 88. The distribution of the colors in aggregates between crossed nicols generally indicates the structure of the aggregates, a granular, fibrous, or scaly aggregate structure corresponding to a speckled, striped, or flaked change of colors.

Serpentine.

Literature.

R. von DRASCHE, Ueber Serpentin und serpentinähnliche Gesteine. T. M. M. 1871. 1.

E. HUSSAK, Ueber einige alpine Serpentine. T. M. P. M. 1882. V. 61–81.

G. TSCHERMAK, Ueber Serpentinbildung. S. W. A. 1867. July No. LVI.

M. WEBSKY, Ueber die Krystallstructur des Serpentins und einiger demselben zuzurechnenden Fossilien. Z. D. G. G. 1856. X. 277.

B. WEIGAND, Die Serpentine der Vogesen. T. M. M. 1875. 183–206.

F. J. WIIK, Mineralogiska och petrografiska meddelanden. Finska Vet. Soc. Förhandl. Helsingfors. 1875.

Serpentine has a fibrous or apparently laminated structure according to the parent mineral from which it originated; still the apparent scales may represent bundles of parallel fibres. The arrangement of the fibres varies greatly, sometimes parallel, at other times confusedly felty, the optical phenomena between crossed nicols changing with the arrangements and dimensions of the fibres. In the parallel aggregates, which are not too finely fibrous, it is evident that they are biaxial with very large axial angle, whose negative bisectrix is normal to the axis of the fibres, which is also the axis of least elasticity. These fibres have a low index of refraction (very nearly the same as that of Canada balsam), and not inconsiderable double refraction. Chrysotile shows these relations very distinctly. In the fine, confusedly fibrous aggregates there may exist such a perfect compensation that they often appear isotropic.

The mineral from which serpentine is most frequently derived is olivine. On page 216 this alteration is described in its incipiency, as well as the development of a peculiar net-like structure. When serpentine is typically developed three forms may be quite distinctly made out. First, dark veins and bands (Pl. XXVI. Fig. 4) of a deep-colored leek-green and blue-green serpentine substance, which is often opaque from metallic oxides. These bands, generally keeping the same direction for some distance, evidently correspond to the first cracks and cleavage in the olivine, from which the whole process started. They sometimes have a laminated structure, each layer being fibrous cross-wise. Within the meshes of the large net a smaller net is found to some extent, made up of grass-green serpentine veins generally crooked and intersecting, which by higher powers, especially in polarized light, exhibit a fibrous structure at right angles to the length of the veins; these bands do not carry metallic oxides. These are also absent from the yellowish green serpentine substance, which fills the meshes of the small nets and evidently corresponds to the olivine grains which were metamorphosed last. Here the structure is very finely scaly and scaly fibrous.

In the alteration of hornblende and actinolite into serpentine the cleavage and transverse parting of amphibole is clearly brought out by the arrangement of the serpentine fibres. Parallel fibrous aggregates of serpentine stand normal to the cleavage of the hornblende, while the spaces within these cleavage cracks are confusedly fibrous. Between crossed nicols the parallel fibrous lines, which partly run parallel to one another, partly intersect at 124°–125°, or make rhombs with other angles and also rectangles, stand out brightly colored from the dark ground of the confusedly fibrous spaces. This gives rise to a highly characteristic structure which Weigand has called *grating* or "*window*"-*structure* (Pl. XXVI. Fig. 5).

Other serpentines which have a laminated structure macroscopically, consist of microscopically laminated masses crossing at right angles and exhibiting a knitted structure (Pl. XXVI. Fig. 4). Scales which can be loosened give in convergent light the interference figure of a biaxial mineral with small axial angle about a negative bisectrix normal to the face of the plate. Thus they behave very nearly or exactly the same as many bastites. These serpentines appear to have been produced from monoclinic pyroxenes, as Hussak and others have shown.

The fibrous serpentines or chrysotile serpentines, which arise from olivine and hornblende, could be placed optically parallel with the micaceous serpentines, bastite or antigorite serpentines, by assuming that each leaf of the latter is built up of parallel fibres, whose longer axis lies parallel to the vertical axis of the pyroxene. But the axial angle for the chrysotile serpentines is considerably greater than for the antigorite serpentines. It appears to grow so large that the axis of smallest elasticity, lying along the axis of the fibre, becomes the acute bisectrix, as in metaxite, from Schwarzenberg in Saxony, according to Websky's observation.

All serpentines are transparent and greenish, bluish green or yellowish brown, often nearly colorless, seldom rust-red. They frequently contain the interpositions belonging to the parent mineral, as well as unaltered remnants of the latter. They are often permeated with opal and with carbonates.

The different varieties of serpentine, as metaxite, picrolite, marmolite, retinalite, jenkinsite, vorhanserite, picrosmine, schwartzerite, etc., have been described by Websky (l. c.), Wiik (l. c.), Des Cloizeaux, v. Drasche (T. M. M. 1871. 57), and Fischer (Kritische mikroskopisch-mineralogische Studien. 1. Fortsetzung. Freiburg i. B. 1871. 31. 47).

Sp. gr. = 2.5–2.7. Chemical composition = $2H_2O, 3MgO, 2SiO_2$. It is attacked quite vigorously by hydrochloric acid, especially at high temperatures, with the separation of gelatinous silica. Sulphuric acid acts more energetically than hydrochloric.

Delessite.

Delessite forms aggregates mostly with divergent fibrous structure, which may take the shape of very perfect spherulites. When it fills amygdaloidal cavities it is in layers or bands parallel to the rock boundary, each layer consisting of fibres standing normal to the rock walls. The different layers correspond to as many interruptions in its growth. Delessite becomes transparent and green or yellowish brown. Index of refraction and double refraction small. Extinction apparently parallel and normal to the axis of the fibres, which is the axis of least elasticity. Pleochroism of varying intensity; rays vibrating parallel to the axis of the fibre are greenish; those at right angles yellowish, greenish white to nearly colorless.

Sp. gr. = 2.5–2.6. Chemical composition not known exactly—a hydrous aluminous silicate of iron and magnesia. Easily decomposed by acids with the separation of gelatinous silica; when heated to redness it becomes opaque brownish black to black.

Delessite forms pseudomorphs after pyroxene and amphibole, or it fills amygdaloidal cavities in basic rocks in combination with carbonates and epidote. Grengesite is identical with delessite both in structure and physical behavior.

Kaolin.

Literature.

A. Knor, Beiträge zur Kentniss der Steinkohlenformation und des Rothliegenden im Erzgebirgischen Bassin. N. J. B. 1859. 593–594.

E. E. Schmid, Die Kaoline des thüringischen Buntsandsteins. Z. D. G. G. 1876. XXVIII. 87–111.

Kaolin forms loose earthy aggregates, which are produced by the weathering of feldspar, elæolite, scapolite, and other minerals. Isolated and loosened in water, these aggregates are found to consist of extremely fine, irregularly bounded plates, which are rarely hexagonal, and are completely colorless. In the large-leaved varieties, known as nakrite or pholerite, rhombic or hexagonal forms have been recognized, which would correspond to a combination of a prism of 120° with a brachypinacoid and a basal plane. The plates in polarized light prove to be partly crossed trillings parallel to a prism face. Fibrous structure which has been mentioned by some observers is probably brought about by a rosette-like arrangement of the plates.

Loose scales of kaolin are transparent and colorless. The index of refraction is about the same as that of Canada balsam; the double refraction is strong. A negative bisectrix emerges from the face of the plate, the axial plane bisects the acute prism angle. The optical behavior is therefore very similar to that of muscovite. Aggregates of kaolin are cloudy and scarcely translucent.

Sp. gr. = 2.2–2.65. Chemical composition = $2H_2O, Al_2O_3, 2SiO_2$. Is not acted on by hydrochloric acid. Is decomposed by boiling sulphuric acid. It can only be distinguished with certainty from colorless mica by chemical reaction, by proving the absence of alkali; its specific gravity cannot be used to advantage because of the micaceous form of both minerals.

INDEX.

EXPLANATION OF PLATES.

PLATE I.

FIG. 1. Solution of sulphur in a mixture of carbon bisulphide and Canada balsam. Globulites and crystals have formed with a halo free from globulites. × 200.

FIG. 2. The same. About the larger globulites have been formed halos free from globulites; about these as well as about gas-bubbles diffusion-streams may be recognized. × 200.

FIG. 3. The same. Globulites and longulites have formed. × 200.

FIG. 4. Globulites in basalt-glass from Hawaii, Sandwich Islands. × 250.

FIG. 5. Longulites and cumulites of sulphur in a solution of sulphur in a mixture of carbon bisulphide and Canada balsam. × 200.

FIG. 6. Globospherite and globulite in the same solution. × 200.

PLATE II.

FIG. 1. Margarites in obsidian. Clear Lake, Cal. × 250.

FIG. 2. Trichites in obsidian. Mexico. × 200.

FIG. 3. Microlites and trichites in schillerizing obsidian. Transcaucasus. × 225.

FIG. 4. Spherulites (*sphærocrystallæ*) in obsidian. Lipari. × 15.

FIG. 5. Cumulite in felsite-pitchstone. Buschbad in Triebischthal, Saxony. × 450.

FIG. 6. Augite with colorless crystallization-halo in trachytic pitchstone. Hammers-fjord, Iceland. × 25.

PLATE III.

FIG. 1. Skeleton crystal of olivine. Palma. × 100.

FIG. 2. Skeleton crystal of magnetite in glassy basalt. Schatung, China. × 90.

FIG. 3. Skeleton crystal of augite in pitchstone. Arran. × 90.

FIG. 4. Spherical aggregates and bundles of feldspar crystals in trachyte. Caucasus. × 60.

FIG. 5. Broken feldspar crystals in augite andesite. Grad-Jakan, Java. × 33.

FIG. 6. Bent and frayed-out mica in augite minette. Fuchmühle near Weinheim on the Bergstrasse. × 42.

PLATE IV.

FIG. 1. Shattered garnet in mica schist. Brixen, Tyrol. × 10.

FIG. 2. Marginally fractured feldspar in anorthite rock. Chicontrini, Quebec, Canada. × 10.

FIG. 3. Albite, fractured and dislocated. In ordinary light. From phillite-gneiss. Allen's Creek, Victoria, Australia. × 10.

FIG. 4. The same between crossed nicols.

FIG. 5. Crystal of biotite, corroded and with pressure figures, from porphyrite. Ilfeld, Hartz. × 22.

FIG. 6. Plagioclase with bent lamellæ, between crossed nicols; from olivine gabbro. Store Bekkafjord, Norway. × 45.

328 EXPLANAT INDEX PLATES.

PLATE V.

Fig. 1. Corroded quartz crystal from quartz porphyry from Scharfenstein, Münster-thal, Black Forest. × 25.
Fig. 2. Corroded nosean crystal from leucitophyre from Burgberg near Rieden. × 12.
Fig. 3. Zonal structure in plagioclase in melaphyre. Bufaure, Fassathal. × 36.
Fig. 4. Zonal structure in melanite in phonolite. Steinriesenweg near Oberbergen, Kaiserstuhl. × 60.
Fig. 5. Zonal structure in augite in leucitite. Kreuzle near Rothweil, Kaiserstuhl. × 21.
Fig. 6. Hour glass-like zonal structure in augite in nephelinite. Eichberg near Rothweil, Kaiserstuhl. × 45.

PLATE VI.

Fig. 1. Gas inclusions in obsidian. Mexico. × 80.
Fig. 2. Fluid inclusions in apatite. Pfitsch, Tyrol. × 150.
Fig. 3. Fluid inclusions in rock salt. Friedrichshall, Würtemberg. × 200.
Fig. 4. Two fluids, which do not mix, in one inclusion in smoky quartz. Branch-ville, Conn. × 100.
Fig. 5. Fluid inclusions with separated crystal in quartz from granite porphyry. Cornwall. × 210.
Fig. 6. Fluid inclusions, which do not wet the walls of the cavity, in topaz. Schneckenstein, Saxony. × 84.

PLATE VII.

Fig. 1. Glass inclusions in labradorite from Monte Pilieri, Etna. × 50.
Fig. 2. Glass inclusions in quartz, dihexahedral, from quartz porphyry from Dos-senheim on the Bergstrasse. × 72.
Fig. 3. Glass inclusions with several bubbles in oligoclase. Pantelleria. × 100.
Fig. 4. Quartz inclusions in heulandite. Färoe. × 30.
Fig. 5. Microlite inclusions in hypersthene. St. Paul's Island. × 30.
Fig. 6. Central accumulation of inclusions in feldspar from trachyte from Monte Olebano near Pozzuoli, Naples. × 5.

PLATE VIII.

Fig. 1. Peripheral accumulation of inclusions in feldspar from hornblende andesite. South Siberia. × 54.
Fig. 2. Zonal arrangement of inclusions in leucite from Vesuvian lava. Monte Somma. × 13.
Fig. 3. Interpenetration of quartz and orthoclase in granophyre. Sperberbächel near Hohwald, Vosges. × 80.
Fig. 4. Aggregate of quartz grains in ordinary light, in granite-porphyry. Gross-sachsener Thal, Odenwald. × 24.
Fig. 5. The same between crossed nicols. The figure should be turned about 90° to the left.
Fig. 6. Penetration of augite by plagioclase, magnetite, and augite from basalt. Löwenburg, Siebengebirge. × 30.

PLATE IX.

Fig. 1. Spherulites (sphærocrystalle) of chalcedony between crossed nicols, from dia-base-porphyrite. Pfalz. × 50.

Fig. 2. Grains of oölite between crossed nicols. From a coral reef, Bahama Islands. × 20.

Fig. 3. Granosphorite of quartz between crossed nicols from Eisenkiesel. Stifts-buckel, Heidelberg. × 25.

Fig. 4. Bertrand's interference crosses on sphærosiderite from Steinheim, Wetterau. × 35.

Fig. 5. Octahedral cleavage; section parallel to O (111) on fluorite. Markirch. × 42.

Fig. 6. Prismatic cleavage; section parallel to oP (001) on scapolite. Oedegarden near Bamle, Norway. × 150.

PLATE X.

Fig. 1. Rhombohedral cleavage; section parallel R (10$\bar{1}$1); calcite. Auerbach on the Bergstrasse. × 30.

Fig. 2. Cleavage parallel to the prism and two vertical pinacoids in sections parallel to oP (001); hypersthene. St. Paul's Island. × 24.

Fig. 3. Pyramidal cleavage, section parallel to oP (001); in anatase. Oisans, Dauphiné. × 24.

Fig. 4. Prismatic cleavage, section at right angles to c, in augite from nepheline tephrite. Neunlinden, Kaiserstuhl. × 39.

Fig. 5. Prismatic cleavage, section at right angles to c, in hornblende from dacite. Timokthal, Servia. × 36.

Fig. 6. Pinacoidal cleavage in a section at right angles to it; mica from granitite. Grasstein near Mauls, Tyrol. × 30.

PLATE XI.

Fig. 1. Cleavage parallel to oP(001) and $\infty P\bar{\infty}$ (010) in a section parallel to $\infty P\bar{\infty}$ (100) in orthoclase from augite syenite. Laurvig, Norway. × 27.

Fig. 2. Cleavage parallel to oP (001) and $\infty P\bar{\infty}$ (100) in a section parallel to $\infty P\bar{\infty}$ (010) in epidote from epidote rock from Auerbach on the Bergstrasse. × 60.

Fig. 3. Cleavage parallel to ∞P(110) in a section at right angles to c in titanite from syenite from Löhrbach, Odenwald. × 75.

Fig. 4. Crystals of sodium fluosilicate. × 72.

Fig. 5. Crystals of sodium fluosilicate and amorphous aluminium fluosilicate from sodalite. Vesuvius. × 27.

Fig. 6. The same. × 160, × 100, and × 140.

PLATE XII.

Fig. 1. Crystals of potassium fluosilicate from apophyllite. Fassathal. × 130.

Fig. 2. Crystals of potassium fluosilicate from sanidine. Wehr. × 140.

Fig. 3. Crystals of lithium fluosilicate and aluminium fluosilicate from Zinnwaldite. × 100.

Fig. 4. Crystals of calcium fluosilicate. × 45.

Fig. 5. Crystals of calcium fluosilicate from apophyllite. × 42.

Fig. 6. Crystals of magnesium fluosilicate from biotite. × 30.

PLATE XIII.

Fig. 1. Gypsum crystals. × 20.

Fig. 2. Crystals of cæsium alum. × 20.

Fig. 3. Crystals of ammonium magnesium phosphate (struvite). × 10.

FIG. 4. The same from very dilute solution. × 30.
FIG. 5. Ammonium-molybdenum phosphate in crystals. × 140.
FIG. 6. Crystals of zirconia. × 120.

PLATE XIV.

FIG. 1. Hornblende with magnetite border and wreath of augite and magnetite, from nepheline tephrite from Gran Canaria. × 130.
FIG. 2. Double refraction in garnet, in a section parallel to ∞O (110). Peru. × 9.
FIG. 3. Interpositions in garnet, arranged along the axial planes; quartzite from Libramont, Luxemburg. × 15 and × 24.
FIG. 4. Garnet with kelyphite rim; olivine rock. Karlstetten, Lower Austria. × 15.
FIG. 5. Leucite with zonal alternation of different inclusions; Vesuvian lava. × 165.
FIG. 6. Leucite in striated melilite, from leucitite from Capo di Bove, near Rome. × 48.

PLATE XV.

FIG. 1. Leucite surrounded by augite in the form of a wreath, from leucitophyre. Olbrück, Upper Brohlthal. × 80.
FIG. 2. Perofskite in melilite basalt. Spitzberg near Wartenberg, Bohemia. × 170.
FIG. 3. Titanite after rutile in amphibole gneiss. Oetzthal, Tyrol. × 100.
FIG. 4. Rutile from clay slate from Kautenbach in Luxemburg (isolated). × 240. And in clay slate from Hahnenbach near Kirn on the Nahe. × 250.
FIG. 5. Zircon crystals out of granitite from Streblen, Silesia (isolated). × 160 and × 140.
FIG. 6. Melilite with peg-structure in nepheline basalt from Oahu, Sandwich Islands. × 66.

PLATE XVI.

FIG. 1. Ilmenite altered peripherally into titanite (leucoxene). Alpbachthal near Brixlegg, Tyrol. × 24.
FIG. 2. Ilmenite almost completely altered into leucoxene; "augite-propylite." Schemnitz. × 18.
FIG. 3. Corundum in norite. Wolfsgrube near Klausen, Tyrol. × 72.
FIG. 4. Tridymite in trachyte from Pomasqui, N. Quito, Equador. × 84.
FIG. 5. Calcite with twin lamellæ parallel to $-\frac{1}{2}$ between crossed nicols. × 45.
FIG. 6. Basal section through nepheline in leucitophyre. Olbrück, Upper Brohlthal. × 190.

PLATE XVII.

FIG. 1. Vertical section through nepheline in leucitophyre from Olbrück, Upper Brohlthal. × 190.
FIG. 2. Basal and vertical sections of andalusite; andalusite hornstone. Andlau, Vosges. × 25.
FIG. 3. Chiastolite in sections parallel and inclined to the base; chiastolite schist. Pyrenees. × 18.
FIG. 4. Sillimanite in quartz; out of gneiss from Freiberg, Erzgebirge. × 25.
FIG. 5. Vertical section of bronzite. Kupferberg, Silesia. × 57.
FIG. 6. Section parallel to (010) through a parallel intergrowth of enstatite and diallage. Gröditzberg near Liegnitz, Silesia × 36.

PLATE XVIII.

FIG. 1. Enstatite altered into bastite, melaphyre. Hohenstein near Ilfeld, Hartz. × 45.

FIG. 2. Olivine with symmetrically formed and arranged glass inclusions, out of glassy basalt. Mauna Loa, Sandwich Islands. × 190.

FIG. 3. Twin-growths of olivine crystals parallel to $P\infty$ (011), out of nepheline basalt from Randen, Hegau. × 75.

FIG. 4. Rough surface of olivine in Canada balsam, basalt. Steinschönau, Bohemia. × 57.

FIG. 5. Olivine with inclusion of groundmass in the form of its host; basalt. Siebenbürgen. × 87.

FIG. 6. Olivine in advanced serpentinization; olivine norite. Obere Baste, Harzburg. × 27.

PLATE XIX.

FIG. 1. Olivine (hyalosiderite) with broad, marginal secretion of iron oxide, limburgite. Limburg near Sasbach, Kaiserstuhl. × 42.

FIG. 2. Olivine altered into hornblende (pilite); kersantite. Marbach, Lower Austria. × 42.

FIG. 3. Penetration trilling of cordierite, between crossed nicols. Asama Yama, Japan. × 140.

FIG. 4. Zoisite in longitudinal and cross sections; amphibolite from Zamborinho near Macedo, Portugal. × 25.

FIG. 5. Augite twinned parallel to $\infty P\infty$ (100) between crossed nicols; palatinite from Martinstein near Kreuznach. × 22.

FIG. 6. Augite with twin lamellæ parallel to oP (001), diabase from New Haven, Conn. × 96.

PLATE XX.

FIG. 1. Intergrowth of augite crystals, limburgite. Limburg, Sasbach; Kaiserstuhl. × 18.

FIG. 2. Form of growth in augite; Vesuvian lava, Monte Somma. × 60.

FIG. 3. Forms of growth in augite; felsite pitchstone from Corriegills on Arran. × 96 and × 130.

FIG. 4. Augite with corroded centre; nephelinite. Herberg near Oberbergen, Kaiserstuhl. × 33.

FIG. 5. Cleavage of augite in sections parallel to c; leucite basalt. Vormberg near Ihringen, Kaiserstuhl. × 33.

FIG. 6. Cleavage of diallage parallel to ∞P (110) and $\infty P\infty$ (100) in a section at right angles to c. Olivine-gabbro. Hausdorf, Silesia. × 30.

PLATE XXI.

FIG. 1. Parallel intergrowth of augite and hornblende. Picrite from Heim near Oberdieten, Nassau. × 42.

FIG. 2. Alteration of augite into chlorite. Proterobase from Stiebitz near Bautzen, Saxony. × 30.

FIG. 3. Hornblende twinned parallel to $\infty P\infty$ (100) in basal section between crossed nicols. Out of amphibole granite from Pré de Fouchon near Gerardmer, Vosges. × 80.

Fig. 4. Zonal structure in hornblende in basal section, quartz diorite. Little Falk
Minnesota. × 80.

Fig. 5. Uralite in basal section, uralite porphyrite. Minsk × 30.

Fig. 6. Uralite in vertical section, uralite porphyrite. Predazzo, Tyrol. × 9.

PLATE XXII.

Fig. 1. Biotite with rutile needles. Out of diorite-porphyrite from Lippenhof near
Triberg, Black Forest. × 120.

Fig. 2. Biotite altered to chlorite. Bodethal, Hartz. × 27.

Fig. 3. Biotite altered to chlorite and epidote, granite-porphyry. Etival, Vosges.
× 15.

Fig. 4. Ottrelite in basal section with zonal structure, phyllite. Harvey Hills.
Leeds, Canada. × 56.

Fig. 5. Twinning in ottrelite in vertical section between crossed nicols, phyllite.
Ottrez, Belgium. × 24.

Fig. 6. Zonal structure in the form of an hour-glass in ottrelite in vertical section.
From the same locality. × 27.

PLATE XXIII.

Fig. 1. Twinned titanite between crossed nicols, eleolite syenite. Foya, Portugal.
× 66.

Fig. 2. Carlsbad twin of sanidine in clinodiagonal section, phonolite. Wolf's Rock,
Land's End, England. × 15.

Fig 3. Baveno twin of sanidine in leucitophyre from Engler Kopf near Rieden; in
polarized light. × 52.

Fig. 4. Zonal structure in orthoclase, brought out by weathering, in a section
parallel to $\infty P \check{\infty}$ (010), amphibole granitite. Val d'Ajol, Vosges. × 8.

Fig. 5. Broken sanidine crystal in phonolite. Oberbergen, Kaiserstuhl. × 24.

Fig. 6. Transverse parting in sanidine, phonolite. Hohenkrähen, Hegau. × 21.

PLATE XXIV.

Fig. 1. Parallel intergrowth of sanidine and plagioclase between crossed nicols,
trachyte. Mont Dore, Auvergne. × 48.

Fig. 2. Zeolitization of sanidine in phonolite from Hohentwiel, Hegau. × 25.

Fig. 3. Interpenetration of orthoclase and plagioclase, between crossed nicols;
augite gneiss. Seyberer Berg, Lower Austria. × 24.

Fig. 4. Microperthitic intergrowth of orthoclase and albite in a vertical section,
granitite. Moslawina, Croatia. × 75.

Fig. 5. The same in different sections, gneiss. Chicontrini, Quebec, Canada. × 40.

Fig. 6. Orthoclase altered to muscovite. Granite-porphyry. Erdmannsdorf,
Silesia. × 33.

PLATE XXV.

Fig. 1. Microcline, section parallel to oP (001), between crossed nicols. Arendal.
× 21.

Fig. 2. Parallel intergrowth of microcline and albite, section parallel to $\infty P \check{\infty}$
(010), between crossed nicols. Unterflockenbach, Odenwald. ∠ 12.

Fig. 3. Lath-shaped plagioclase, nepheline basanite. Palma, Canary Islands. × 87.

Fig. 4. Jagged outline of plagioclase, between crossed nicols; basalt. Same locality ✕ 24.

Fig. 5. Zonal structure of plagioclase with different optical orientation in the separate zones, between crossed nicols; felsite-pitchstone. Cumardo near Lugano. ✕ 39.

Fig. 6. Twin striation of plagioclase according to the albite law, between crossed nicols; diabase. Biella, Piedmont. ✕ 45.

PLATE XXVI.

Fig. 1. Twin lamination in plagioclase according to the albite and pericline law, between crossed nicols; olivine gabbro. Le Prese, Veltlin. ✕ 12.

Fig. 2. Baveno twin of plagioclase, between crossed nicols. Vesuvian lava. Torre dell'Annunziata. 1734. ✕ 45.

Fig. 3. Net-like intergrowth of plagioclase with glas inclusions; augite andesite. Tokayer Bahnhof, Hungary. ✕ 57.

Fig. 4. Serpentine derived from olivine, with mesh structure. Schweidnitz, Silesia. ✕ 24.

Fig. 5. Serpentine derived from amphibole, with grating structure, between crossed nicols. Rauenthal near Markirch, Vosges. ✕ 45.

Fig. 6. Serpentine derived from augite with bar structure, between crossed nicols. Sprechenstein near Sterzing, Tyrol. ✕ 45.

ERRATA.

Errors in Crystallographic Symbols.

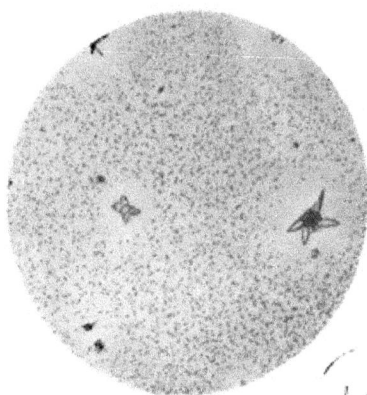

1

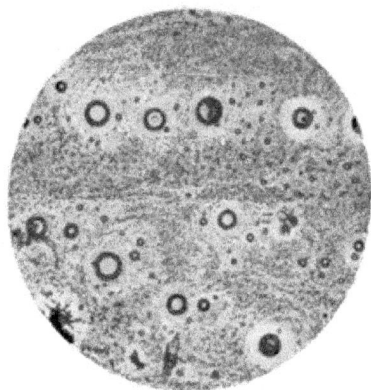

2

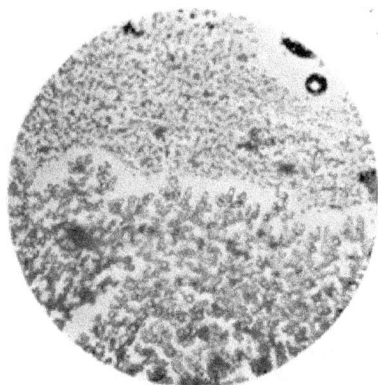

3

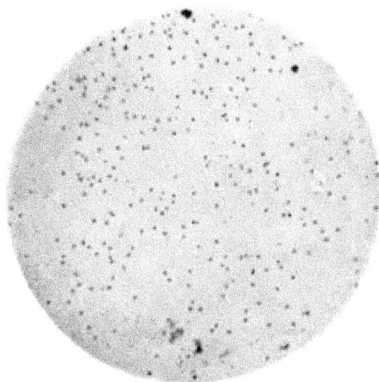

4

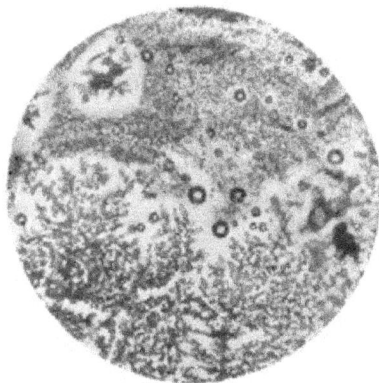

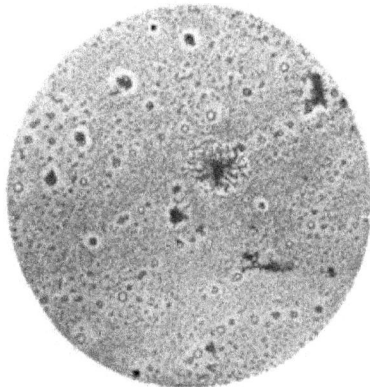

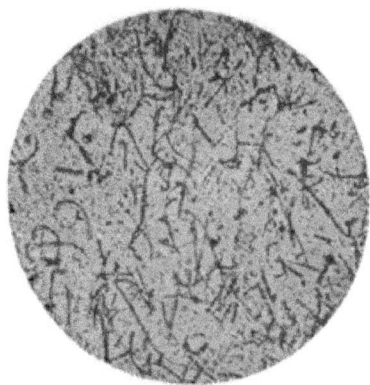

1

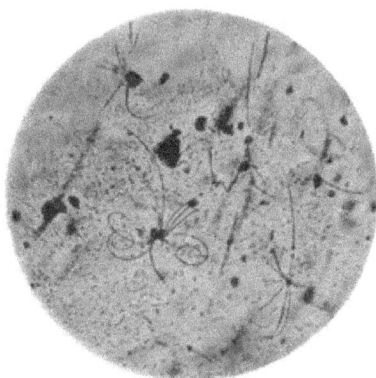

2

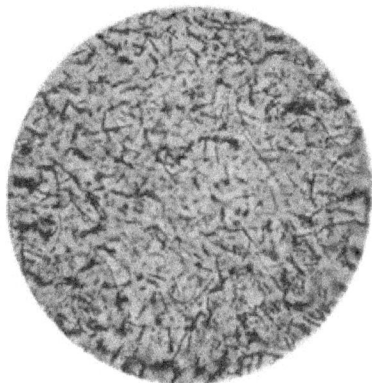

3

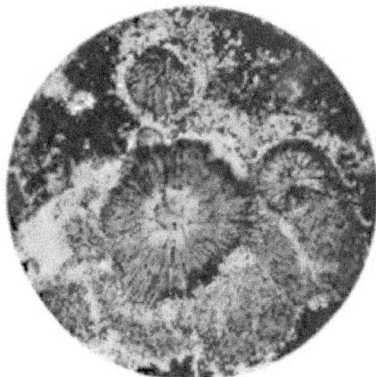

4

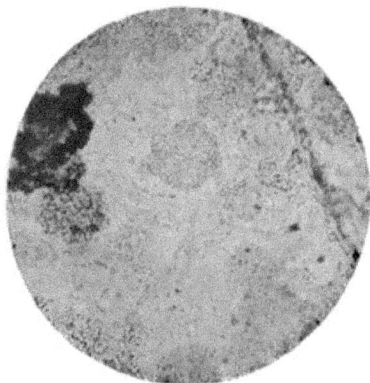

5

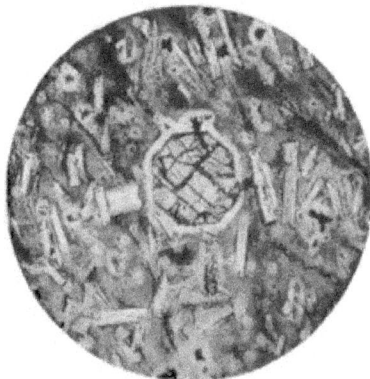

6

1

2

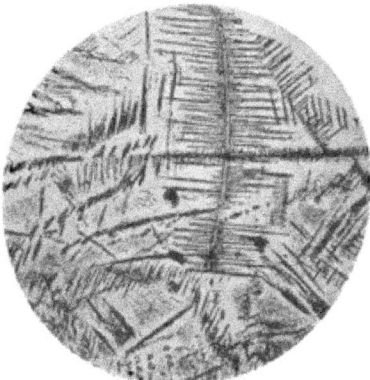

3

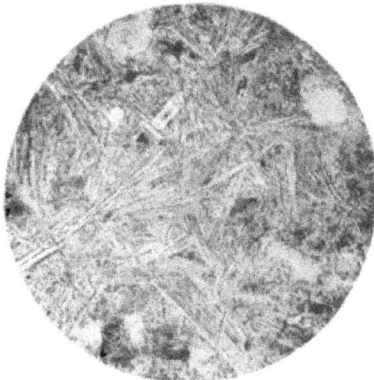

4

5

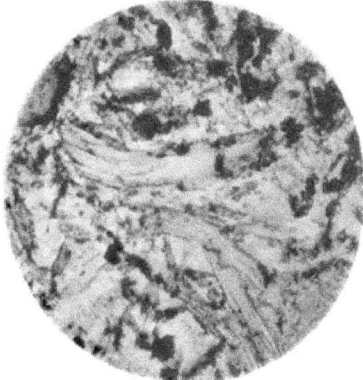

6

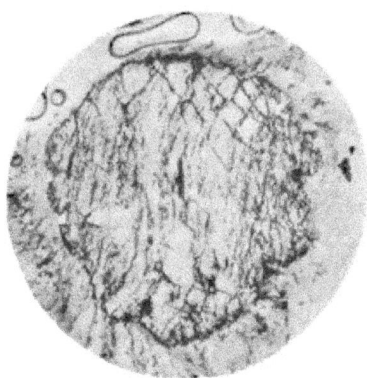

1

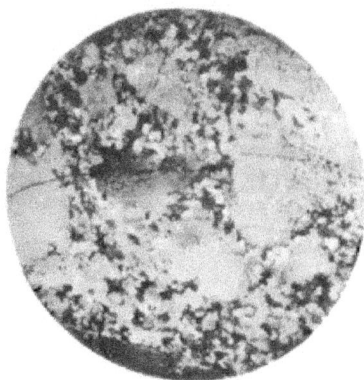

2

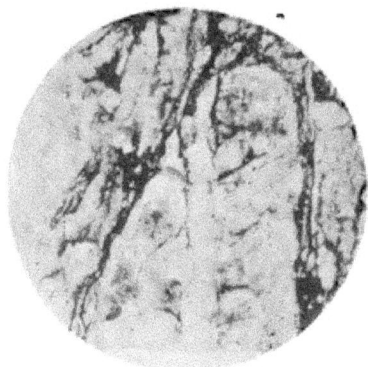

3

4

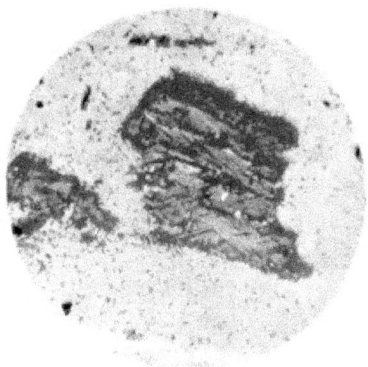

5

6

1

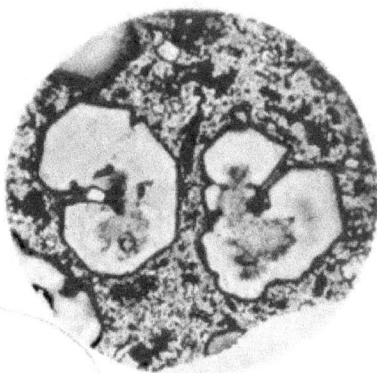

2

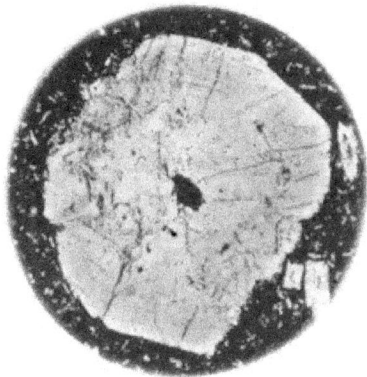

3

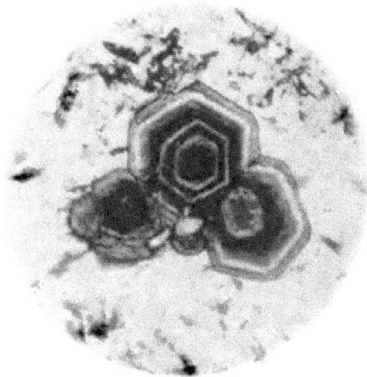

4

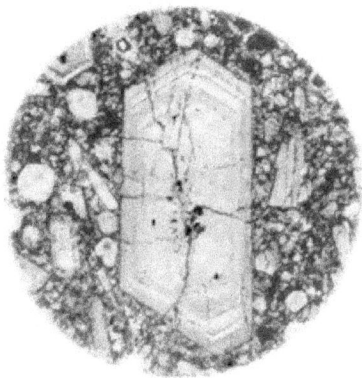

5

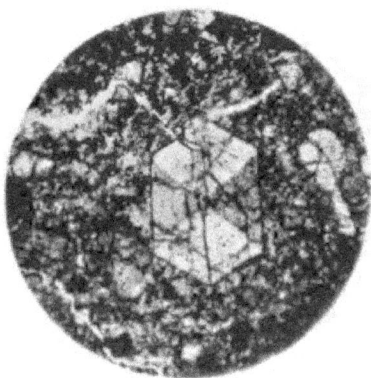

6

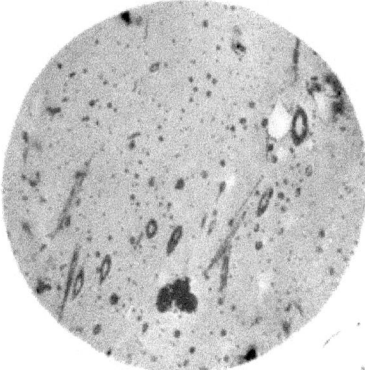

1

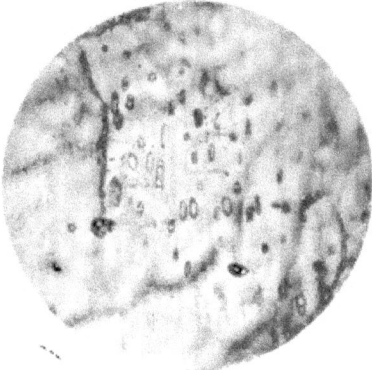

2

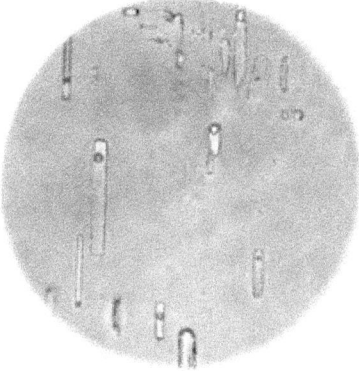

3

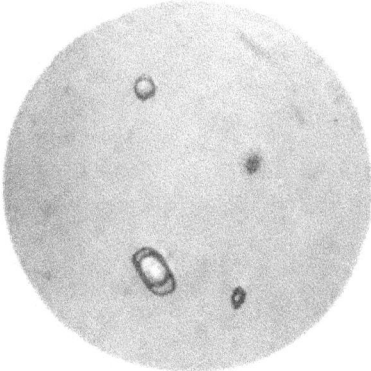

4

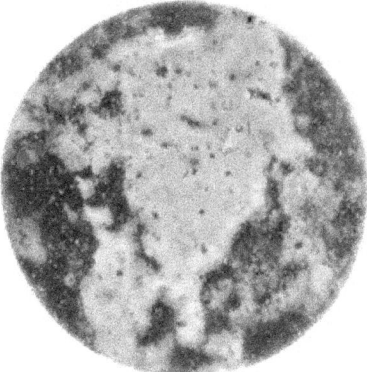

5

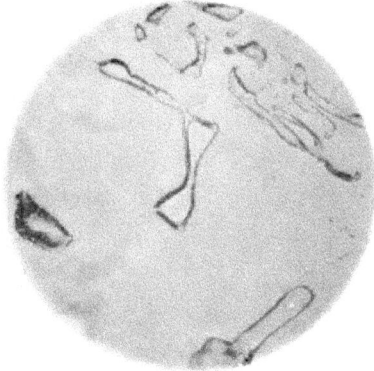

6

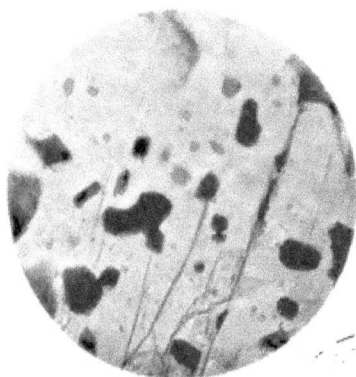

1

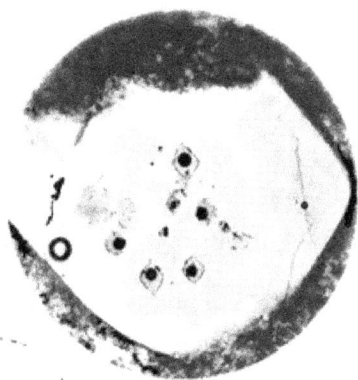

2

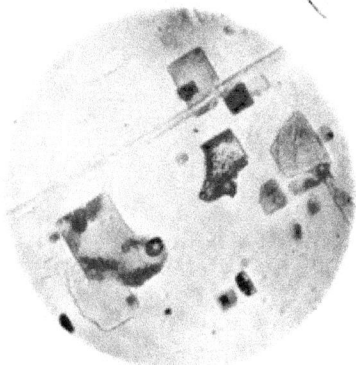

3

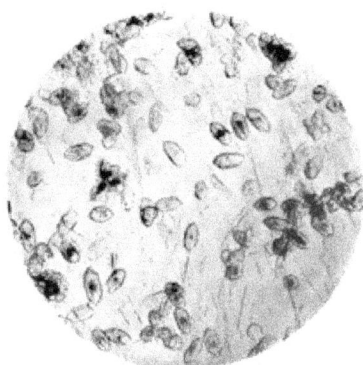

4

5

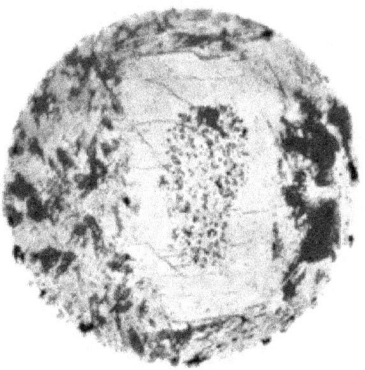

6

Plate VIII.

1

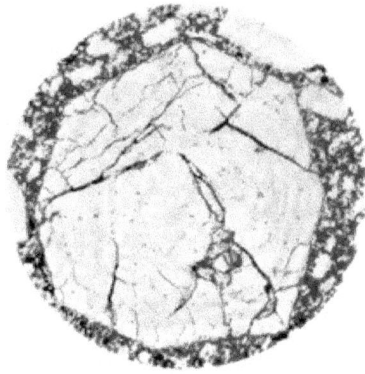

2

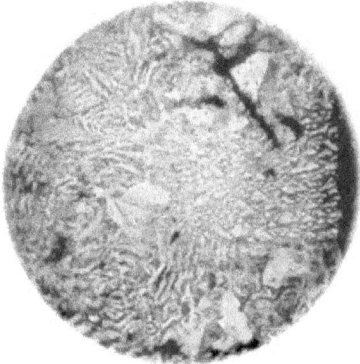

3

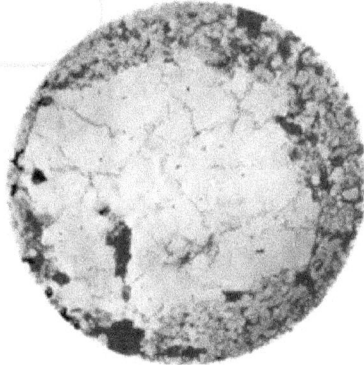

4

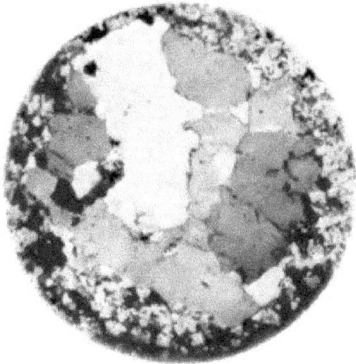

5

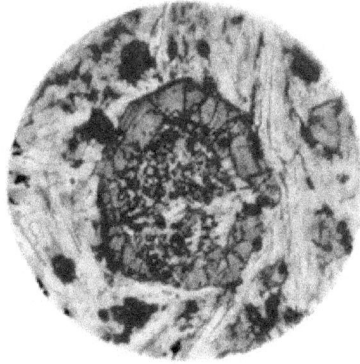

6

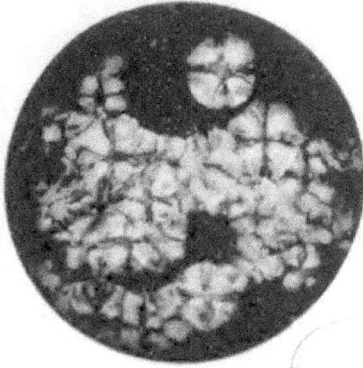

1

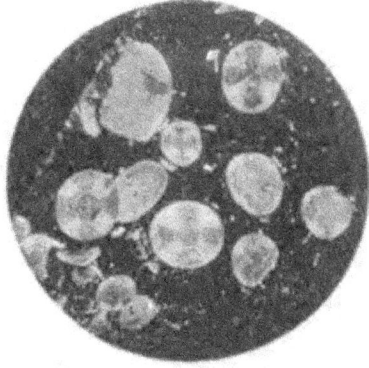

2

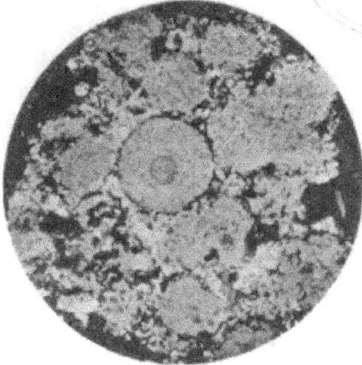

3

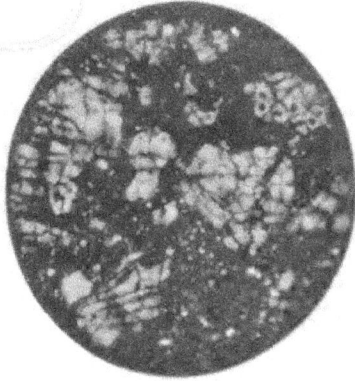

4

5

6

1

2

3

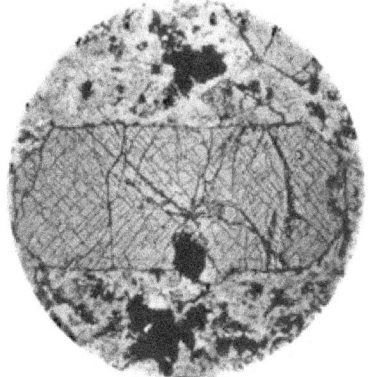

4

5

6

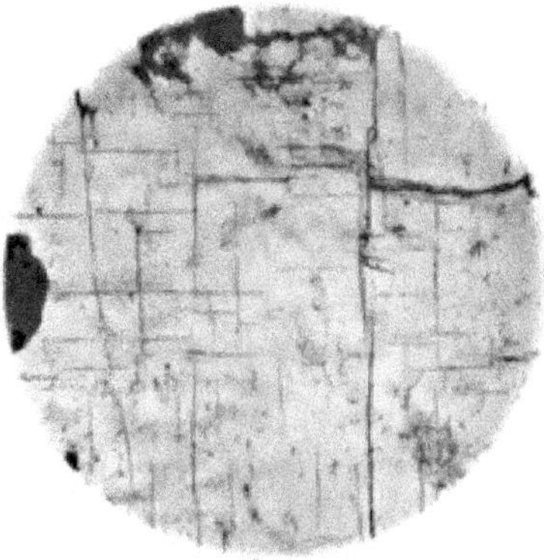

1

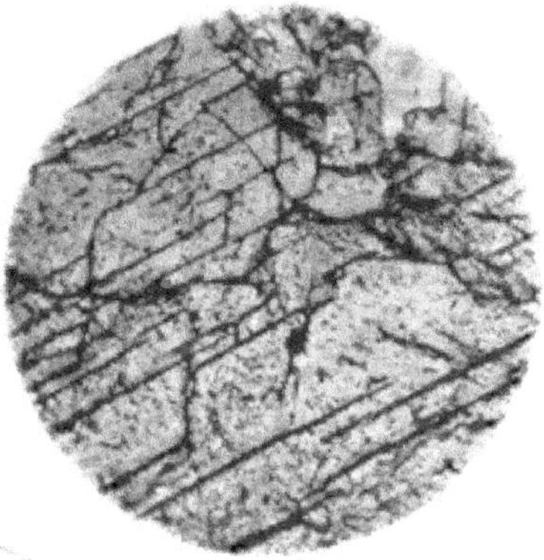

2

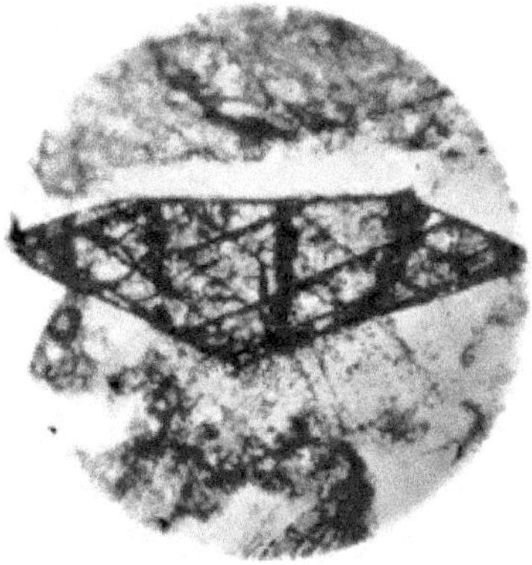

3

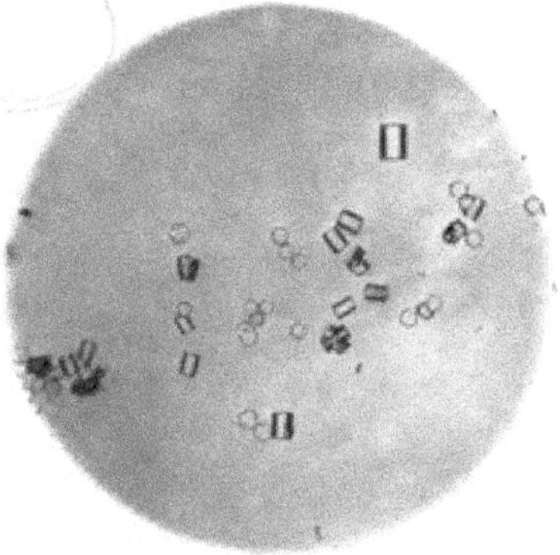

4

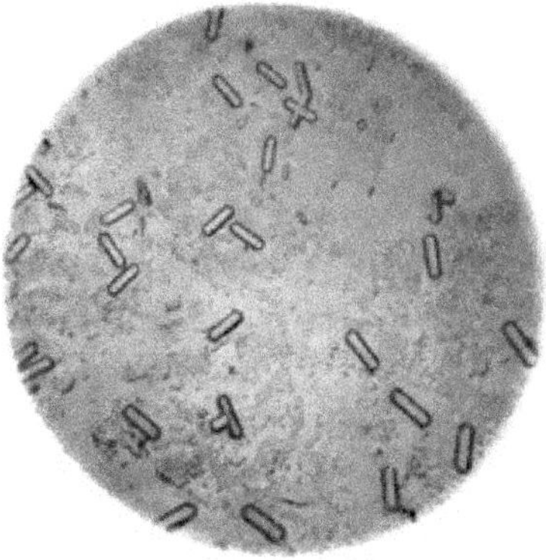

5

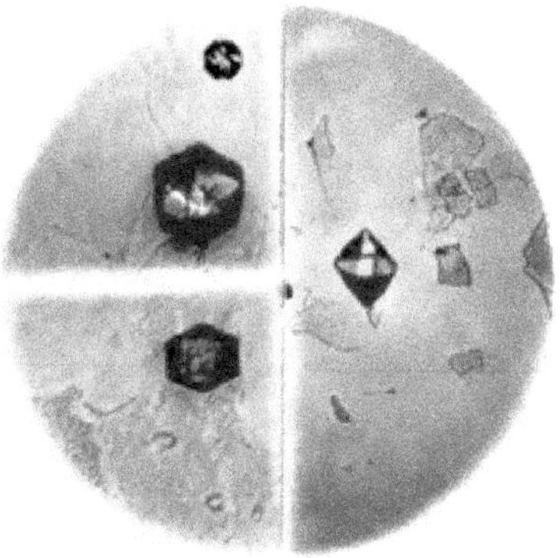

6

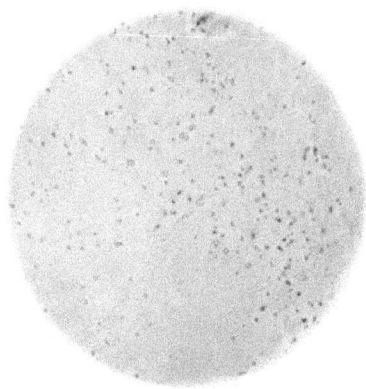

1

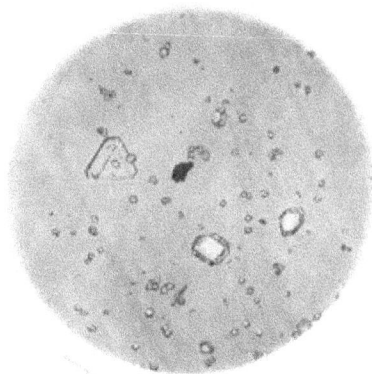

2

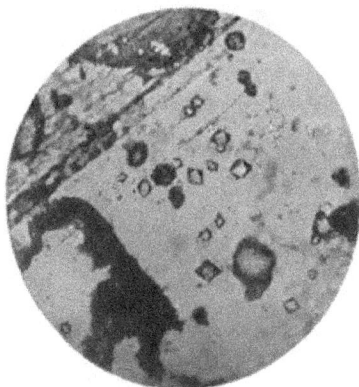

3

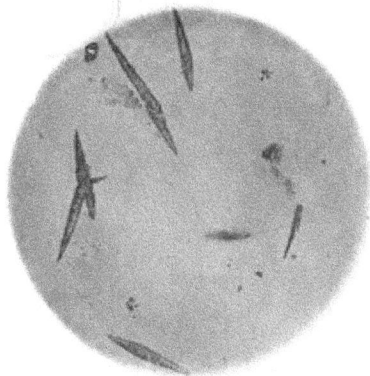

4

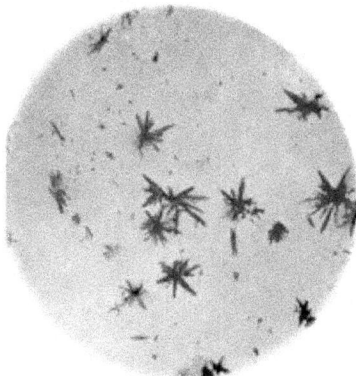

5

6

Plate XIII.

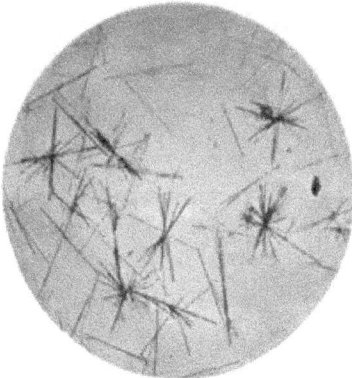

1

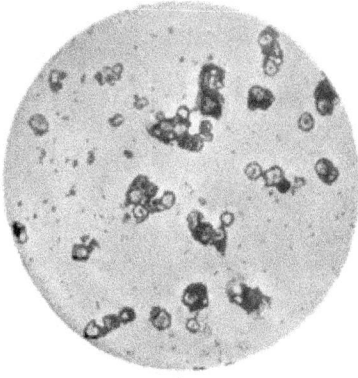

2

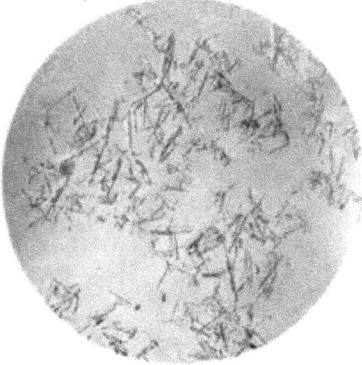

3

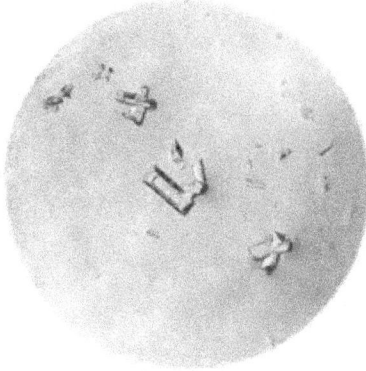

4

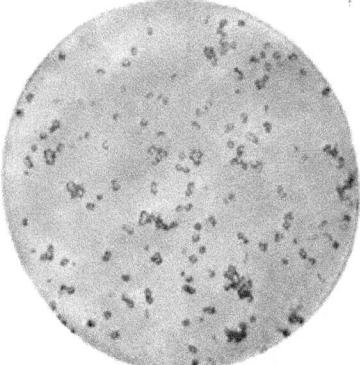

5

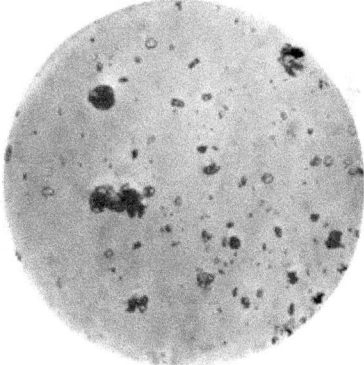

6

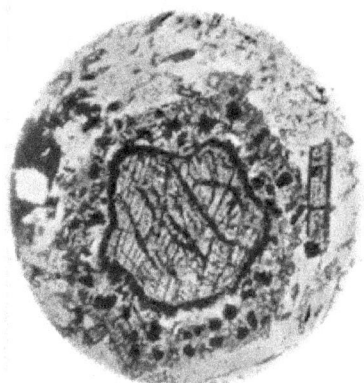

1

2

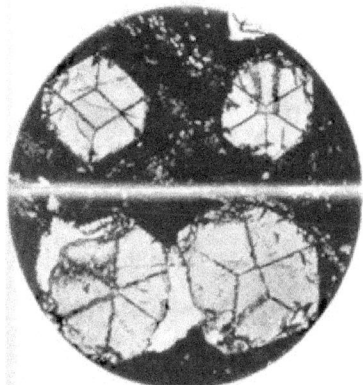

3

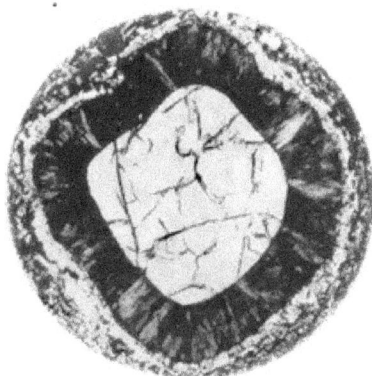

4

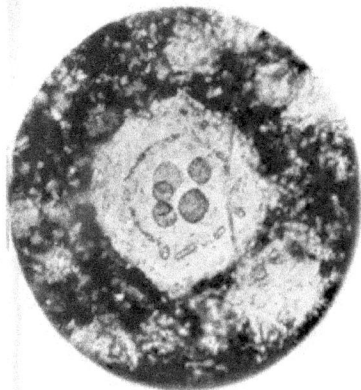

5

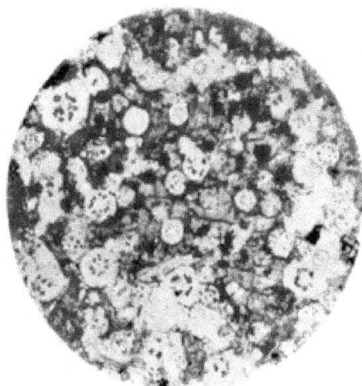

6

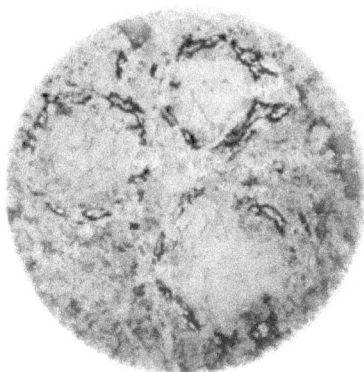

1

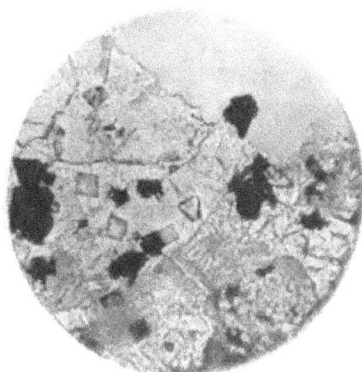

2

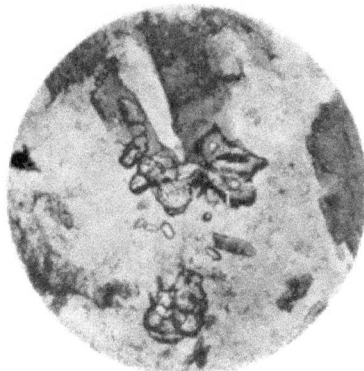

3

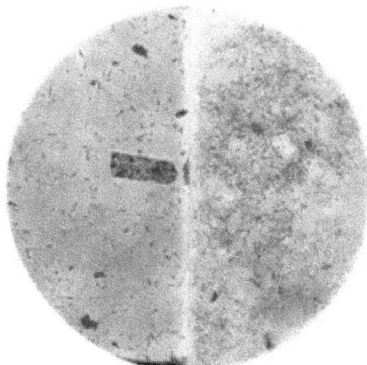

4

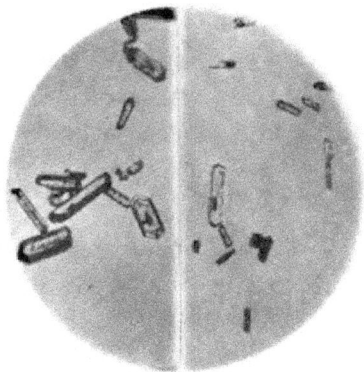

5

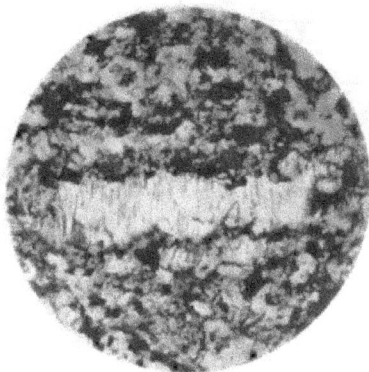

6

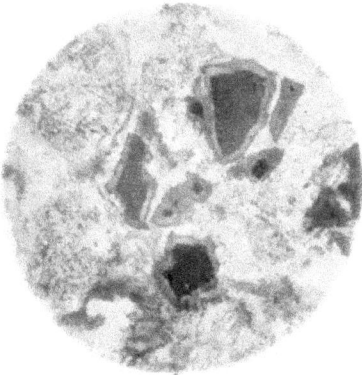

1

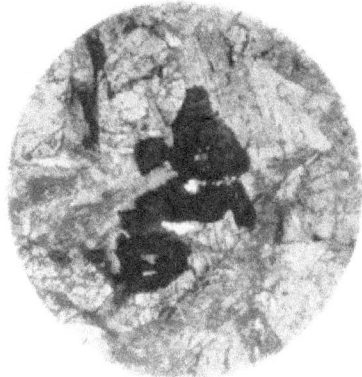

2

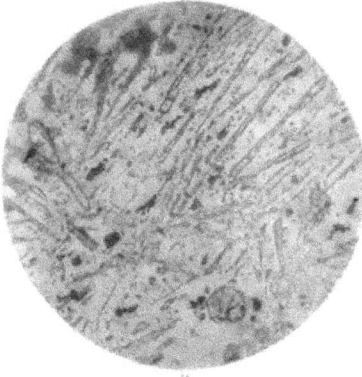

3

4

5

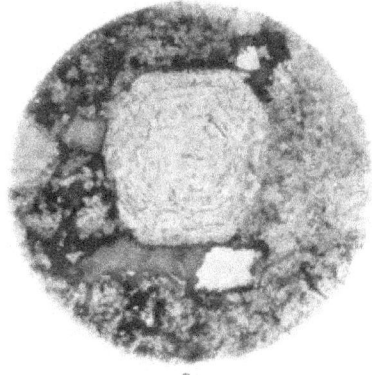

6

1

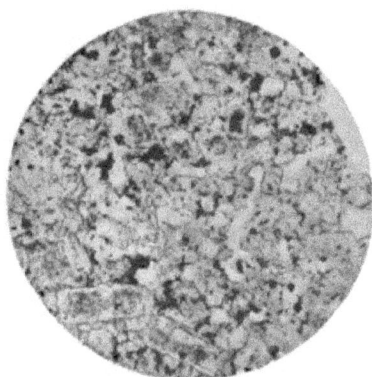

2

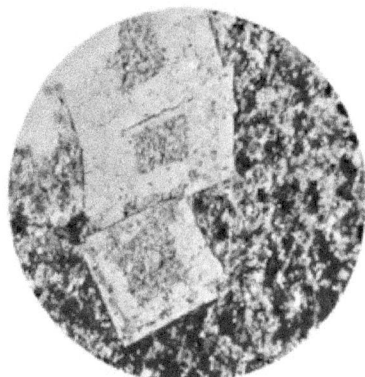

3

4

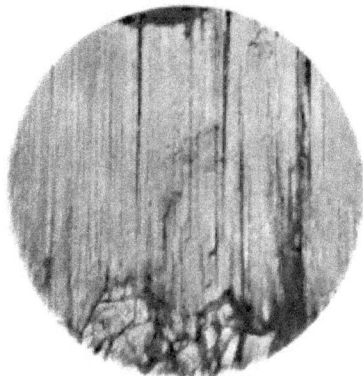

5

6

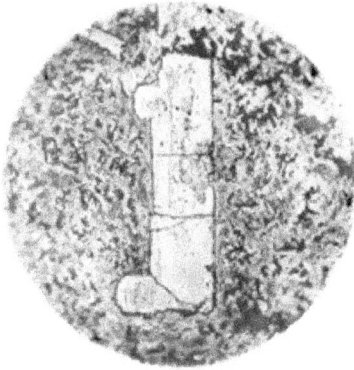

1

2

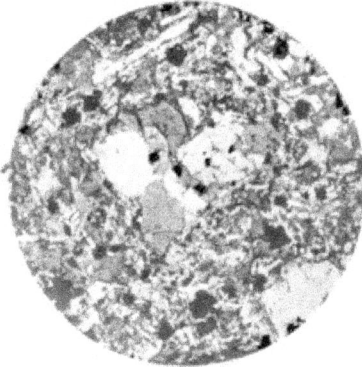

3

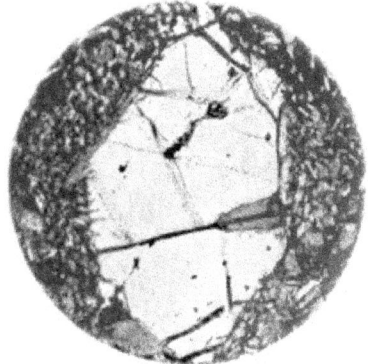

4

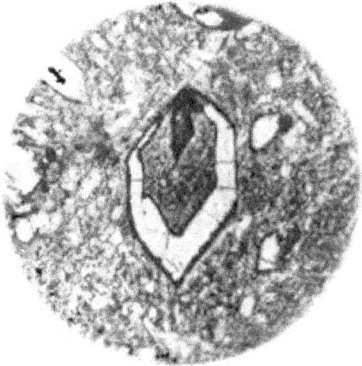

5

6

1

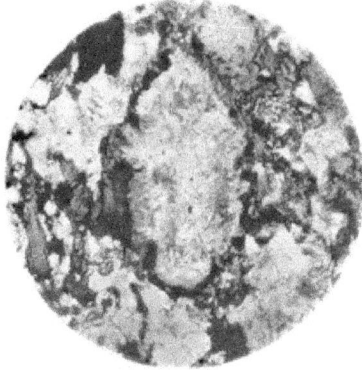

2

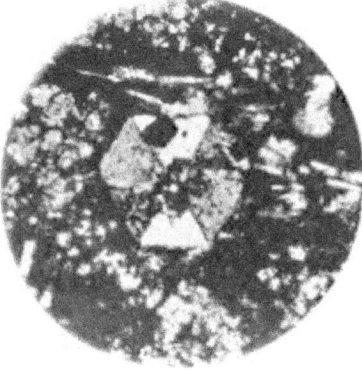

3

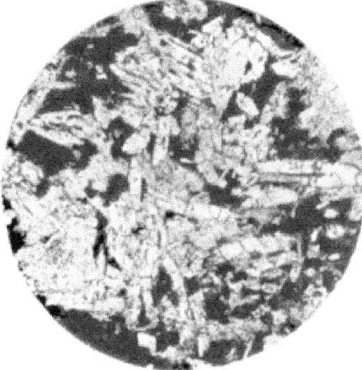

4

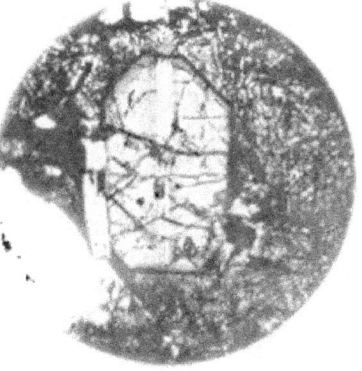

5

6

1

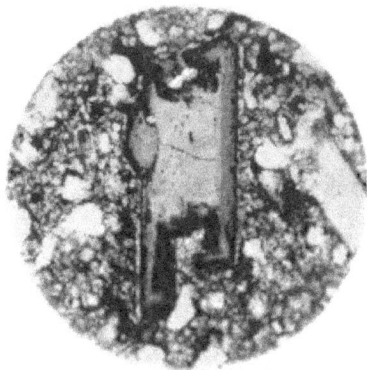

2

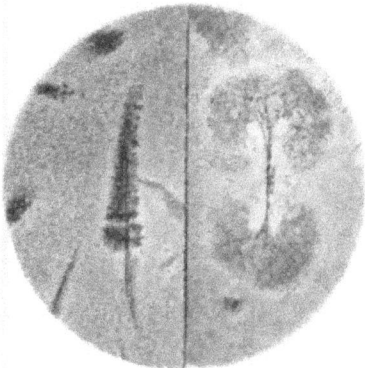

3

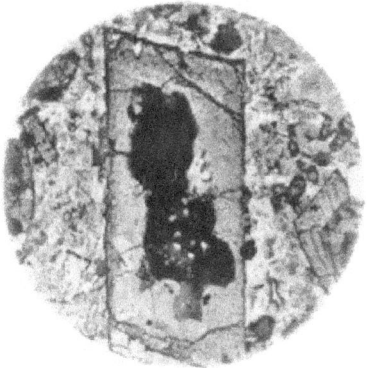

4

5

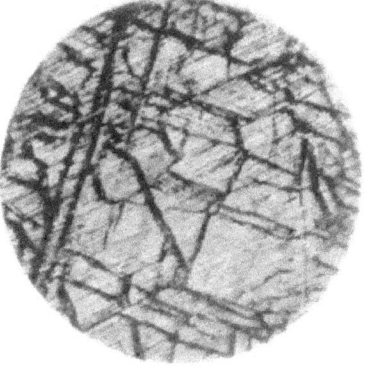

6

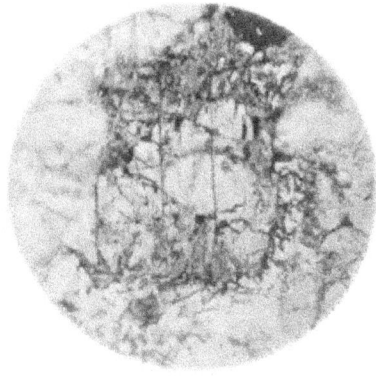

1   2

3   4

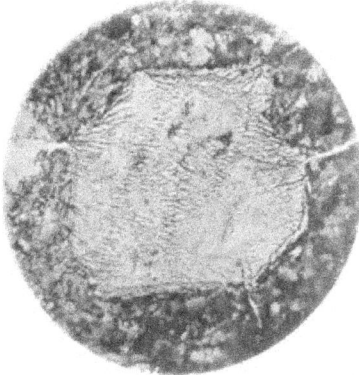
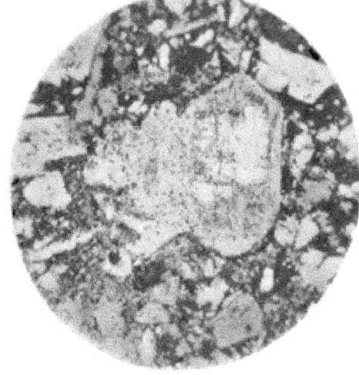

5   6

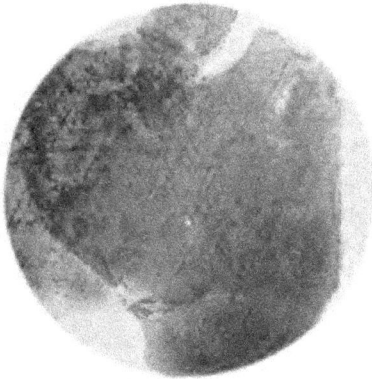

1

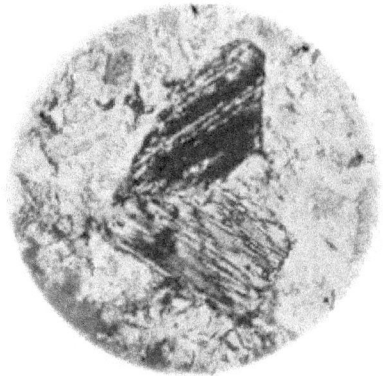

2

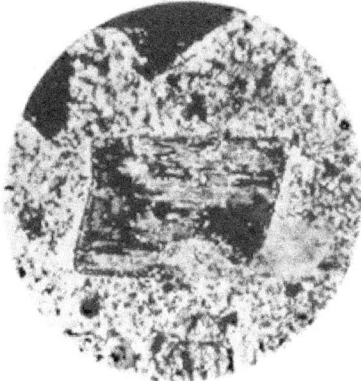

3

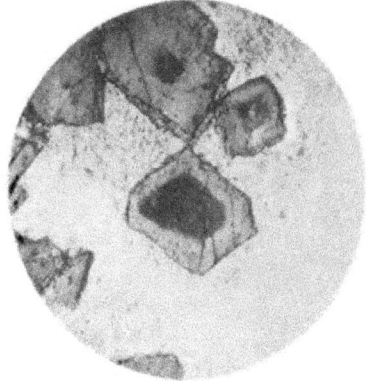

4

5

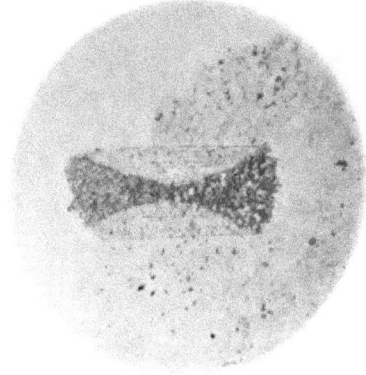

6

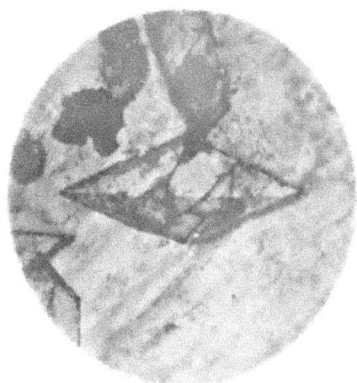

1

2

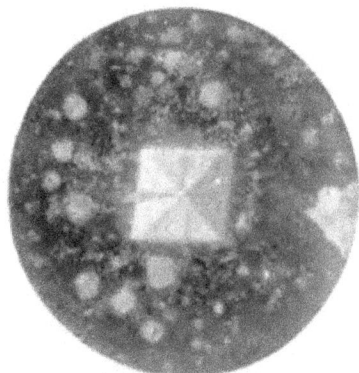

3

4

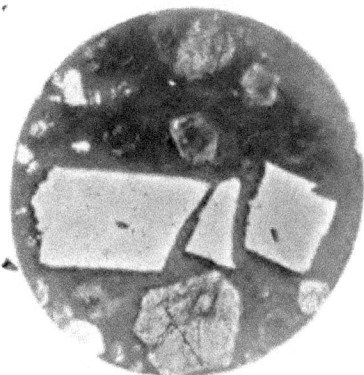

5

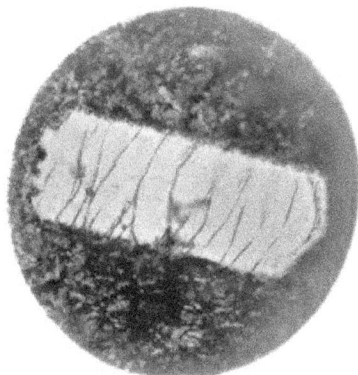

6

1

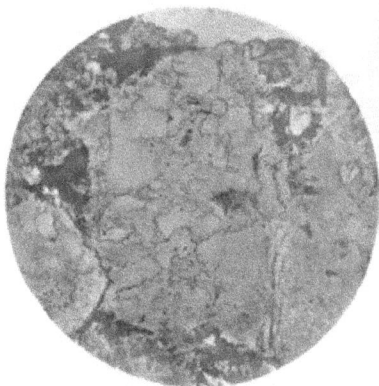

2

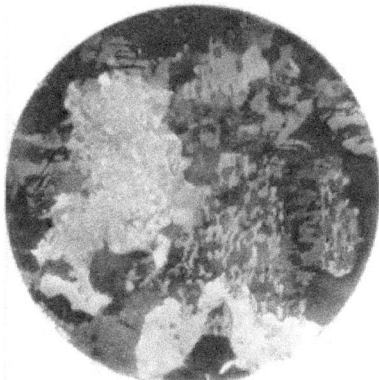

3

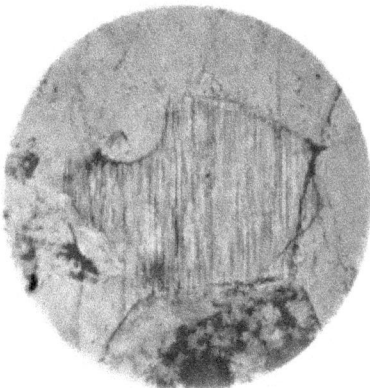

4

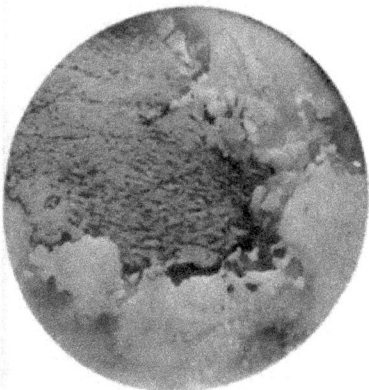

5

6

1

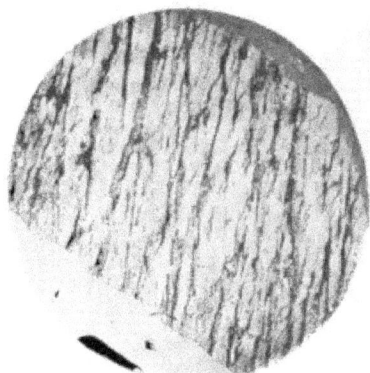

2

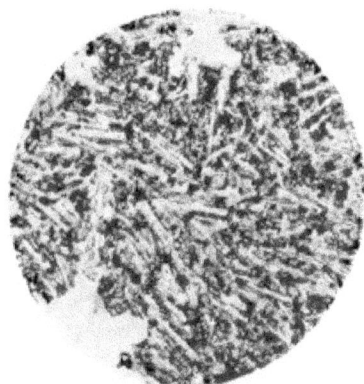

3

4

5

6

Plate XXVI.

1

2

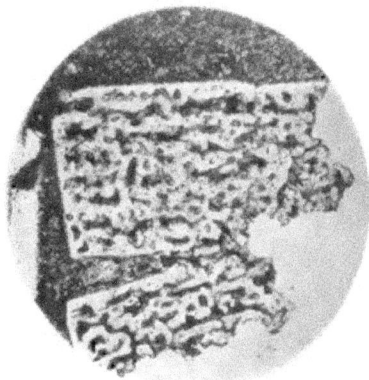

3

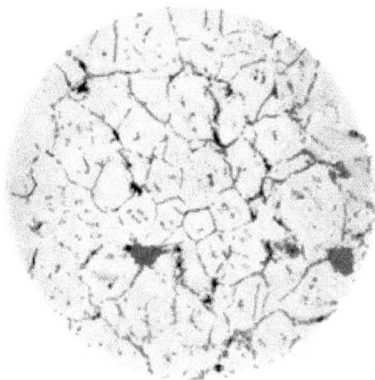

4

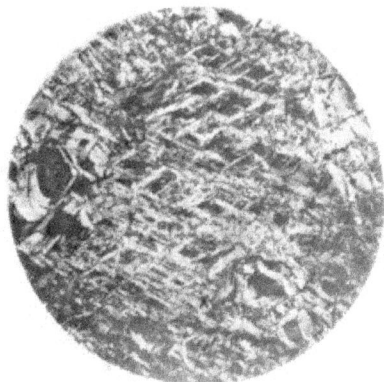

5

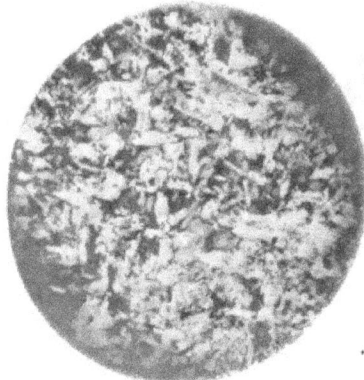

6